Table of Atomic Weights

Element	Symbol	Atomic Number	Atomic Weight	Element	Symbol	Atomic Number	Atomic Weight
Actinium	Ac	89	(227)[a]	Manganese	Mn	25	54.9380
Aluminum	Al	13	26.98154	Mendelevium	Md	101	(258)
Americium	Am	95	(243)	Mercury	Hg	80	200.59
Antimony	Sb	51	121.75	Molybdenum	Mo	42	95.94
Argon	Ar	18	39.948	Neodymium	Nd	60	144.24
Arsenic	As	33	74.9216	Neon	Ne	10	20.179
Astatine	At	85	(210)	Neptunium	Np	93	237.0482
Barium	Ba	56	137.34	Nickel	Ni	28	58.71
Berkelium	Bk	97	(247)	Niobium	Nb	41	92.9064
Beryllium	Be	4	9.01218	Nitrogen	N	7	14.0067
Bismuth	Bi	83	208.9804	Nobelium	No	102	(255)
Boron	B	5	10.81	Osmium	Os	76	190.2
Bromine	Br	35	79.904	Oxygen	O	8	15.9994
Cadmium	Cd	48	112.40	Palladium	Pd	46	106.4
Calcium	Ca	20	40.08	Phosphorus	P	15	30.97376
Californium	Cf	98	(251)	Platinum	Pt	78	195.09
Carbon	C	6	12.01115	Plutonium	Pu	94	(244)
Cerium	Ce	58	140.12	Polonium	Po	84	(210)
Cesium	Cs	55	132.9054	Potassium	K	19	39.098
Chlorine	Cl	17	35.453	Praseodymium	Pr	59	140.9077
Chromium	Cr	24	51.996	Promethium	Pm	61	(147)
Cobalt	Co	27	58.9332	Protactinium	Pa	91	231.0359
Copper	Cu	29	63.546	Radium	Ra	88	226.0254
Curium	Cm	96	(247)	Radon	Rn	86	(222)
Dysprosium	Dy	66	162.50	Rhenium	Re	75	186.2
Einsteinium	Es	99	(254)	Rhodium	Rh	45	102.9055
Erbium	Er	68	167.26	Rubidium	Rb	37	85.4678
Europium	Eu	63	151.96	Ruthenium	Ru	44	101.07
Fermium	Fm	100	(257)	Samarium	Sm	62	150.4
Fluorime	F	9	18.99840	Scandium	Sc	21	44.9559
Francium	Fr	87	(223)	Selenium	Se	34	78.96
Gadolinium	Gd	64	157.25	Silicon	Si	14	28.086
Gallium	Ga	31	69.72	Silver	Ag	47	107.868
Germanium	Ge	32	72.59	Sodium	Na	11	22.98977
Gold	Au	79	196.9665	Strontium	Sr	38	87.62
Hafnium	Hf	72	178.49	Sulfur	S	16	32.06
Hahnium	Ha	105	(260)[b]	Tantalum	Ta	73	180.9479
Helium	He	2	4.00260	Technetium	Tc	43	98.9062
Holmium	Ho	67	164.9304	Tellurium	Te	52	127.60
Hydrogen	H	1	1.00797	Terbium	Tb	65	158.9254
Indium	In	49	114.82	Thallium	Tl	81	204.37
Iodine	I	53	126.9045	Thorium	Th	90	232.0381
Iridium	Ir	77	192.22	Thulium	Tm	69	168.9342
Iron	Fe	26	55.847	Tin	Sn	50	118.69
Krypton	Kr	36	83.80	Titanium	Ti	22	47.90
Kurchatovium	Ku	104	(260)[c]	Tungsten	W	74	183.85
Lanthanum	La	57	138.9055	Uranium	U	92	238.029
Lawrencium	Lr	103	(256)	Vanadium	V	23	50.9414
Lead	Pb	82	207.19	Xenon	Xe	54	131.30
Lithium	Li	3	6.941	Ytterbium	Yb	70	173.04
Lutetium	Lu	71	174.97	Yttrium	Y	39	88.9059
Magnesium	Mg	12	24.305	Zinc	Zn	30	65.38
				Zirconium	Zr	40	91.22

[a] Value in parentheses is the mass number of the most stable or best-known isotope.
[b] Suggested by American workers but not yet accepted internationally.
[c] Suggested by Russian workers. American workers have suggested the name Rutherfordium.

BASIC CONCEPTS
OF CHEMISTRY

BASIC CONCEPTS
OF CHEMISTRY

Leo J. Malone/Saint Louis University—St. Louis, Missouri

JOHN WILEY & SONS,
New York • Chichester • Brisbane • Toronto • Singapore

Library of Congress Cataloging in Publication Data:

Malone, Leo J   1938–
  Basic concepts of chemistry.

  Includes index.
  1. Chemistry. I. Title.
QD31.2.M344   540   80-19501
ISBN 0-471-06381-9

Printed in the United States of America

10 9 8 7 6 5 4 3 2

Preface

Basic concepts of chemistry is an introduction to chemical concepts and to the way chemistry is studied and mastered. The level, content, and sequence of topics have been chosen with a sensitivity to students who have had little or no background in chemistry or who have had a significant interruption of their studies since high school. The text is primarily for students who wish to obtain the background and confidence needed to pursue a main sequence general chemistry course.

The style of the text is conversational yet concise. There is a traditional sequence of topics that provides a step-by-step construction of the science, with one topic building logically on the previous one. Another reason for the chosen sequence that makes it particularly appropriate for a course using this text is that many, if not most, students need some mathematical preparation or review before encountering extensive quantitative concepts applied to chemical systems. The mathematical tools of measurement and conversions are introduced in Chapter 2, supplemented by detailed reviews in the appendixes. The students, with the direction of the instructor, should improve their mathematical, algebraic, and problem-solving skills while important topics that are not primarily quantitative are introduced (that is, electron structure, the periodic table, bonding, and nomenclature). There are separate appendixes on basic math (Appendix A), basic algebra (Appendix B), scientific notation (Appendix C), logarithms (Appendix E), and graphs (Appendix F). The appendixes include discussion, worked-out examples, and drill problems (with answers provided). Appendixes A, B, and C also include self-diagnostic tests so that students can quickly determine whether they need the full review.

Almost all introductory and general chemistry texts now use the unit-factor (dimensional analysis) method of problem solving. I feel, however, that the typical student needs more than a one- or two-page introduction in order to apply this important and useful tool consistently and confidently. In Chapter 2, there is a detailed introduction to the topic, *supplemented* by an extensive appendix that provides additional explanations, solved examples, and exercise problems (with answers). Diligent students can therefore become comfortable with this important tool by the time it is applied extensively to chemical systems.

Besides the reviews in the appendixes, the following features will help students throughout the course.

1. Simple analogies are used that relate the concrete to the abstract. Analogies that are easily understood themselves can be helpful in making new concepts understandable.
2. Introductions to each chapter discuss the general objectives of the chapter,

how it follows from the previous discussions and, most important, what specific topics previously discussed should be reviewed before the chapter is covered. The approach is to help counter the view that chemistry is a study of isolated topics.

3. Numerous example problems are worked out in the text, step by step. There are usually two or three examples of each type of problem.

4. End-of-chapter problems are assigned in the margins of the text after a particular topic has been discussed and examples shown. This is designed to give students direction for immediate reinforcement of a concept without affecting the continuity of the discussion.

5. End-of-chapter problems are numerous and of varying difficulty. They are categorized by topic. About two-thirds of the answers are provided in Appendix H (many of the quantitative problems include solutions).

6. Questions are provided at the end of the chapter that can serve as a checklist of important concepts and definitions learned.

7. New terms are introduced in the text in boldface type. The definitions of the terms are in italics.

8. A comprehensive glossary of terms provides easy reference to definitions used throughout the text.

There is obviously more material in this text than can be easily covered in a one-semester course. Generally, Chapters 1 to 12 or 13 would be covered. Chapters 14 to 16 may be covered for special purposes or may be used for future reference. The sequence of topics may be altered, particularly if parts of Chapters 8 and 9 (moles and stoichiometry) need to be covered early. The nuclear discussion in Chapter 3 is completely optional; discussion and problems are separated from the rest of the chapter so that they can be included at any time. It is included here instead of in a later chapter because it follows logically from the discussion of the makeup of the nucleus, it is current, and it can be used early to build interest. It can be omitted or included later. The discussions of orbitals and electron spin in Chapter 4 are also optional and may be deleted without prejudice in later discussions.

I hope you find the study or the teaching of this course rewarding. It is a fascinating discipline.

Leo J. Malone

Acknowledgments

 Writing a text such as this is never accomplished without a great deal of help. First, I thank my colleagues at Saint Louis University who helped me by commenting on early drafts of the manuscript, especially Dr. Judith Durham, who read and commented on the entire project. I also thank my editor, Gary Carlson, who guided this project with his encouragement and suggestions. Finally, the following people reviewed the manuscript and offered many useful comments and suggestions on the preliminary drafts: Thomas Beall, Brevard Community College; Robert Coley, Montgomery Community College; Jo Beran, Texas A and I University; Helen Hauer, Delaware Technical and Community College, Stanton Campus; Lavier Lokke, San Diego Mesa College; Amel Anderson, University of Maryland; and John Chrysochoos, University of Toledo.

L.J.M.

Table of Contents

CHAPTER 1 • Chemistry—Matter, Changes, and Energy   **1**

1-1 The Study of Chemistry   3
1-2 Chemistry and the Nature of Matter   6
1-3 Classification of Homogeneous Matter   8
1-4 Classification of Pure Substances   10
1-5 The Names and Symbols of the Elements   11
1-6 Properties of Matter   11
1-7 Physical and Chemical Changes   13
1-8 The Conservation of Mass   14
1-9 Energy Changes in Chemical Reactions   16
1-10 The Relationship of Matter and Energy   16
Review Questions   16
Problems

CHAPTER 2 • Math and Measurements in Chemistry   **18**

2-1 Significant Figures   19
2-2 Significant Figures in Mathematical Operations   21
2-3 Scientific Notation   23
2-4 Length, Volume, and Mass   25
2-5 Conversion of Units by the Unit-Factor Method   28
2-6 Density   35
2-7 Temperature   38
Review Questions   40
Problems   41

CHAPTER 3 • The Structure of Matter; The Nucleus, and Nuclear Reactions   **45**

3-1 The Basic Structures of the Elements   46
3-2 Compounds and Formulas   47
3-3 Ions and Ionic Compounds   48
3-4 The Structure of the Atom   49
3-5 Atomic Number, Mass Number, and Atomic Weight   51
3-6 Unstable Nuclei   54
3-7 Rates of Decay of Radioactive Isotopes   56
3-8 The Effects of Radiation   57

3-9   Nuclear Reactions   59
3-10 Nuclear Fission   62
3-11 Nuclear Fusion   65
Review Questions   66
Problems   67

CHAPTER 4 · The Nature of the Electrons in the Atom   69

4-1   The Periodic Table of the Elements   70
4-2   The Emission Spectra of the Elements   71
4-3   A Model for the Electrons in the Atom   74
4-4   Shells   76
4-5   Subshells   77
4-6   Electron Notation of the Elements   78
4-7   Orbitals   84
4-8   Electron Spin   85
4-9   Paramagnetism and Diamagnetism   85
4-10 The Shapes of Orbitals   87
Review Questions   88
Problems   89

CHAPTER 5 · The Periodic Nature of the Elements   91

5-1   The Physical Properties of the Elements   92
5-2   Periods   93
5-3   Groups   93
5-4   Periodic Trends: Atomic Radius   100
5-5   Periodic Trends: Ionization Energy   103
5-6   Periodic Trends: Electron Affinity   105
Review Questions   106
Problems   107

CHAPTER 6 · The Nature of Bonding   109

6-1   Bond Formation and the Representative Elements   110
6-2   Lewis Dot Structures   110
6-3   The Formation of Ions   111
6-4   Formulas of Binary Ionic Compounds   113
6-5   The Covalent Bond   115
6-6   Simple Binary Molecules   118
6-7   The Multiple Covalent Bond   120
6-8   Polyatomic Ions   121
6-9   Writing Lewis Structures   121
6-10 Resonance Hybrids   125
6-11 Polarity and Electronegativity   126

6-12 Molecular Polarity   129
Review Questions   131
Problems   132

CHAPTER 7 · The Naming of Compounds   134

7-1   Oxidation States   135
7-2   Naming Metal–Nonmetal Binary Compounds   137
7-3   Naming Compounds with Polyatomic Ions   138
7-4   Naming Nonmetal–Nonmetal Binary Compounds   140
7-5   Naming Acids   141
Review Questions   143
Problems   143

CHAPTER 8 · Quantitative Relationships: The Mole   145

8-1   The Mole   146
8-2   The Molar Mass of the Elements   146
8-3   The Molar Mass of Compounds   151
8-4   Percent Composition, Empirical and
Molecular Formulas   153
8-5   The Use of Empirical and Molecular Formulas   157
Review Questions   159
Problems   159

**CHAPTER 9 · Quantitative Relationships: The
Chemical Equation   162**

9-1   Chemical Equations   163
9-2   Types of Chemical Reactions   166
9-3   Stoichiometry   168
9-4   Percent Yield   173
9-5   Limiting Reactant   175
Review Questions   177
Problems   178

CHAPTER 10 · The Nature of Gases   181

10-1   The Pressure of a Gas   182
10-2   Volume and Pressure   184
10-3   Volume and Temperature   187
10-4   The Combined Gas Law   189
10-5   The Nature of Gases   192
10-6   Temperature and Velocity   194
10-7   The Nature of Mixtures of Gases   195

10-8   Volume and Quantity   198
10-9   The Ideal Gas Law   199
10-10 Molar Volume   202
10-11 Stoichiometry Involving Gases   204
Review Questions   206
Problems   206

CHAPTER 11 · The Nature of Water and Aqueous Solutions   211

11-1   The Conductivity of Aqueous Solutions   212
11-2   Water as a Solvent   214
11-3   Ionic Equations and Double Displacement
Reactions   218
11-4   Solubility of Ionic Compounds   220
11-5   Concentration: Percent by Weight   222
11-6   Concentration: Molarity   223
11-7   Stoichiometry Involving Solutions   227
Review Questions   229
Problems   230

CHAPTER 12 · Acids, Bases, and Salts   232

12-1   Early Criteria for Acids and Bases   233
12-2   The Nature of Acids and Bases   234
12-3   Neutralization and Salts   236
12-4   The Strengths of Acids   239
12-5   The Strengths of Bases   242
12-6   Brønsted-Lowry Acids and Bases   242
12-7   Oxides as Acids and Bases   247
Review Questions   249
Problems   250

CHAPTER 13 · Oxidation–Reduction Reactions   252

13-1   The Nature of Oxidation and Reduction   253
13-2   Balancing Redox Equations: Oxidation
State Method   255
13-3   Balancing Redox Equations: The Ion–Electron
Method   258
13-4   Spontaneous Redox Reactions   262
13-5   Voltaic Cells   267
13-6   Electrolytic Cells   270
Review Questions   272
Problems   273

CHAPTER 14 · Reaction Kinetics and Equilibrium   **276**

14-1   The Mechanism of a Reversible Reaction   277
14-2   Examples of Reactions at Equilibrium   279
14-3   Reaction Rates and the Equilibrium Constant   281
14-4   The Effect of Stress and Catalysts on Equilibrium   288
Review Questions   292
Problems   292

CHAPTER 15 · Aqueous Acid–Base Equilibria   **295**

15-1   Equilibrium in Water   296
15-2   pH and pOH   299
15-3   Equilibria of Weak Acids and Bases in Water   301
15-4   Ions as Acids and Bases: Hydrolysis   306
15-5   Buffer Solutions   308
Review Questions   311
Problems   311

CHAPTER 16 · Organic Chemistry   **314**

16-1   Carbon and Its Chemical Bonds   315
16-2   Alkanes   317
16-3   Alkenes   323
16-4   Alkynes   326
16-5   Aromatic Compounds   328
16-6   Organic Functional Groups   329
16-7   Alcohols   330
16-8   Ethers   333
16-9   Amines   333
16-10 Aldehydes and Ketones   335
16-11 Carboxylic Acids, Esters, and Amides   337
Review Questions   341
Problems   342

Forward to the Appendixes   **344**
Appendix A · Basic Mathematics   **346**

A. Test of Basic Mathematics   346
B. A Review of Basic Mathematics   347
    1. Addition and Subtraction   348
    2. Multiplication   349
    3. Division and Fractions   350
    4. Multiplication and Division of Fractions   351
    5. Roots of Numbers   353

    6. Construction of Fractions   354
    7. Fractions Expressed as Percent   355

APPENDIX B · Basic Algebra   **357**

    A. Test of Basic Algebra   357
    B. A Review of Basic Algebra   359
       1. Operations on Algebra Equations   359
       2. Word Problems   363
       3. Direct and Inverse Proportionalities   369

APPENDIX C · Scientific Notation   **372**

    A. Test of Scientific Notation   372
    B. A Review of Scientific Notation   373
       1. Expressing Numbers in Standard Scientific
          Notation   373
       2. Addition and Subtraction   375
       3. Multiplication and Division   376
       4. Powers and Roots   378

APPENDIX D · Problem Solving by the Unit-Factor Method   **380**

    A. Conversion Factors   381
    B. Manipulation of Units   383
    C. One-Step Conversions   385
    D. Two-Step or More Conversions   390

APPENDIX E · Logarithms   **397**

    A. Logs of Numbers Between 1 and 10   397
    B. Logs of Numbers Less than 1 or Greater Than 10   398
    C. The Antilog of a Positive Logarithm   399
    D. The Antilog of a Negative Logarithm   400
       Log Tables   402

APPENDIX F · Graphs   **404**

    A. Direct Linear Relationships   404
    B. Nonlinear Relationships   406

APPENDIX G · Glossary   **409**

APPENDIX H · Answers to Problems   **417**

Index   **445**

BASIC CONCEPTS
OF CHEMISTRY

Chemistry—Matter, Changes, and Energy

An open pit copper ore mine. Copper occurs in nature as both an ore (chemically combined) and as a metal. One of the first important chemical achievements of the human race was the conversion of copper ore to the metal.

In Greek mythology there existed a god named Prometheus. Prometheus supposedly gave to animals the special tools they needed to survive in a hostile world. In the eyes of the other gods, however, he went too far when he got to humans, because he gave them a tool reserved for the gods themselves. This tool was fire.

In reality, when prehistoric cavedwellers controlled their natural fear of fire enough to discover that it could heat a cave and make food easier to eat and digest, they did prove that they were distinct and indeed superior animals. Other species can communicate with each other, and some can even use primitive tools. Only humans have dared to use fire.

Until about 1800, fire itself was considered by scientists to be one of the four basic forms of matter called elements (the others were earth, air, and water). We now know that fire results from hot gases and the energy associated with the occurrence of a fundamental change in substances called a *chemical reaction.* It could thus be argued that we proved to be the superior species when we first put to use the science of chemistry in the form of fire.

Throughout the millennia since the days of the cavepeople, many of our significant advances have been based on the observance and use of a chemical phenomenon. At first people used only stone tools, but around 5000 B.C. it was discovered that a certain naturally occurring substance was different. This material could be pounded into various shapes and sharpened to a much finer edge than the stone chips that had been used. The metal copper had been discovered. Still, this metal was relatively rare in its natural state so it could not be utilized to a large extent. What was probably an accidental discovery opened the way for the copper age. The discovery that copper could be extracted from an ore probably occurred somewhere in the Sinai Desert. Someone discovered that a beautiful green rock (now known as malachite) could be heated in an earthen furnace with glowing coals and converted to copper. Imagine the commotion that must have occurred as some prehistoric citizen insisted that one could change a common and useless rock into a valuable metal! The conversion of ores to metals is now a well-known chemical process called *metallurgy.* When first discovered, however, it must have been considered pure magic.

The evidence that chemistry has been a major factor in our progress is impressive. In about 3000 B.C. the Egyptians could dye cloth and embalm their dead. They were so good at these chemical processes that we can still tell what caused the death of some of the mummies from these times and even what diseases they may have had during their lifetimes. Despite their use of chemistry, however, the early Egyptians actually had no idea of why any of these procedures occurred.

Around 400 B.C. the Greeks went in the other direction. They thought and argued a great deal about why things occurred, but they didn't really do much to check out their ideas by experimentation. Still, some of their thoughts have proved to be consistent with modern developments.

The middle ages (500–1600 A.D.) are usually looked upon as the dark ages because of the retreat from the arts and literature, the decline of central governments, and the general loss of so much of the civilization that Egypt, Greece, and Rome had previously built. In the case of the science of chemistry, however, considerable advancement occurred. The chemists of the time were known as alchemists and were thought to mix in a little magic along with some solid chemical knowledge. Among other things, these people were looking for ways to convert cheaper metals such as lead and zinc into gold, which was thought to be the perfect metal. Although they never accomplished their goal, they did establish certain important chemical procedures, such as distillation and crystallization. They also discovered and prepared many previously unknown chemicals now called elements and compounds.

Modern chemistry had its foundation in the late 1700s when the use of the balance became widespread. At that time chemistry became a quantitative science, and from then on theories used to describe chemistry had to be checked and correlated with the results of experiments performed in the laboratory. From this came the modern atomic theory, first proposed by John Dalton around 1803. This theory, which in a slightly modified form is still accepted today, gave chemistry the solid base from which it could grow to serve humanity on such an impressive scale. Since most of our actual knowledge of chemistry is comparatively recent (the past 100 years), chemistry actually doubles as the oldest and, in a way, the youngest of sciences.

1-1 The Study of Chemistry

Curiosity may have killed the cat, but it is also responsible for the overwhelming impact of science on our daily lives. Airplanes and automobiles, plastics, synthetic fibers (orlon, dacron, and nylon), chemicals that halt cancer and cure diseases all resulted from someone's curiosity. Chemistry, like other sciences, attracts the profoundly curious person, one who never outgrew the childhood tendency to constantly ask the question "why?" As children, most scientists probably drove their parents nuts with their continual questions about nature.

Why study chemistry? The obvious answer is because it is required. But maybe you can appreciate *why* it is required. Chemistry is *the* fundamental natural science. Biology, physics, geology, as well as all branches of engineering and medicine, are based on an understanding of the basic chemical substances of which matter is composed. It is the beginning point in the course of studies that eventually produces scientists, engineers, and physicians.

Besides scientists and engineers, all intelligent citizens need a foundation of scientific knowledge in order to make intelligent decisions as to the future of our complex environment and the need for energy. What is air and water pollution (see Figure 1-1)? How did it get there and how do we get it out? How does pollution relate to the energy crisis? These questions and others can't be discussed intelligently without some knowledge about the nature of the matter involved. Neither should these questions be left solely to scientists. Everyone has a big stake in the answers. Thus it seems reasonable to expect that a fundamental knowledge of chemistry is a prerequisite not only for graduation but for anyone living in confusing and complicated times.

Chemistry can be studied and appreciated by the average person. It need not be feared. Chemistry is orderly, predictable, and entirely reasonable, which gives the study of the science an advantage over the social sciences. Psychology, for example, deals with the whims of human nature. The nature of humans is constantly changing. On the other hand, once we know the

Figure 1-1 SMOG. To fully appreciate the origin and control of pollution, citizens need a basic understanding of chemistry.

facts about chemicals, we can be assured that the nature of the matter that we have studied will never change.

If you enjoy mastering concepts in your mind or find a sense of satisfaction in solving a problem successfully, then you will find the study of chemistry challenging. Keep an open mind; it may be fun. The study of chemistry is much like the study of a language. As such, it requires special study habits and academic self-discipline. What follows is a checklist of questions dealing with the academic self-discipline needed in the study of this science.

1   *Can you budget time on a regular basis to study and work problems?*

Like playing basketball or the piano, chemistry requires practice to master. The practice should be done very soon after the topic is covered and not right before the test. One wouldn't wait until the night before the big game to first practice jump shots or the night before the recital to first practice the sonata. Starting with Chapter 2, this text uses the margins to assign problems that emphasize a certain concept. It is wise to stop at that point and practice before you proceed. Each chapter and each topic within a chapter is like constructing a building. You can't put up the second floor until you have built a firm foundation with the first. The night before a test should be reserved for *review* and a good night's sleep—it is no time to break new ground.

2 *Can you read chapters assigned* **before** *the lectures?*

Believe it or not, this practice saves time. Just being aware of some of the concepts you don't understand puts you ahead. You're ready for the lecture, and you'll listen more intently. You'll also save time on your note taking, since you'll know that certain definitions and tables are available in the book.

3 *Can you attend lectures regularly and take good notes?*

As you proceed in the study of chemistry you will notice that the successful students (besides yourself) miss few classes. Regular and punctual attendance means you are always up to date. This text supplements your instructor—it doesn't replace him or her.

Taking good and orderly notes will help you organize the material for future reference. Copy problems in your notes so that you can cover them up later and rework them yourself.

4 *Can you ask questions in class or afterwards?*

You can't let a basic concept go by without owning it. If the class is such that asking questions during class is impossible (or embarrassing), ask later. Your instructor is available, and you owe yourself the answers. Step forward; this is your life, and your question is not "dumb."

5 *Can you keep at it even if you are disappointed on the first test?*

An "A" is a great way to start, but what if that doesn't happen? In that case you'll have to reanalyze your study habits and try again. Don't be afraid to ask your instructor for advice in this matter. For most people mastering chemistry takes perseverance. Give the course a chance with your best shot. You'll know eventually whether you should "haul up the flag."

6 *Can you memorize?*

Some definitions, names, and formulas have to be memorized. Some people will say, "I don't want to memorize, I want to understand." To be sure, understanding is what you are after, but memorization is often necessary before understanding develops. Chemistry in this respect is like studying a foreign language. Before you can learn to speak or write a new language, such things as vocabulary and verb declensions must be committed to memory.

7 *Can you do problems systematically and neatly?*

It is amazing how the good performances in chemistry courses usually correlate with neat and orderly notes, papers, and tests. Order on paper shows an orderly mind. That helps in chemistry.

If you are able to show by your habits positive answers to these questions, you just may surprise yourself with how well this material comes to you and, as a result, just how fascinating chemistry can be.

1-2 Chemistry and the Nature of Matter

The definition of chemistry at first seems simple. **Chemistry** *is the branch of science that deals with the nature and composition of matter and the changes that matter undergoes.* Actually, there are two key words in this definition that require further discussion. These words are "matter" and "changes." First, we will discuss **matter,** which is defined as *anything that has mass and occupies space.* After this we will discuss changes.

All matter can be classified (arranged systematically) into one of three physically distinct states. *The* **solid state** *has a definite shape and a definite volume. The* **liquid state** *has a definite volume but not a definite shape.* Thus liquids take the shape of the bottom of the container. Finally, *the* **gaseous state** *has neither a definite shape nor a definite volume.* Thus a gas fills a container uniformly. (See Figure 1-2.)

a   *b*   *c*

Figure 1-2 THE THREE PHYSICAL STATES OF MATTER.
(*a*) Solids have a definite shape and a definite volume.
(*b*) Liquids have a definite volume but an indefinite
shape. (*c*) Gases have an indefinite volume and indefinite shape.

In nature we find matter composed of many distinct forms (called substances) that represent all physical states. Rocks, salt, and steel are substances that exist as solids; gasoline, water, and alcohol exist as liquids; air, ammonia, and natural gas are present in nature as gases. All of these substances can be classified further as to the complexity of their nature. To understand this, we will discuss the most complex form of nature first and move step by step toward the most basic and simple form of nature.

In addition to physical state, all matter can be divided into two categories: homogeneous and heterogeneous. **Homogeneous matter** *has the same properties and composition throughout and exists in a single phase. A* **phase** *is a state of matter that has definite and identifiable boundaries.* Water, salt dissolved in water, copper pennies, air, and table sugar are all examples of homogeneous matter. **Heterogeneous matter** *is a nonuniform mixture, since it exists as two or more homogeneous phases with definite boundaries between the phases.* Oil and water (two liquid phases), salt and sand (two solid phases), fog (a liquid and a gas phase), and ice water (a solid and a liquid phase) are all examples of heterogeneous matter. Sometimes it is not easy to tell whether matter is homogeneous or heterogeneous. Creamy salad dressing appears uniform and homogeneous to the naked eye, but putting a sample under a microscope shows that it is heterogeneous with drops of oil suspended in the vinegar. Fog also requires a close look to notice that tiny water droplets are suspended in the gaseous air. (See Figure 1-3.)

Heterogeneous matter is the more complex form of matter and can be separated into two or more homogeneous phases. In Figure 1-4 we have started with a heterogeneous mixture composed of a liquid phase (table salt dissolved in water) and a solid phase (copper powder). By performing a

Figure 1-3
HETEROGENEOUS MIXTURES. A magnification of fog reveals a liquid and a solid phase. A magnification of creamy salad dressing reveals two liquid phases.

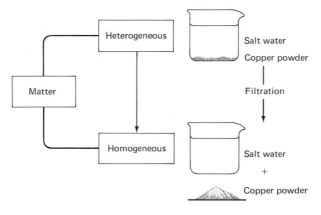

Figure 1-4 CLASSIFICA-TION OF MATTER. A heterogeneous mixture can be separated into homogeneous sub-stances.

simple filtration as shown in Figure 1-5, the mixture can be separated into two homogeneous phases—salt water and copper powder.

Even though salt water is still a mixture of two substances (salt and water), it is homogeneous. Even the most high-powered microscope does not show any evidence of salt crystals suspended in the water.

1-3 Classification of Homogeneous Matter

Homogeneous matter can also be divided into two categories: solutions and pure substances (see Figure 1-6). A **solution** *is the more complex, as it is a mixture of two or more pure substances*. Many solutions can be separated into pure substances by laboratory procedures such as distillation, shown in Figure 1-7. **Pure substances** *cannot be separated into other components by distillation or similar methods*. Thus our original mixture contained three pure substances.

Figure 1-5 FILTRA-TION. This labora-tory procedure is used to separate a solid phase from a liquid phase.

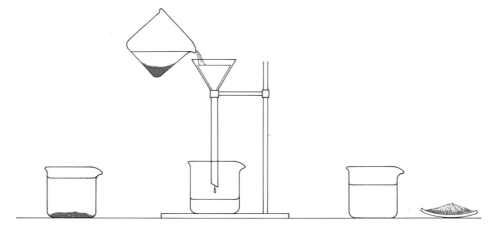

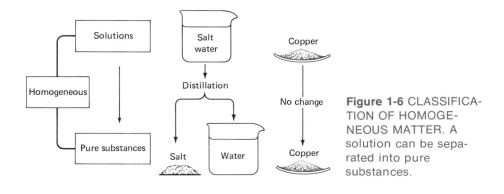

Figure 1-6 CLASSIFICA-
TION OF HOMOGE-
NEOUS MATTER. A
solution can be sepa-
rated into pure
substances.

If solutions and pure substances are both homogeneous, how can you tell the difference? The answer lies in a careful examination of the properties of the two forms of matter. *The* **properties** *of a substance describe its unique, observable characteristics or traits. Pure substances have definite compositions with definite and distinct properties such as melting point and boiling point.* Pure water freezes sharply at 0°C (32°F), no matter where on the face of the earth it is found. Pure table salt also has unique and definite properties that serve as its fingerprint, such as a melting point of 804°C. In contrast to the sharp and distinct properties of pure substances, *mixtures (both heterogeneous and homogeneous) of different pure substances have properties that vary according to the proportions of the mixture.* In our example, although both salt and water have definite melting points, a salt water solution has a melting

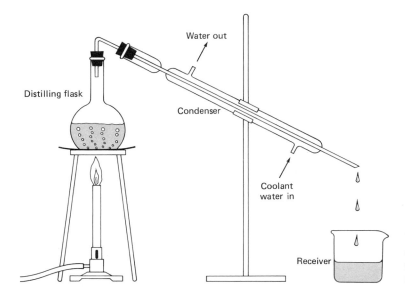

Figure 1-7 DISTILLATION.
This laboratory proce-
dure is used to separate
a solution into pure
substances.

point that varies from about $-18°C$ to just under $0°C$ depending on the amount of salt dissolved in the water. The application of this phenomenon is important in the winter in the northern parts of this country. Salt is spread on ice to lower its freezing point and thus melt the ice from the streets. Within limits, the more salt that is spread, the lower the freezing point of the salt and ice. The whole country takes advantage of this phenomenon to make home-made ice cream. The ice cream is prepared by suspending a container in a mixture of ice, salt, and water that will have a temperature below the freezing point of pure water.

For another example of the varying properties of a solution we can refer to the addition of sugar to coffee or tea. The more sugar we add the sweeter the coffee or tea tastes, although there is a limit to how much sugar can be dissolved in a given amount of water.

1-4 Classification of Pure Substances

Table salt, water, and copper are all pure substances with distinct and un-changing composition and properties. However, pure substances can also be divided into two categories: compounds and elements. **Compounds** *are the more complex, as they can be separated into two or more elements.* **Elements** *can be broken down no further by ordinary chemical means, so they represent the simplest and most basic form of matter.* In our example, two of the three pure substances (water, salt, and copper) can be decomposed further and are thus classified as compounds. The table salt (whose chemical name is sodium chloride) can be divided into the elements sodium and chlorine. Water can be divided into the elements hydrogen and oxygen. The copper powder cannot be separated further, so it is classified as an element. (See Figure 1-8.)

Figure 1-8 CLASS-IFICATION OF PURE SUB-STANCES. A compound can be separated into elements.

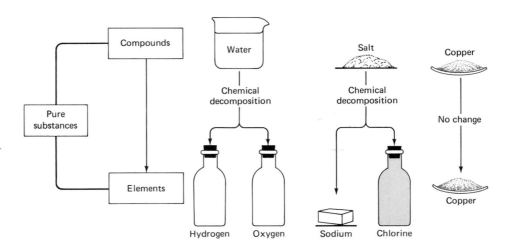

1-5 The Names and Symbols of the Elements

An element is designated by its name and symbol. *In most cases the first one or two letters of the name are used as the element's* **symbol.** When elements have a two-letter symbol, the first is capitalized but the second is not. The names and symbols of all of the 105 named elements are shown in the inside front cover (element 106 has not yet been named). In a chart of the elements called the **periodic table** (which is shown inside the back cover), the symbols of the elements are listed in addition to other important information that will be discussed in Chapter 3. Some common elements and their symbols are shown in Table 1-1.

In the table you probably noticed that some symbols do not seem to relate to their names. Many of these elements, however, derive their symbols from their original Latin or Greek names. For example, the Latin name for sodium is *natrium* (Na); for potassium, *Kalium* (K); for gold, *Aurum* (Au); and for lead, *plumbum* (Pb). In the chapters that follow more and more elements are used in the discussions. When you see an unfamiliar element, note and memorize its name and symbol.

Table 1-1　Some Common Elements[a]

Element	Symbol	Element	Symbol
Hydrogen	H	Sodium	Na
Carbon	C	Sulfur	S
Nitrogen	N	Magnesium	Mg
Oxygen	O	Chlorine	Cl
Aluminum	Al	Potassium	K
Gold	Au	Copper	Cu
Phosphorus	P	Lead	Pb
Calcium	Ca	Iron	Fe

[a] An element that isn't particularly common is Promethium (Pm), named after the giver of fire who was mentioned earlier.

1-6 Properties of Matter

A compound results from a combination of elements in what is called a **chemical reaction.** A compound is not just an intimate mixture of elements. For example, in Figure 1-9 two elements, zinc and sulfur, can be mixed as powders to form an obviously heterogeneous mixture in which both solid phases can be easily seen. Zinc is a shiny, bluish-white metal that melts at 419°C. Sulfur is a soft, yellow solid that melts at 113°C. It is possible to ignite the mixture in a hot flame so that a chemical reaction occurs to form a compound called zinc sulfide. If the proportions are just right, there is no zinc or sulfur remaining. The zinc sulfide formed no longer has properties similar to the zinc and

Zinc and
sulfur mixture

Chemical
reaction

Zinc sulfide
(a compound)

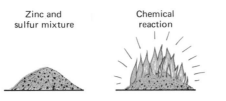

Figure 1-9 FORMATION OF ZINC SULFIDE. In this chemical reaction a mixture of elements is changed into a compound.

sulfur from which it was formed. It has distinct, definite, and uniform properties of its own. Zinc sulfide is a white solid with a melting point of 1850°C.

In our previous discussion we mentioned some of the properties of three pure substances—zinc, sulfur, and the compound zinc sulfide. There are two kinds of properties of matter: physical and chemical. **Physical properties** *can be observed without changing the substance into another substance.* Physical properties of chemical substances are properties such as odor, color, specific gravity, state of matter (solid, liquid, or gas), melting point, and boiling point (see Figure 1-10). Your physical properties are listed on your driver's license, such as your height, weight, sex, and the color of your eyes and hair.

Just as there's more to falling in love with a person than just noticing his or her physical properties (theoretically, anyway, there should be some "chemistry"), there's more to describing a substance than noting physical properties. A substance also has **chemical properties,** *which relate to its ability or tendency to change into other substances by a chemical reaction.* Chemi-

Zinc
1. Solid
2. Melts at 419 °C
3. Boils at 907 °C
4. Bluish-white
5. Metallic taste
6. Specific gravity 7.1

Sulfur
1. Solid
2. Melts at 113 °C
3. Boils at 444 °C
4. Yellow
5. Pungent smell
6. Specific gravity 2.1

Zinc Sulfide
1. Solid
2. Melts at 1850 °C
3. Colorless or white
4. Specific gravity 4.0

Figure 1-10 PHYSICAL PROPERTIES. Zinc, sulfur, and zinc sulfide all have distinct and different physical properties.

cal properties are like a personality. You have to see how the substance acts in the presence of other substances to see what its "chemistry" is all about. Since chemical properties describe chemical changes, examples are given in the next section.

1-7 Physical and Chemical Changes

A **physical change** *in a substance does not involve a change in the composition of the substance but is simply a change in physical state.* When water freezes to form ice or boils to form steam, the liquid state has simply undergone a physical change to another phase. Ice and steam are both forms of water just as much as is the more familiar liquid state. On the other hand, *a* **chemical change** *involves a change of substances into other substances by means of a chemical reaction.* (See Figure 1-11.) As discussed previously, zinc sulfide is fundamentally different than the elements from which it came. The decay of vegetation, the burning of coal, and the growth of a tree all involve chemical changes.

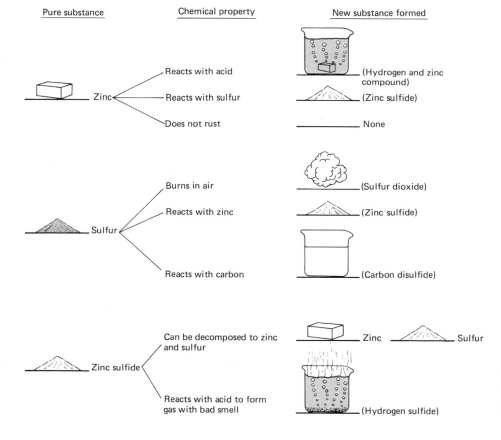

| Pure substance | Chemical property | New substance formed |

Zinc —
- Reacts with acid — (Hydrogen and zinc compound)
- Reacts with sulfur — (Zinc sulfide)
- Does not rust — None

Sulfur —
- Burns in air — (Sulfur dioxide)
- Reacts with zinc — (Zinc sulfide)
- Reacts with carbon — (Carbon disulfide)

Zinc sulfide —
- Can be decomposed to zinc and sulfur — Zinc   Sulfur
- Reacts with acid to form gas with bad smell — (Hydrogen sulfide)

Figure 1-11 CHEMICAL PROPERTIES AND CHANGES OF SUBSTANCES.

1-8 The Conservation of Mass

In Figure 1-9, the zinc and sulfur mixture weighs exactly the same as the zinc sulfide formed (if the violence of the reaction didn't blow any of it away). Not more than 200 years ago, however, scientists were still puzzled when wood burned and most of the mass seemed to disappear. This was explained by various theories, but our understanding of chemical changes is now based on the **law of conservation of mass,** which states that *matter is neither created nor destroyed in a chemical reaction.* Two hundred years ago the involvement of gases in chemical reactions was not fully understood. It is now known that the weight of the wood plus the weight of the oxygen in the air used in combustion equals the weight of the gases formed plus the weight of the ashes (see Figure 1-12). The gases involved are all invisible but still have mass.

In Chapter 8, chemical reactions from a quantitative point of view will be introduced. At that time you will notice that many of the calculations are based on the law of conservation of mass.

Figure 1-12 THE COMBUSTION OF WOOD. In a chemical reaction matter is conserved.

1-9 Energy Changes in Chemical Reactions

One of the more obvious and useful results of a burning log is the large amount of energy in the form of heat released by the combustion process. Chemistry is also vitally concerned with the changes in energy involved in chemical reactions. *Chemical reactions that release energy, such as combustion, are called* **exothermic,** *and those that absorb energy, such as photosynthesis in plants, are called* **endothermic.**

What is energy? When we have no energy left at the end of the day the implication is clear: We have no capacity to do work. That is exactly how energy is defined: **Energy** *is the capacity or the ability to do work.* Energy exists in several forms. Almost all of our energy on earth originates from the sun in the form of *radiant* or *light energy.* (Light energy will be discussed in Chapter 4.) By a process called photosynthesis, a living tree can transform the solar, radiant energy into *chemical energy.* In this process, energy-poor compounds in the environment are transformed into energy-rich compounds in the tree. When logs from the tree burn, the chemical energy is transformed into *heat energy.* Energy-poor compounds are also formed and returned to the environment. If the heat energy is used to turn a turbine in a power plant, it is transformed into *mechanical energy,* which is then converted into *electrical energy.* In a car battery, however, chemical energy is converted directly into elec-

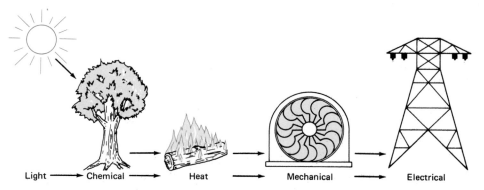

Light ——→ Chemical ——→ Heat ——→ Mechanical ——→ Electrical

Figure 1-13 EN-
ERGY. Energy
is neither created
nor destroyed but
can be transformed.

trical energy. In all of these energy transformations that have been discussed, energy is not gained or lost. It is simply transformed. (See Figure 1-13.) *The* **law of conservation of energy** *states that energy cannot be created or destroyed.*

The energy of motion (mechanical energy) is familiar to us in our everyday world. This energy, however, is of two types depending on whether it is available but not being used or actually in use. For example, a weight suspended above the ground has energy available because of its suspended position. *Energy due to position is called* **potential energy.** Other examples are water stored behind a dam or a coiled spring. In our example, when the weight is released, the energy can be put to use by the downward motion. *Energy due to motion is called* **kinetic energy.** In order to move the weight back into position, energy must be supplied. (See Figure 1-14.)

In future chapters, potential energy and kinetic energy will play a part in discussions of some chemical systems.

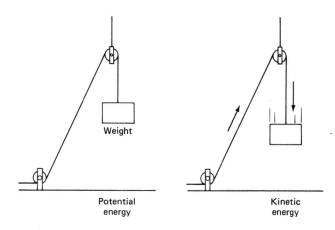

Weight

Potential
energy

Kinetic
energy

Figure 1-14 POTENTIAL
AND KINETIC ENERGY.
Potential energy is
energy of position and
kinetic energy is energy
of motion.

1-10 The Relationship of Matter and Energy

The laws of conservation of mass and energy turn out to be less exact than was originally thought. In 1905 Albert Einstein proposed the now well-known relationship between mass and energy,

$$E = mc^2$$

$$E = \text{energy} \qquad m = \text{mass} \qquad c = \text{the speed of light}$$

This amazing theory tells us that, in fact, energy results from a conversion of mass. Since a small amount of mass produces a tremendous amount of energy, the mass loss or gain in a chemical reaction is far too small to be detectable. For our purposes in chemistry, the laws of conservation of mass and energy remain valid and can be applied to the understanding of the nature of matter and its changes. In nuclear reactions, however, there is a significant mass loss, which results in a vast amount of energy. *Nuclear energy* is another form of energy and is the source of the energy of the sun. Nuclear energy is discussed in Chapter 3.

Review Questions

1 Define chemistry.
2 What is "matter"?
3 Describe the three physical states in nature.
4 What is meant by a phase in the chemical sense?
5 Explain the difference between homogeneous and heterogeneous matter.
6 What is a solution? What is a pure substance?
7 How does the chemist differentiate between mixtures and pure substances?
8 Identify the two forms of pure substances.
9 How is the symbol of an element obtained?
10 What is the difference between a physical and a chemical property?

11 Describe physical and chemical changes.
12 What is the law of conservation of mass?
13 What is meant by an exothermic chemical reaction? An endothermic reaction?
14 Define energy. What are the various forms of energy?
15 How is the law of conservation of energy defined?
16 Describe the difference between potential and kinetic energy.
17 Give the equation that relates matter and energy. Is the conversion of mass to energy important in chemical reactions?

Problems

(*Throughout the text, answers to all problems except those marked* ■ *are provided in Appendix H.*)

1-1 Identify each of the following as homogeneous or heterogeneous matter:
(a) Gasoline
(b) Dirt
(c) Smog
(d) Alcohol

(e) A nail

(f) Vinegar

(g) Ice cubes in water

(h) Aerosol spray

(i) Air

■ (j) Dry Ice

■ (k) Whipped cream

■ (l) Bourbon

■ (m) Natural gas

■ (n) Grapefruit

1-2 Identify each of the following pure, homogeneous substances as either elements or compounds:

(a) Carbon dioxide

(b) Hydrogen gas

(c) Iron

(d) Titanium dioxide

(e) Helium

■ (f) Ammonia

■ (g) Hydrogen peroxide

■ (h) Aspirin

■ (i) Mercury

1-3 Using the table of the elements in the front of the book, determine the symbol for each of the following elements:

(a) Fluorine   (d) Magnesium

(b) Neon   (e) Manganese

(c) Tin   (f) Tungsten

1-4 Using the table of the elements in the front of the book, determine the names of the elements with the following symbols:

(a) Fe   (e) Cm

(b) Bi   (f) Hg

(c) Eu   (g) Ag

(d) U   (h) Sb

1-5 Identify each of the following as either a physical or a chemical property:

(a) Diamond is the hardest substance.

(b) Carbon monoxide is poisonous.

(c) Soap is slippery.

(d) Silver tarnishes.

(e) Gold does not rust.

(f) Carbon dioxide freezes at $-78°C$.

(g) Tin is a shiny, gray metal.

(h) Sulfur burns in air.

■ (i) Sodium burns in the presence of chlorine gas.

■ (j) Mercury is a liquid at room temperature.

■ (k) Water boils at 100°C at sea level.

■ (l) Limestone gives off carbon dioxide when heated.

1-6 Identify each of the following as either physical or chemical change:

(a) The frying of an egg

(b) The vaporization of Dry Ice

(c) The boiling of water

(d) The burning of gasoline

(e) The breaking of glass

■ (f) The tanning of leather

■ (g) The fermentation of apple cider

■ (h) The compression of a spring

2 Math and Measurements in Chemistry

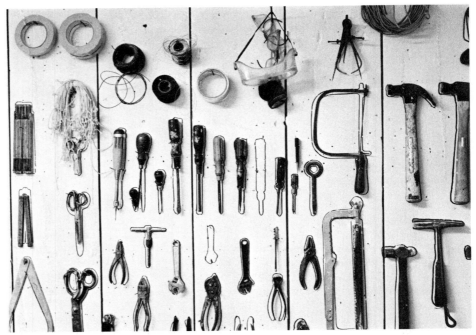

These are the tools of the trade of a carpenter. One of the tools of the trade of a physical scientist is mathematics.

In a popular television series in 1980, an alien from the planet Ork used an undefined unit of time measurement called a "bleem." In fact, modern science on this planet uses plenty of legitimate units that sound even stranger. There are rads, rems, roentgens, and barns; ergs, oersteds, and henrys; ohms, gauss, and newtons. Each of these units is used in some particular way to answer the question, "how much?". The use of units refers, then, to the quantitative aspects of science.

Quantitative questions concern measurements. **Measurements** *determine quantity, dimensions, or extent of something, usually in comparison to some standard.* For example, if you are six feet tall, you are six times the length of a long-dead English king's foot, which is used as the standard. We will see that modern units of measurement have less arbitrary standards than somebody's foot.

The purpose of this chapter is to examine the language the chemist uses to describe measurements. We will look into weight, length, volume, and temperature. In doing this we convert from our familiar English system to the metric system by the unit-factor method, which is introduced in this chapter and used throughout the text.

It is hard to imagine how a mechanic could repair an automobile engine without being able to use tools such as wrenches and screwdrivers. It is also hard to picture how one could grasp the meaning of measurements without the ability to use mathematics. "Why is there so much math in chemistry?" is a question often heard from the student new to this science. The answer is that chemistry, like physics and astronomy, is a **physical science,** that is, *a science concerned with the natural laws of matter other than those laws concerned with life (biology, botany) and the earth (geology).* As we will see throughout this text, many of these natural laws involve quantitative measurements.

The math involved in this course and in most general chemistry courses at the college or university level is not at all awesome. In fact, no more than a year of high school algebra is needed. Still, many students need a review of mathematics. As you know, being in good physical shape allows physical exertion in sports to be relaxing, challenging, and just downright fun. If you're not in shape, exertion is not only painful but unsatisfying. Being in good mathematical shape has a similar effect on the study of chemistry. The normal sequence of chapters in this text will allow you time to review and strengthen your math background before the quantitative aspects of chemistry are introduced in force in Chapter 8. Hopefully, students with specific needs will take full advantage of this time to "get in shape."

One of the main purposes of this text is to help students who require some direction with their mathematical needs. With this in mind, Appendixes A and B contain diagnostic tests in basic mathematical and algebraic calculations. It is advised that you take each self-test, grade it, and evaluate your own strengths and weaknesses in light of the recommendations in the appendixes. If the tests in Appendixes A and B indicate that you are a bit "rusty," work through the examples and assigned problems before proceeding.

Refer to Appendixes A and B for a review of basic math and algebra.

2-1 Significant Figures

In most large cities there is one official who is an expert at estimating the size of a crowd. Let's say that he estimates the size of a certain gathering at 9000 people. Then he notices that eight people leave. Should his new estimate be changed to 8992? No, of course not. The original number wasn't that good. The three zeros in the number 9000 were not really known numbers (as opposed to the zero in 108), but were there merely as fillers to indicate the magnitude of the number or, in other words, to locate the decimal point. In this example, the crowd would have to diminish by at least 500 before the official might feel justified in changing the estimate of the crowd to 8000. In the number 9000 only one of the numbers, the 9, is "significant." *The number of* **significant figures** *in a measurement is simply the number of valid numbers.* Significant figures refer to how precisely a number is known.

The science of chemistry is vitally concerned with the precision of the numbers that we use in measurements. The precision of a measurement is illustrated in Figure 2-1. Big Ben in London measures time to the nearest minute. Notice that even so the actual minute must be estimated. The wrist watch measures time more *precisely*, since the seconds as well as the minute can be read. Again, the last digit of the second hand is estimated. Finally, the stopwatch used to time track events is even more precise, since time can be measured with it to tenths of a second.

In Figure 2-1, the actual time can be compared to the standard, known as Greenwich time, which is established at an observatory in Greenwich, England. In the example above, the Greenwich time is 1:22:45.8. Notice that the stopwatch time is the most precise, since its value has the largest number of significant figures. However, it is the least accurate. **Accuracy** *refers to how close the value of a measurement is to the true value.* The wrist watch is the most accurate. In scientific measurements, however, the most precise number is generally the most accurate.

Since the advent of the inexpensive hand calculator, it has become even more important for the student of chemistry to understand the precision of the answers to quantitative problems. For example, 7.8 divided by 2.3 reads 3.33913043 on a calculator. But if the original numbers represent measurements (e.g., 7.8 in. and 2.3 in.) the calculator's answer is much too precise. The calculator makes no judgment about the precision of the numbers punched into it. Therefore, *we* must know how to report an answer properly despite what the calculator reads. The answer must be expressed to the correct number of significant figures.

As we've mentioned, the number of significant figures in a measurement

Big Ben
Time 1:21

Wrist watch
Time 1:22:46

Stopwatch
Time 1:20:50.4

Figure 2-1 PRECISION. Precision refers to the "exactness" of a measurement.

is simply the number of valid numbers. For example, 76.3 has three significant figures, and 4562 has four significant figures. That would be all there is to it except for the number zero. Zero serves two functions: It can be a legitimate number indicating "none," or simply a filler to locate a decimal point (like the zeros in the crowd of 9000 we had earlier). It is unfortunate that centuries ago when numbers were invented no one invented a symbol that could be used in place of a zero that is not significant. Since we don't have a separate symbol, we have to memorize some rules that tell us whether a zero is a significant figure or not.

1 When a zero is between other digits, it is significant (e.g., 709 has three significant figures).

2 Zeros to the right of a decimal point that are also to the right of a digit are significant (e.g., 8.0 has two significant figures, 5.780 has four significant figures).

3 Zeros to the right of a decimal point but to the left of a digit are not significant (e.g., 0.0078 has two significant figures, 0.0460 has three significant figures).

4 Zeros to the left of a decimal point when no decimal point is shown may or may not be significant. It is usually assumed that they are not (e.g., 9000 has one significant figure and 670 has two). What if a number such as 890 is actually known precisely to three significant figures? That is, the zero is really a zero. This is a tough question. Some texts use a line over the zero to indicate that it is significant (i.e., 890). As we will see shortly, scientific notation gives us a way to solve the problem. *In most calculations in this text it is assumed that the numbers are known to three significant figures.*

See Problems 2-1, 2-2

Example 2-1
How many significant figures are in the following measurements?

(a) 1508 cm (b) 300.0 ft (c) 20.003 lb (d) 0.00705 gal (e) 36 g (f) 20,000 in.^{2}

Answers:
(a) 4 (b) 4 (c) 5 (d) 3 (e) 2 (f) 1

2-2 Significant Figures in Mathematical Operations

The rules for manipulation of significant figures are different for addition and subtraction than for multiplication and division. In addition and subtraction, the answer is limited by the term with the least number of places to the right of the decimal. This rule is illustrated by the following summation:

$$
\begin{array}{r}
10.68 \\
0.473 \\
\underline{1.32} \\
12.473 = 12.47
\end{array}
$$

Notice that the "3" in the addition cannot be expressed in the answer, since the other two numbers are not known to that many decimal places. In fact, since the last significant figure in a measured number is estimated, it is assumed that there is an uncertainty of ± 1 in that number. Therefore, the number above is expressed as 12.47 although it may actually range from 12.46 to 12.48.

In addition and subtraction we must also be careful with manipulations involving numbers containing nonsignificant zeros. The final answer must be rounded off to show the same number of nonsignificant zeros as in the original number. This can be illustrated with the crowd estimated at 9000 people discussed earlier.

$$
\begin{array}{r}
9000 \\
- \quad 8 \\
\hline
8992 = \underline{9000}
\end{array}
\qquad
\begin{array}{r}
9000 \\
- \quad 80 \\
\hline
8920 = \underline{9000}
\end{array}
\qquad
\begin{array}{r}
9000 \\
- \quad 800 \\
\hline
8200 = \underline{8000}
\end{array}
$$

The rules for rounding off are as follows:

1 If the digit to be dropped is 4 or less, simply drop that digit (e.g., 12.472 rounds off to 12.47).

2 If the digit to be dropped is greater than 5, increase the preceeding digit by one (e.g., 1.577 rounds off to 1.58, 3.452 rounds off to 3.5).

3 If the digit to be dropped is exactly 5, increase the preceeding digit by one if odd or leave unchanged if even (e.g., 5.45 rounds off to 5.4, 6.715 rounds off to 6.72).

See Problems 2-3 through 2-8

Example 2-2
Add the following:

$$
\begin{array}{r}
7.56 \\
0.375 \\
\underline{14.2203} \\
22.1553 = 22.16
\end{array}
$$

In multiplication and division there is no concern as to the location of the decimal point. The rule is simply that the answer must be expressed with the same amount of significant figures as the number with the least amount of significant figures. To use an old cliche, the chain (the answer) is only as strong as its weakest link.

Example 2-3

Multiply the following: 2.34 in. × 3.225 in.

    The answer on a calculator is 7.5465. Since the first multiplier has three significant figures and the second has four, the answer should be expressed to three figures. The answer rounds off to <u>7.55 in.^{2}</u>.

Divide 11.688 ft by 4.0 sec.

    The answer, 2.922, should round off to two significant figures. The answer is <u>2.9 ft/sec</u>.

    Before leaving the topic of significant figures, we should mention the effect of *exact* numbers on a calculation. An exact number is one that is either defined exactly or originates from a count. For example, 1 gal = 4 qt is an exactly defined relationship. As such it has unlimited significant figures (i.e., 1.00 . . . etc. gal = 4.00 . . . etc. qt). Counted numbers are also exact and have unlimited significant figures. For example, if there are 32 students in a certain classroom, the number 32 is considered exact (i.e., 32.00 . . . etc.). As we will see in this chapter, an exact relationship does not limit or affect the number of significant figures in a calculation. Relationships that originate from a measurement are not exact and limit the number of significant figures in a calculation as described. You will become more familiar with exact numbers as we proceed.

2-3 Scientific Notation

Chemistry requires the use of some very large and some very small numbers. For example, numbers such as

$$602{,}000{,}000{,}000{,}000{,}000{,}000{,}000$$

are used in Chapter 8. Obviously, this number as written is extremely awkward. A much more convenient way of both writing and reading the above number is

    6.02×10^{23} ("six point oh two times ten to the twenty-third")

The latter number is expressed in scientific notation. **Scientific notation** *expresses zeros used to locate the decimal point in powers of 10*. A number expressed in scientific notation has two parts: *the* **coefficient** *is the number that is multiplied by the power of 10, and the* **exponent** *is the power to which 10 is raised*.

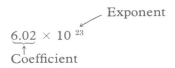

The following reviews some powers of 10 and their equivalent numbers:

$10^0 = 1$ $10^{-1} = \dfrac{1}{10^1} = 0.1$

$10^1 = 10$ $10^{-2} = \dfrac{1}{10^2} = 0.01$

$10^2 = 10 \times 10 = 100$ $10^{-3} = \dfrac{1}{10^3} = 0.001$

$10^3 = 10 \times 10 \times 10 = 1000$ $10^{-4} = \dfrac{1}{10^4} = 0.0001$

$10^4 = 10 \times 10 \times 10 \times 10 = 10,000$ etc.

etc.

The *standard* method of expressing numbers in scientific notation is with one digit to the left of the decimal point (e.g., 4.56×10^6 rather than 45.6×10^5).

Example 2-4
Express each of the following numbers in scientific notation with one digit to the left of the decimal point:

(a) 47,500 (b) 5,030,000 (c) 0.0023 (d) 0.0000470

Solution:
(a) The number 47,500 can be factored as $4.75 \times 10,000$. Since $10,000 = 10^4$, the number can be expressed as

$$4.75 \times 10^4$$

A more practical way to transform this number to scientific notation is to count to the left from the *old* decimal point to where you wish to put the new decimal point. The number of places counted to the left will be the positive exponent of 10.

4 3 2 1

$4\,7\,5\,0\,0 = 4.75 \times 10^4$

(b)

6 5 4 3 2 1

$5,0\,3\,0,0\,0\,0 = 5.03 \times 10^6$

(c) 0.0023 can be factored into 2.3 × 0.001. Since 0.001 = 10^{-3}, the number can be expressed as

$$2.3 \times 10^{-3}$$

A more practical way is to count to the right from the *old* decimal point to where you wish the new decimal point. The number of places counted to the *right* will be the negative exponent of 10.

$$0.\underset{1\ \ 2\ \ 3}{0\ 0\ 2}3 = 2.3 \times 10^{-3}$$

(d)

$$0.\underset{1\ 2\ 3\ 4\ 5}{0\ 0\ 0\ 0\ 4}7\ 0 = 4.70 \times 10^{-5}$$

The use of scientific notation can help remove the ambiguity of numbers containing zero that may or may not be significant. For example, the number 9000 can have from one to four significant figures as written. The following expressions, however, are clear as to the number of significant figures:

9×10^3 has one significant figure

9.0×10^3 has two significant figures

9.00×10^3 has three significant figures

9.000×10^3 has four significant figures

The preceding discussion is supplemented in Appendix C. This appendix includes a diagnostic test and a discussion of the mathematical manipulation of numbers expressed in scientific notation. You are urged to take the short test at this time and work through the appendix if practice is indicated by the results of the test.

Refer to Appendix C for further review.
See Problems 2-9, 2-10, and 2-11

2-4 Length, Volume, and Mass

There are many observable physical properties of a portion of matter. There is color, temperature, density, physical state, the length of its dimensions, volume, and mass. In this section we are concerned with length, volume, and mass (see Figure 2-2). Later in this chapter we will discuss density. There is often confusion between the terms "mass" and "weight." **Mass** *is the quantity of matter that a substance contains. It is the same anywhere in the universe.* **Weight** *is a measure of the attraction of gravity for the substance.* An astronaut

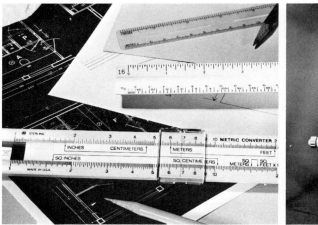
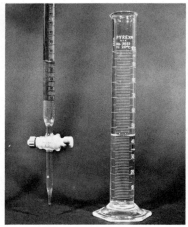

a   b

Figure 2-2
LENGTH, VOL-
UME, MASS. These
properties of a
quantity of matter
are measured with
common laboratory
equipment. (a)
Length. Metric
rulers. (b) Volume.
A buret and a
graduate cylinder.
(c) Mass. A pan
balance and an
electric balance.

c

has the same mass on the moon as on earth. Since the gravity of the moon is much less than earth, however, his or her weight is much less on the moon. You may weigh less on the moon, but you'll be just as big (or small). The terms mass and weight are often used interchangeably, but mass is correct.

As mentioned in the introduction to this chapter, the United States inherited a rather strange system of measurement for length, volume, and mass from England. Some of the units of length use a King's anatomy as a standard. The major problem, however, is the lack of any systematic relation between units. For example, the relations of length are 12 in. = 1 ft; 3 ft =

Table 2-1 Metric or SI Units

Measurement	Unit	Symbol
Mass	Gram	g
Length	Meter	m
Volume	Liter	l
Time	Second	s
Temperature	Degree Celsius	°C
	Kelvin	K
Quantity	Mole	mol
Energy	Kilojoule	kJ
Pressure	Kilopascal	kPa

1 yd; and 1760 yd = 1 mile. As a result, two units are usually needed to express one measurement (e.g., 5 ft 11 in. rather than 5.92 ft). Our monetary system makes more sense (cents?) because our currency is based on multiples of 10. Therefore, only one unit is needed to show a typical student's dismal cash position (e.g., $11.98). The British system still requires the use of two units (e.g., 2 pounds, 3 shillings).

The rest of the world uses a system that also has mass, volume and length units based on multiples of 10 with a convenient relation between mass and volume. This is, of course, the metric system, which this country is slowly adopting although it has long been used in science. In slightly modified form the metric system is also known as the SI system after the French words for International System. The basic SI or metric units with which we will be concerned are shown in Table 2-1.

There are many other SI units that designate quantities that are not used in this text. Most SI units have very precisely defined standards based on certain precisely known properties of matter.

Use of the SI or metric units is simplified by their *exact relationships by*

Table 2-2 Prefixes Used in the Metric System

Prefix	Symbol	Relation to Basic Unit
Mega-	M	10^6
Kilo-	k	10^3
Deci-	d	10^{-1}
Centi-	c	10^{-2}
Milli-	m	10^{-3}
Micro-	μ^a	10^{-6}
Nano-	n	10^{-9}
Pico-	p	10^{-12}

[a] A Greek letter.

Table 2-3 Relationships Among Metric Units

Mass Unit	Symbol	Relation to Basic Unit	Volume Unit	Symbol
Kilogram	kg	10^3 g	Kiloliter	kl
Decigram	dg	10^{-1} g	Deciliter	dl
Centigram	cg	10^{-2} g	Centiliter	cl
Milligram	mg	10^{-3} g	Milliliter	ml
Microgram	μg	10^{-6} g		

powers of 10. This is illustrated in Table 2-3 by use of the more common prefixes listed in Table 2-2.

In the metric system, 1 liter is the volume occupied by 1 cubic decimeter (i.e., 1 liter = 1 dm³). Thus 1 ml is the volume occupied by 1 cubic centimeter (i.e., 1 ml = 1 cm³ = 1 cc). The units milliliter (ml) and cubic centimeter (cm³ or cc) can be used interchangeably when expressing volume. (See Figure 2-3.)

In Table 2-4 the relations of metric units to English units (expressed to three significant figures) are given. (See also Figure 2-4.)

2-5 Conversion of Units by the Unit-Factor Method

The majority of problems with which we will be concerned in this text are basically of the same type. They simply involve the conversion of one unit of measurement to another. To do this we use the unit-factor method, which is

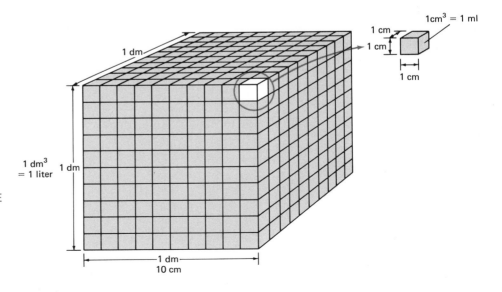

Figure 2-3 VOLUME AND LENGTH. One milliliter is the volume occupied by one cubic centimeter.

Relation to Basic Unit	Length Unit	Symbol	Relation to Basic Unit
10^3 liter	Kilometer	km	10^3 m
10^{-1} liter	Decimeter	dm	10^{-1} m
10^{-2} liter	Centimeter	cm	10^{-2} m
10^{-3} liter	Millimeter	mm	10^{-3} m

Figure 2-4 METRIC UNITS. The use of metric units in this country is becoming increasingly evident.

Table 2-4   The Relationship Between English and Metric Units

	English	Metric Equivalent
Length	1.00 in	2.54 cm
	0.621 mile	1.00 km
Mass	1.00 lb	454 g
	2.20 lb	1.00 kg
Volume	1.00 qt	0.943 liter
	1.06 qt	1.00 liter
Time	1.00 sec	1.00 sec

also known as dimensional analysis. A brief introduction is included in this chapter, with supplemental explanations and exercises presented in Appendix D.

The **unit-factor method** *converts from one unit to another by use of conversion factors. A* **conversion factor** *is an equality or equivalency relationship between units or quantities expressed in fractional form.* For example, in the metric system we have the exact relationship

$$10^3 \text{ m} = 1 \text{ km}$$

This can be expressed in fractional or factor form as

$$\frac{1 \text{ km}}{10^3 \text{ m}} \quad \text{or the reciprocal } \frac{10^3 \text{ m}}{1 \text{ km}} \quad \left(\text{or simply } \frac{10^3 \text{ m}}{\text{km}}\right)$$

This is read as "1 kilometer per 10^3 meters" or "10^3 meters per one kilometer." The latter factor is usually simplified to "10^3 meters per kilometer." When just a unit (no number) is read or written in the denominator, the number is assumed to be "one."

Conversion factors are also constructed from the equalities between the English and metric units. For example, the equality

$$1.00 \text{ in.} = 2.54 \text{ cm}$$

can be expressed in factor form as

$$\frac{1.00 \text{ in.}}{2.54 \text{ cm}} \quad \text{or} \quad \frac{2.54 \text{ cm}}{1.00 \text{ in.}}$$

Factors such as the examples above are sometimes referred to as **unity factors,** since *they relate a quantity to "one" of another.* As such, the factors are often written in a somewhat simplified form as follows:

$$\frac{1 \text{ in.}}{2.54 \text{ cm}} \quad \text{or} \quad \frac{2.54 \text{ cm}}{\text{in.}}$$

Unity factors are written as such in this text. *It should be understood, however, that "one" (as written in the numerator or implied in the denominator) is known to as many significant figures as the other number.*

These factors can be used to convert a measurement in one unit to the other. In the unit-factor method all units of the numbers are maintained in the calculation. Like the numbers themselves, the units are multiplied, divided, and cancelled in the course of the calculation. This adds a little time to the calculation, but the payoff is *correct* answers. Students who consistently use this method are often amazed at the orderliness and logic of chemistry. In more complex conversions, where several conversion factors are required, the student can plan the calculation in a stepwise and orderly manner from what is given to what is requested.

The general procedure of a one-step conversion by the unit-factor method is as follows:

$$[\text{What's given}] \times [\text{conversion factor}] = [\text{what's requested}]$$

In most calculations, the proper conversion factor that converts "given" to "requested" has the unit of what is given in the denominator (the old unit) and the unit of what is requested in the numerator (the new unit). In this way, the old unit *cancels* as would an identical numerical quantity. This leaves the new unit in the numerator.

$$[\text{Given}]\ \cancel{\text{old unit}} \times \left[\begin{array}{c}\text{conversion} \\ \text{factor}\end{array}\right] \frac{\text{new unit}}{\cancel{\text{old unit}}} = [\text{requested}]\ \text{new unit}$$

The key, then, is to select the conversion factor that does the right job. As we will see, many conversions require several steps and thus need more than one conversion factor to convert from what's given to what's requested. Proper organization and planning help avoid difficulties for these problems. The following examples illustrate the use of the unit-factor method in one-step conversion between units.

Example 2-5

Convert 0.468 m to (a) kilometers and (b) millimeters.

a

(1) Given: 0.468 m.   Requested: ____?____ km
(2) Procedure: Use a conversion factor that cancels m and leaves km in the numerator. In shorthand our procedure is

$$\text{m} \longrightarrow \text{km}$$

(3) Relationship: 10^3 m = 1 km (from Table 2-3).
(4) Conversion factor: Of the two possible conversion factors that originate from the above relationship, we choose the one with km (requested) in the numerator and m (given) in the denominator. This is 1 km/10^3 m.
(5) Solution:

$$0.468\ \cancel{\text{m}} \times \frac{1\ \text{km}}{10^3\ \cancel{\text{m}}} = 0.468 \times 10^{-3}\ \text{km} = \underline{\underline{4.68 \times 10^{-4}\ \text{km}}}$$

b

(1) Given: 0.468 m.   Requested: ____?____ mm.
(2) Procedure: m → mm
(3) Relationship: 1 mm = 10^{-3} m.
(4) Conversion factor: 1 mm/10^{-3} m.
(5) Solution:

$$0.468 \; \cancel{m} \times \frac{1 \; mm}{10^{-3} \; \cancel{m}} = 0.468 \times 10^3 \; mm = \underline{\underline{468 \; mm}}$$

Example 2-6
Convert 825 cm to inches.

(1) Given: 825 cm.   Requested: ___?___ in.
(2) Procedure: cm → in.
(3) Relationship: 1.00 in. = 2.54 cm (from Table 2-4).
(4) Conversion factor: 1 in./2.54 cm.
(5) Solution:

$$825 \; \cancel{cm} \times \frac{1 \; in.}{2.54 \; \cancel{cm}} = \underline{\underline{325 \; in.}}$$

Example 2-7
Convert 6.85 qt to liters.

(1) Given: 6.85 qt.   Requested: ___?___ liters.
(2) Procedure: qt → liters
(3) Relationship: 1.00 liter = 1.06 qt.
(4) Factor: 1 liter/1.06 qt.
(5) Solution:

$$6.85 \; \cancel{qt} \times \frac{1 \; liter}{1.06 \; \cancel{qt}} = \underline{\underline{6.46 \; liters}}$$

What would have happened if we had multiplied instead of divided? In that case, the units would have served as a red flag indicating that a mistake had been made. The units of the answer would have been qt²/liter, which is obviously not correct.

$$6.85 \; qt \times \frac{1.06 \; qt}{1 \; liter} = 7.26 \; \frac{qt^2}{liter} \qquad ??????$$

The one-step conversions that have been worked so far are analogous to a direct, nonstop airline flight between your home city and your destination. The problems that follow are like a flight between cities where it is necessary to switch planes at least once between your point of origin and your destination. In such a case you would plan your journey carefully from city to city. Likewise, for multistep conversions you should plan your path from your origin (what's given) to your destination (what's requested). In this case each step along the way requires one conversion factor.

Example 2-8
Convert 4978 mg to kilograms.

(1) Given 4978 mg.   Requested: ___?___ kg
(2) Procedure: In Table 2-3, notice that one relationship between milli-

grams and kilograms is not directly available. However, notice that we can take a two-step journey by (a) converting milligrams to grams and then (b) converting grams to kilograms. In shorthand the procedure is

$$mg \xrightarrow{(a)} g \xrightarrow{(b)} kg$$

(3) Relationships: 1 mg = 10^{-3} g; 1 kg = 10^3 g.

(4) Conversion factors: Treat each step as a distinct operation or problem. In step (a) we need g in the numerator and mg in the denominator. In step (b) we need kg in the numerator and g in the denominator.

$$\text{(a) } 10^{-3} \text{ g/mg} \qquad \text{(b) } 1 \text{ kg/}10^3 \text{ g}$$

(5) Solution:

$$\text{(a)} \qquad\qquad \text{(b)}$$

$$4978 \text{ mg} \times \frac{10^{-3} \text{ g}}{mg} \times \frac{1 \text{ kg}}{10^3 \text{ g}} = 4978 \times 10^{-6} \text{ kg} = \underline{\underline{4.978 \times 10^{-3} \text{ kg}}}$$

Example 2-9
Convert 9.85 liters to gallons.

(1) Given: 9.85 liters.   Requested ___?___ gal.

(2) Procedure:

$$\text{liters} \xrightarrow{(a)} \text{qt} \xrightarrow{(b)} \text{gal}$$

(3) Relationships: 1.06 qt = 1.00 liter; 4 qt = 1 gal.

(4) Factors:

$$\text{(a) } \frac{1.06 \text{ qt}}{1 \text{ liter}} \qquad \text{(b) } \frac{1 \text{ gal}}{4 \text{ qt}}$$

(5) Solution:

$$\text{(a)} \qquad\qquad \text{(b)}$$

$$9.85 \text{ liters} \times \frac{1.06 \text{ qt}}{\text{liter}} \times \frac{1 \text{ gal}\star}{4 \text{ qt}} = \underline{\underline{2.61 \text{ gal}}}$$

(⋆ An exact number does not limit the number of significant figures in the answer.)

Example 2-10
Convert 55 miles/hr to meters per minute.

(1) Given: 55 miles/hr.   Requested: ___?___ m/min.

(2) Procedure: Notice in this problem that both the numerator and denominator must be converted to other units. The first two steps convert the numerator and the third the denominator.

$$\text{Numerator:} \qquad \text{miles} \xrightarrow{(a)} \text{km} \xrightarrow{(b)} \text{m}$$

$$\text{Denominator:} \qquad \text{hr} \xrightarrow{(c)} \text{min}$$

(3) Relationships: 0.621 mile = 1.00 km; 10^3 m = 1 km;* 60 min = 1 hr.*
(* Exact numbers.)

(4) Conversion factors:

$$\text{(a)} \ \frac{1 \text{ km}}{0.621 \text{ mile}} \qquad \text{(b)} \ \frac{10^3 \text{ m}}{\text{km}}$$

To convert a denominator to another unit, notice that the conversion factor must be set up with what's given in the *numerator* and what's requested in the *denominator*.

$$\text{(c)} \ \frac{1 \text{ hr}}{60 \text{ min}}$$

(5) Solution:

$$\qquad \qquad \text{(a)} \qquad \quad \text{(b)} \qquad \text{(c)}$$

$$55 \ \frac{\cancel{\text{miles}}}{\cancel{\text{hr}}} \times \frac{1 \ \cancel{\text{km}}}{0.621 \ \cancel{\text{mile}}} \times \frac{10^3 \text{ m}}{\cancel{\text{km}}} \times \frac{1 \ \cancel{\text{hr}}}{60 \text{ min}} = \underline{\underline{1500 \text{ m/min}}}$$

The general procedure for working conversion problems can be summarized as follows:

1   Write down what's given and what's requested.

2   Outline a procedure in shorthand for a path from what's given to what's requested. Determine whether one or more steps are required to solve the problem.

3   Write down the relationship for each step in the conversion. Consult the appropriate tables if necessary.

4   Determine the appropriate conversion factor from the relationships in step 3. *For each step*, the unit of what's wanted should be in the numerator and that of what's given in the denominator. (An exception to this is where a problem requires a change in a denominator such as miles per hour to miles per minute.)

See Problems 2-12 through 2-32

5   Put the equation together. Make sure that the proper units cancel and that you are left with only the unit or units requested. *Carefully* do the math.

Refer to Appendix D for additional discussion on problem solving by the unit-factor method.

   If these examples and the problems at the end of this chapter cause you any difficulties, you are strongly urged to work through Appendix D. There you will find an extensive supplement including solved problems and exercises employing familiar English units, imaginary units, and actual chemical units. It is designed to increase your proficiency in the mechanics of the unit-factor method so that it can serve as your "backup system" for obtaining correct answers. (Your primary system should eventually be a "feeling" for the type and general magnitude of the answer you are seeking.)

2-6 Density

Styrofoam is usually considered "light" and the metal lead "heavy." However, since a large volume of Styrofoam can also be "heavy," we must use equal volumes of both substances to actually compare the two. What we wish to compare is the densities of the two substances. *The density of a substance is the ratio of the mass of a substance (in grams) to the volume (in milliliters or liters)*. Both mass and volume are variable properties that depend on amount. The density (the ratio of the two) is the *same* for any amount of a pure substance *under the same conditions of pressure and temperature*. Density is thus a physical property that is characteristic of a pure substance. The densities of several liquids and solids are listed in Table 2-5. The densities of gases are discussed in Chapter 10.

The following examples illustrate how density is calculated and how it is used to identify pure substances.

Example 2-11
A young lady was interested in purchasing a sample of pure gold weighing 8.99 g. Being wise, she wished to confirm that it was actually gold before she paid for it. With a quick test using a graduated cylinder like that shown in Figure 2-5, she found that the "gold" had a volume of 0.796 ml. Was the substance gold?

Solution:
From the volume and the mass, the density can be calculated and compared to that of pure gold.

$$\text{Density} = \frac{8.99 \text{ g}}{0.796 \text{ ml}} = 11.3 \text{ g/ml}$$

The sample was *not* gold. It apparently was lead that was dipped in gold paint.

Example 2-12
A sample of a pure substance was weighed and found to have a mass of 47.5 g. As shown in Figure 2-5, a quantity of water has a volume of 12.5 ml. When the substance is added to the water the volume reads 31.8 ml. The difference in volume is the volume of the substance. What is the density?

Solution:
Refer to Figure 2-5.

In the previous examples, density was calculated from a given weight and volume. Density itself is used as a conversion factor to:

1   Convert a given weight to an equivalent volume and

2   Convert a given volume to an equivalent weight.

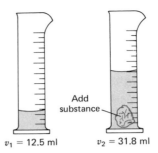

Figure 2-5 DEN-
SITY. To determine
the density, both
the mass and the
volume of a sub-
stance are
measured.

Substance   Mass =
              47.5 g

Add
substance

$v_1 = 12.5$ ml $v_2 = 31.8$ ml

Volume of substance $= 31.8 - 12.5 = 19.3$ ml

Density $= \dfrac{47.5 \text{ g}}{19.3 \text{ ml}} = 2.46$ g/ml

---

Example 2-13
Using Table 2-5, determine the volume occupied by 485 g of table salt.

Given: 485 g of salt.   Requested: ___?___ ml of salt.
Procedure: g → ml.
Relationship: 1.00 ml = 2.16 g.
Conversion factor: 1 ml/2.16 g.
Solution:

$$485 \ \cancel{g} \times \frac{1 \text{ ml}}{2.16 \ \cancel{g}} = \underline{\underline{225 \text{ ml}}}$$

Example 2-14
What is the weight of 1.52 liters of kerosine?

Given: 1.52 liters.   Requested: _____ g of kerosine.

Procedure:

$$\text{liters} \xrightarrow{\text{(a)}} \text{ml} \xrightarrow{\text{(b)}} \text{g.}$$

Relationships: 1 ml = 10^{-3} liter; 0.82 g = 1.0 ml.
Conversion factors: (a) 1 ml/10^{-3} liter; (b) 0.82 g/ml.
Solution:

$$1.52 \ \cancel{\text{liters}} \times \frac{1 \ \cancel{\text{ml}}}{10^{-3} \ \cancel{\text{liter}}} \times \frac{0.82 \text{ g}}{\cancel{\text{ml}}} = \underline{\underline{1200 \text{ g}}}$$

Assuming that there is no reaction or mixing, *a substance with a density
lower than a certain liquid floats or is* **buoyant** *in that liquid.* In the case of
water, anything with a density less than 1.00 g/ml floats. Notice that gasoline
and ice float on water but most solids and carbon tetrachloride sink.

Table 2-5 Density (at 20°C)

Substance (Liquid)	Density (g/ml)	Substance (Solid)	Density (g/ml)
Ethyl alcohol	0.790	Aluminum	2.70
Gasoline (a mixture)	~0.67 (variable)	Gold	19.3
Carbon tetrachloride	1.60	Ice	0.92 (0°C)
Kerosine	0.82	Lead	11.3
Water	1.00	Lithium	0.53
Mercury	13.6	Magnesium	1.74
		Table salt	2.16

Liquids can be mixed in such proportions to provide a range of densities. Gemologists use this principle to determine the authenticity of gemstones. For example, liquids can be mixed whereby an emerald (density 2.70 g/ml) floats in one mixture but sinks in another. Fakes do not have the same density as the authentic stone and can be discovered as shown in Figure 2-6.

The density of a liquid can be determined by a device called a **hydrometer.** The hydrometer tube, as shown in Figure 2-7, is exactly balanced with weights in the bottom so that its level in pure water is exactly at the 1.00 mark. When immersed in liquids of other densities, it becomes more or less buoyant. The scale is calibrated (the divisions previously determined and checked) in such a manner that the density of the liquid is read directly from the scale. A hydrometer is used to measure the density of the acid in a car battery because this indicates its charge condition.

Specific gravity is a term sometimes used in place of density. **Specific**

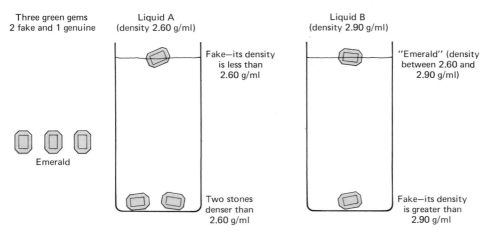

Three green gems
2 fake and 1 genuine

Emerald

Liquid A
(density 2.60 g/ml)

Fake—its density
is less than
2.60 g/ml

Two stones
denser than
2.60 g/ml

Liquid B
(density 2.90 g/ml)

"Emerald" (density
between 2.60 and
2.90 g/ml)

Fake—its density
is greater than
2.90 g/ml

Figure 2-6 BUOYANCY OF AN EMERALD. A stone sinks in a liquid that is less dense than the stone but floats in a more dense liquid.

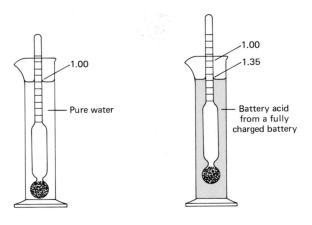

Figure 2-7 THE HYDRO-
METER. This device is
used to measure the density
of a liquid.

gravity *is a comparison of the density of a substance to that of water.* Since the
density of water is 1.00 g/ml, specific gravity is simply the density expressed
without units.

See Problems 2-33
through 2-44

$$\text{Density of Hg} = 13.6 \text{ g/ml}$$

$$\text{Density of water} = 1.00 \text{ g/ml}$$

$$\text{Specific gravity} = \frac{\text{density of mercury}}{\text{density of water}} = \frac{13.6 \text{ g/ml}}{1.00 \text{ g/ml}} = \underline{\underline{13.6}}$$

2-7 Temperature

Another important measurement is temperature. **Temperature** *is a measurement of the heat intensity of a substance and relates to the average kinetic energy of the system. A* **thermometer** *is a device that measures temperature.* The
thermometer scale with which we are most familiar is the Fahrenheit Scale
(°F), but the Celsius scale (°C) is used in most of the rest of the world and in
science. As you've probably noticed on the TV news, most weather reports
now give the temperature readings in both scales as this country slowly
switches to use of the Celsius scale (see Figure 2-8).

We learned in Chapter 1 that the boiling and melting points of pure
water (at sea level) are constant and definite properties. We can take advantage of this fact to compare the two temperature scales and establish a relationship between them. In Figure 2-9 the temperature of an ice and water
mixture is shown to be exactly 0°C. This temperature was originally established by definition. This corresponds to exactly 32°F on the Fahrenheit
thermometer. The boiling point of pure water is exactly 100°C, which corresponds to 212°F.

On the Celsius scale there are 100 equal divisions between these two
temperatures, whereas on the Fahrenheit scale there are 212 − 32 = 180

Figure 2-8 CELSIUS TEM-
PERATURE. Thanks to the
evening news, Americans
are becoming more familiar
with the Celsius scale.

equal divisions between the two temperatures. Thus we have the following
relation between scale divisions:

$$100 \text{ C div.} = 180 \text{ F div.}$$

$$1 \text{ C div.} = \frac{180 \text{ F div.}}{100} = \frac{9}{5} \text{ F div.}$$

$$1 \text{ F div.} = \frac{100 \text{ C div.}}{180} = \frac{5}{9} \text{ C div.}$$

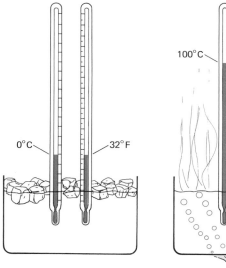

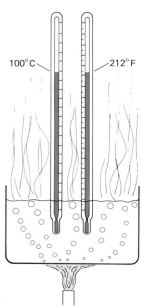

Figure 2-9 THE TEMPER-
ATURE SCALES. The
freezing and boiling
points of water are used
to calibrate the tempera-
ture scales.

You can see that a change of 1 Celsius degree is equivalent to a change of 1.8 Fahrenheit degrees.

To convert from Celsius to Fahrenheit:

1 Convert the Celsius degrees to the equivalent number of Fahrenheit divisions.

$$\text{F div.} = \tfrac{9}{5} \, C°$$

2 Add 32 to this number, since zero on the Fahrenheit scale is not the same as zero on the Celsius scale.

$$°F = \tfrac{9}{5} \, °C + 32$$

To convert from Fahrenheit temperature to Celsius:

1 Subtract 32 from the Fahrenheit reading so that both scales start at zero.

2 Multiply that number by $\tfrac{5}{9}$.

$$°C = \tfrac{5}{9}(°F - 32)$$

See Problems 2-45 through 2-50

Example 2-15
A person with a cold has a fever of 102°F. What would be the reading on a Celsius thermometer?

$$°C = \tfrac{5}{9}(°F - 32)$$
$$= \tfrac{5}{9}(102 - 32) = \tfrac{5}{9}(70) = \underline{39°C}$$

Example 2-16
On a cold winter day the temperature is −10.0°C. What is the reading on the Fahrenheit scale?

$$°F = \tfrac{9}{5}°C + 32$$
$$= \tfrac{9}{5}(-10.0) + 32 = -18.0 + 32 = \underline{14.0 \ °F}$$

Review Questions

1 Define a measurement.
2 What is meant by precision? How does it differ from accuracy?
3 What is a significant figure?
4 Give the rules regarding zero as a significant figure?
5 What is the rule regarding the number of decimal places in an answer from addition and subtraction?

6 How is a number rounded off?
7 What is the rule regarding the number of significant figures in an answer from multiplication and division?
8 When is a number "exact"? How many significant figures are in an exact number?
9 How is a number expressed in scientific notation?
10 Define mass. How does it differ from weight?

11 Give the basic metric units for mass, length, and volume.

12 What is a relation between metric and the English system for mass? Length? Volume?

13 What is a conversion factor? What is its origin?

14 Discuss the advantages of the unit-factor method of solving problems.

15 Define density. How can it be determined experimentally?

16 When is a solid buoyant in a certain liquid?

17 How does a hydrometer work?

18 What is temperature? What is a thermometer?

19 Give the relationships between the Fahrenheit and the Celsius scales.

Problems

(More difficult problems are marked by an asterisk.)

Significant Figures

2-1 How many significant figures are in each of the following measurements?
(a) 7030 g (b) 4.0 kg (c) 4.01 lb (d) 0.01 ft (e) 4002 m (f) 0.060 hr (g) 8200 km (h) 0.00705 ton

■ 2-2 How many significant figures are in each of the following measurements?
(a) 10.070 in. (b) 0.0023 mm (c) 0.606 ml (d) 0.300 sec (e) 3000 lb (f) 100.0 cm (g) 1.002 kl (h) 2060 yr

Math and Rounding Off

2-3 Round off each of the following numbers to three significant figures:
(a) 15.9994 (b) 1.0080 (c) 0.6654 (d) 4885 (e) 87,550 (f) 0.027225 (g) 301.4

■ 2-4 Round off each of the following numbers to two significant figures:
(a) 18.998 (b) 10.81 (c) 3,650,000 (d) 0.07482 (e) 444 (f) 9750 (g) 6.02 (h) 9500

2-5 Carry out each of the following operations. Express your answer to the proper decimal place.
(a) 14.72 + 0.611 + 173
(b) 0.062 + 11.38 + 1.4578
(c) 1600 − 4 + 700 (d) 47 + 0.91 − 0.286
(e) 0.125 + 0.71

■ 2-6 Carry out each of the following operations. Express your answer to the proper decimal place.
(a) 0.0128 + 0.00102 + 0.00416 (b) 173 + 150 + 122 (c) 15.3 − 1.12 + 3.377 (d) 9300 + 1100 − 188

2-7 Carry out the following calculations. Express your answer to the proper number of significant figures.
(a) 40.0 cm × 3.0 cm (b) 179 ft × 2.20 lb (c) 4.386 cm² ÷ 2 cm (d) (14.65 in. × 0.32 in.) ÷ 2.00 in.

■ 2-8 Carry out the following calculations. Express your answer to the proper number of significant figures.
(a) 1400 m ÷ 0.6 m (b) 0.565 mm³ ÷ (1.62 mm × 0.50 mm) (c) 106.0 ft × 3.0 ft (d) (14.72 cm² × 2 cm) ÷ 5.678 cm

Scientific Notation

2-9 Express the following numbers in scientific notation with one digit to the left of the decimal point in the coefficient.
(a) 157 (b) 0.157 (c) 0.0300 (d) 40,000,000 (two significant figures) (e) 0.0349 (f) 32,000 (g) 32 billion (h) 0.000771 (i) 2340

2-10 Express the following as numbers without powers of 10.
(a) 4.76×10^{-4} (b) 6.55×10^3 (c) 788×10^{-5} (d) 0.489×10^5 (e) 475×10^{-2} (f) 0.0034×10^{-3}

■ 2-11 Change the following numbers to scientific notation with one digit to the left of the decimal point in the coefficient.
(a) 489×10^{-6} (b) 0.456×10^{-4} (c) 0.0078×10^6 (d) 571×10^{-4} (e) 4975×10^5 (f) 0.030×10^{-2}

Conversions Within the Metric System

2-12 Complete the following table:

	mm	cm	m	km
(Example)	108	10.8	0.108	1.08×10^{-4}
(a)	7200			
(b)			56.4	
(c)				0.250

2-13 Complete the following table:

	mg	g	kg
(a)	8900		
(b)		25.7	
(c)			1.25

■ **2-14** Complete the following table:

	ml	Liters	kl
(a)			0.0976
(b)		432	
(c)	45,800		

Conversion Between the English and Metric Systems

2-15 Complete the following table:

	Miles	Feet	m	km
(a)			7800	
(b)	0.450			
(c)		8980		
(d)				6.78

2-16 Complete the following table:

	Gallons	Quarts	Liters
(a)	6.78		
(b)		670	
(c)			7.68×10^3

▪ **2-17**   Complete the following table:

	lb	g	kg
(a)	_____	1980	_____
(b)	178	_____	_____
(c)	_____	_____	4.40

2-18   A punter on a professional football team averaged 28.0 m per kick. What is his average in yards? Should he be kept on the team?

▪ **2-19**   A prospective basketball player is 2.11 m tall and weighs 98,800 g. What are his dimensions in feet and pounds? Will he make the team?

2-20   Gasoline is sold by the liter in Europe. How many gallons does a 55.0-liter gas tank hold?

▪ **2-21**   When the United States changes to the metric system, a football field will be 100 m long. Is this longer or shorter than the 100-yd field? How many yards would a first-and-ten be on the metric field?

2-22   The speed limit is 55 miles/hr. What is the speed limit in kilometers per hour?

2-23   If gold costs \$285/oz, what is the cost of 1.00 kg of gold?

▪ **2-24**   It is 245 miles from St. Louis to Kansas City. How far is this in kilometers?

2-25   The planet Jupiter is about 4.0×10^8 miles from Earth. If radio signals travel at the speed of light, which is 3.0×10^{10} cm/sec, how long does it take a radio command from Earth to reach a Voyager spacecraft passing Jupiter?

2-26   Gasoline sold for \$0.989/gal in 1979. What was the cost per liter? What did it cost to fill an 80.0-liter tank?

★▪ **2-27**   If grapes sell for \$1.15/lb and there are 255 grapes per pound, how many grapes can you buy for \$5.15?

2-28   Using the price of gas from Problem 2-26, how much does it cost to drive 551 miles if your car averages 21.0 miles/gal? How much does it cost to drive 482 km?

2-29   An aspirin contains 0.324 g (5.00 grains) of aspirin. How many pounds of aspirin are in a 500-aspirin bottle?

▪ **2-30**   A certain type of nail costs \$0.95/lb. If there are 145 nails per pound, how many nails can you purchase for \$2.50?

2-31   Another type of nail costs \$0.92/lb, and there are 185 nails per pound. What is the cost of 5670 nails?

★▪ **2-32**   At a speed of 35 miles/hr, how many centimeters will you travel per second?

Density

2-33   A handful of sand weighs 208 g and displaces a volume of 80.0 ml. What is its density?

▪ **2-34**   A 45.5-g quantity of iron has a volume of 5.76 ml. What is its density?

2-35   Referring to Table 2-4, what is the weight of 1.00 liter of gasoline?

2-36   What is the volume in milliliters occupied by 1.00 kg of carbon tetrachloride?

2-37   Pumice is a volcanic rock that contains many trapped air bubbles in the rock. A 155-g sample is found to have a volume of 163 ml. What is the density of pumice? What is the volume of a 4.56-kg sample? Will pumice float or sink in water? In ethyl alcohol?

▪ **2-38**   The density of diamond is 3.51 g/ml. What is the volume of the Hope diamond if it weighs 44.0 carats (1 carat = 0.200 g)?

2-39   A small box is filled with liquid mercury. The dimensions of the box are 3.00 cm wide, 8.50 cm long, and 6.00 cm high. What is the weight of the mercury in the box? (1.00 ml = 1.00 cm³.)

▪ **2-40**   Which has a greater volume, 1 kg of lead or 1 kg of gold?

2-41   Which has a greater weight, 1 liter of gasoline or 1 liter of water?

▪ **2-42**   What is the weight of 1 gal of gasoline in grams? In pounds?

*2-43   Calculate the density of water in pounds per cubic foot (lb/ft³)?

*2-44   In certain stars matter is tremendously compressed. In some cases the density is as high as 2.0 × 10⁷ g/ml. A tablespoon full of this matter is about 4.5 ml. What is this weight in pounds?

Temperature

2-45   The temperature of the water around a nuclear reactor core is about 300°C. What is this temperature in degrees Fahrenheit?

2-46   The temperature on a comfortable day is 76°F. What is this temperature in degrees Celsius?

■ 2-47   The lowest possible temperature is −273°C and is called absolute zero. What is absolute zero on the Fahrenheit scale?

2-48   Mercury thermometers cannot be used in cold arctic climates because mercury freezes at −39°C. What is this temperature in degrees Fahrenheit?

*2-49   At what temperature are the Fahrenheit and Celsius scales equal?

■ 2-50   The coldest temperature recorded on earth was −110°F. What is this temperature in degrees Celsius?

The Structure of Matter, The Nucleus, and Nuclear Reactions

The most basic unit of matter is the atom. The core or nucleus of the atom is a source of energy that can be converted to electricity.

Suppose we took a sample of the element copper and divided it into smaller and smaller pieces? Before the 1800s it was thought that matter was continuous, meaning that it could be divided into infinitely smaller pieces without changing the nature of the element. Around 1803, however, the theory of an English scientist named John Dalton (1766–1844) gained acceptance. The nature of matter and the manner in which elements combined to form compounds suggested that there was a limit to how far an element could be subdivided. We now accept that if we divide our sample of copper into smaller and smaller pieces eventually we would find a basic unit that could not be divided further without changing the nature of the element. This basic unit is called the atom. *An* **atom** *is the smallest particle of an element that can exist and still have the properties of that element.*

In this chapter we have three goals. First, we will examine the nature of the atom as it exists in elements and compounds. Then we will take a closer look at the atom itself in order to understand its internal structure. Finally (in an optional section), we will discuss how the interior of the atom, known as its nucleus, undergoes changes that are important yet controversial in our modern world.

The first thing we must realize is that atoms are extremely small. Since the diameter of an atom is on the order of 10^{-8} cm, it would take 100 million atoms in a line to extend for 1 cm (less than half an inch). Even the most powerful microscope now available can produce only fuzzy images of some of the larger atoms. However, let's not let the present limits of technology confine our vision. Let's assume that we have for our exclusive use an "atomscope" from the distant future. This "atomscope" magnifies samples of pure matter so that atoms can be seen as small spheres. Later we will turn up the power of the "atomscope" so that we can look inside the atom itself and see why the atoms of one element differ from those of another.

3-1 The Basic Structures of the Elements

The element that we discussed in the introduction to this chapter was copper. Let's take a look at a sample of copper wire under our "atomscope" from the future. To the naked eye, copper wire appears shiny, solid, and continuous. Under the "atomscope" a different picture appears. The element is indeed composed of small spherical atoms all of which appear to be identical. This phenomenon is much like a brick wall—it looks completely featureless and continuous from a distance, but up close it is obvious that the wall is constructed of basic units (bricks). In the wire, the atoms of copper are firmly packed together as would be any hard spheres of the same size such as marbles. (See Figure 3-1.)

All elements do not exist in nature as simple congregations of individual

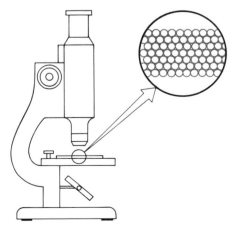

Figure 3-1 THE BASIC STRUCTURE OF COPPER. Most elements exist in nature as closely packed atoms.

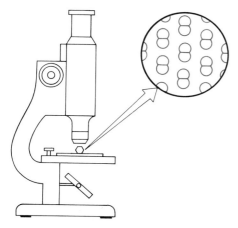

Figure 3-2 THE BASIC STRUCTURE OF IODINE. Some elements exist in nature as two or more atoms joined together.

atoms as in copper. In some elements the basic unit is composed of molecules. *A* **molecule** *is a group of two or more atoms held together by a force called a* **chemical bond.**

Iodine is an example of an element whose basic structure is composed of molecules. In Figure 3-2 we have placed a sample of solid iodine under the "atomscope." We see that the atoms of iodine exist in pairs, with the two spherical atoms overlapping to some extent. The two atoms in the molecule are obviously more intimately connected than the atoms between separate molecules that just touch. This overlap of atoms is the result of a chemical bond between the two atoms. We will have much more to say about the nature of the chemical bond in Chapter 6.

Examples of other elements that exist as diatomic (two-atom) molecules under normal conditions such as are found on the surface of the earth are oxygen, nitrogen, hydrogen, fluorine, chlorine, and bromine. Phosphorus consists of molecules composed of four atoms, and sulfur consists of molecules composed of eight atoms.

3-2 Compounds and Formulas

The molecules of an element are composed of identical atoms. Of far more significance to the chemist is the fact that atoms of different elements can also combine to form molecules. *The basic structure of most compounds is that of molecules made up of atoms of two or more different elements.* For example, water is a compound composed of molecules. Each water molecule is composed of two atoms of hydrogen joined to one atom of oxygen by chemical bonds (see Figure 3-3).

A molecule is symbolized by the atoms of which it is composed. This is

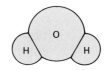

Figure 3-3 A WATER MOLECULE. A water molecule is composed of two hydrogen atoms bound to one oxygen atom.

called the **formula** *of the compound or element.* The familiar formula for water is therefore

$$H_2O$$

This is, of course, pronounced H–two–O. You should notice that the "2" is written as a *subscript,* indicating that the molecule has two hydrogen atoms. When there is only one atom of a certain element present (e.g., oxygen), no subscript is written after the element.

Each unique chemical compound has a unique formula or arrangement of atoms in the molecule. For example, another compound exists whose molecules are composed of only hydrogen and oxygen and is called hydrogen peroxide. Since the molecules of hydrogen peroxide are composed of two hydrogen atoms and two oxygen atoms, its formula is H_2O_2 and its properties are distinctly different than those of water (H_2O). In the previous section, elemental iodine has the formula I_2. Ordinary table sugar is a compound whose molecules are composed of 45 atoms. The formula for table sugar (sucrose) is $C_{12}H_{22}O_{11}$.★

See Problems 3-1, 3-2

3-3 Ions and Ionic Compounds

Ordinary table salt is composed almost entirely of sodium chloride, which has the formula NaCl. This compound has a basic unit that is profoundly different from the example of H_2O. In this case, the atoms in the compound have an electrical charge. *Atoms or groups of atoms that have an electrical charge are known as* **ions.** *Positive ions are known as* **cations** *and negative ions are known as* **anions.** In NaCl the sodium exists as a cation with a single positive charge and the chlorine as an anion with a single negative charge. This is illustrated below with a + or − as a superscript to the right of the element.

$$Na^+Cl^-$$

Let's look at a sample of NaCl under the "atomscope." This is shown in Figure 3-4.

In this case, notice that there are spheres of two different sizes. The spheres do not overlap as in the case of molecules like H_2O and I_2. In fact, the structure is held together because the Na^+ ions (the small spheres) are surrounded by Cl^- ions (the large spheres) and vice versa. Since charges of opposite sign attract, the compound is held firmly together. *Forces of attraction*

★ Molecules can be very complex; some are composed of millions of atoms. The atoms in these molecules are sometimes arranged in continuous chains, as in DNA or in plastics such as polyethylene. In other substances, such as quartz, silicon and oxygen atoms are bound together in a continuous three-dimensional arrangement that extends throughout the whole crystal.

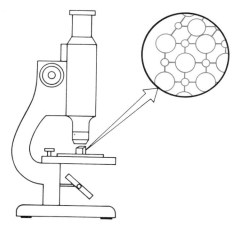

Figure 3-4 THE BASIC STRUCTURE OF SODIUM CHLORIDE. An ionic compound exists as charged atoms or groups of atoms.

(between opposite charges) and forces of repulsion (between like charges) are referred to as **electrostatic forces.** *Compounds such as NaCl in which ions are present are referred to as* **ionic compounds.**

Certain formulas at first may appear more complex than necessary. For example, barium nitrite has the formula

$$Ba(NO_2)_2$$

instead of the simpler BaN_2O_4. In this case, the NO_2^- exists as an ion and is a single entity. The formula tells us that two NO_2^- ions are present for each Ba^{2+} ion. The NO_2^- ion is an example of a group of atoms that form an ion.

In Figure 3-5, the discussion in Sections 3-1 through 3-3 has been summarized. Under normal conditions found on the earth, matter is composed of either *atoms, molecules, or ions. The formula of a compound represents either a discrete molecular unit in the case of molecules or a ratio of ions in the case of ionic compounds.* Eventually, we will discuss why some elements are composed of molecules and why one compound is molecular whereas another is ionic. This is the subject of Chapter 6.

3-4 The Structure of the Atom

From Earth with the naked eye, the moon looks like a solid, round sphere. Through a telescope, however, the moon is seen to be complex, with mountains, valleys, and large craters. In the previous section the atom was also pictured as a featureless sphere. To see the details of the atom, let's again employ the "atomscope" with the magnification turned up as high as it will go. What follows is a simplification of how scientists currently view the atom. How this picture evolved is an intriguing story beginning in the late 1800s and continuing to the present day. Many famous scientists were involved in this story. Scientists by the names of Thomson, Rutherford, Becquerel,

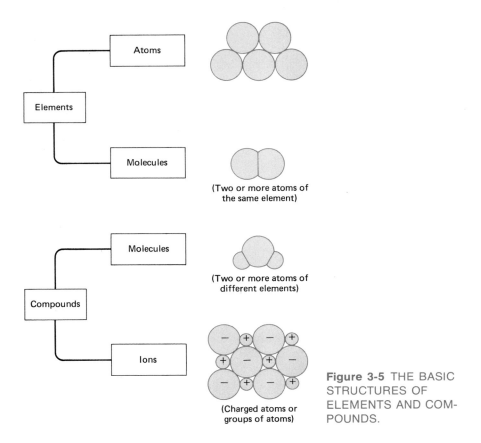

Figure 3-5 THE BASIC STRUCTURES OF ELEMENTS AND COMPOUNDS.

Curie, and Roentgen, among others, all had a hand in the development of this model known as the nuclear concept of the atom.

In a close look into the atom, the first thing to be noticed is that the atom is not solid as it first appeared. In fact, the atom is mostly empty space. Most of this vast space of the atom is the home of small particles called electrons. Electrons were the first particles in the atom to be identified. *In 1897, J. J. Thomson characterized the* **electron** *by noting that it had a negative charge and was common to the atoms of all elements.* In 1911 the English scientist, Lord Ernest Rutherford, proved by his experiments that *these electrons exist outside of a small dense core at the center of the atom called the* **nucleus.** The nucleus is so small compared to the total volume of a typical atom that if it were expanded to the size of a softball, the radius of the atom would extend for about 1 mile. Despite the small size of the nucleus, it contains almost all of the mass of the atom. *The nucleus is composed of particles called* **nucleons.** *There are two types of nucleons,* **protons,** *which have a positive charge, and* **neutrons,** *which do not carry a charge.* (See Figure 3-6.)

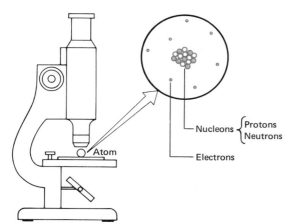

Figure 3-6 THE BASIC STRUC-
TURE OF THE ATOM. The
atom is composed of electrons,
neutrons, and protons.

The data on the particles in the atom is summarized in Table 3-1. *The unit of weight,* **amu,** *is an abbreviation of* **atomic mass units** *and is approximately the weight of one proton (or one neutron).* The origin of this unit of weight will be discussed later.

At this point in the discussion it would appear that the atom is composed of just these three particles. In recent years, however, the picture of the atom has become much more complex, as physicists are now looking into the basic structure of neutrons and protons. The search for what is considered to be *the* fundamental particles of nature continues with the discovery of nuclear particles with names such as quarks, pions, and gluons. With each discovery the picture of the nucleus becomes increasingly complex. Fortunately, the three-particle model of the atom still meets the needs of the chemist.

3-5 Atomic Number, Mass Number, and Atomic Weight

The simplest of all atoms is a hydrogen atom. Most neutral hydrogen atoms are made up of a nucleus with one proton (with a positive charge of $+1$); this is balanced by one electron outside of the nucleus with a charge of -1. *The number of protons in the nucleus of an atom is referred to as its* **atomic number.** *The atomic number is what distinguishes the atoms of one element from another.* The atomic number of hydrogen is therefore 1.

Table 3-1 Atomic Particles

Name	Symbol	Electrical Charge	Mass (amu)	Mass (g)
Electron	e	-1	0.000549	9.110×10^{-28}
Proton	p	$+1$	1.00728	1.673×10^{-24}
Neutron	n	0	1.00867	1.675×10^{-24}

Another form of hydrogen (called deuterium) exists in nature that contains one neutron in addition to the proton. Obviously, the mass of deuterium, with two nucleons, is different than that of the common form of hydrogen. *Atoms having the same atomic number (number of protons) but different* **mass numbers** *(number of nucleons) are known as* **isotopes.** A third isotope of hydrogen (called tritium) exists that contains two neutrons as well as the one proton. Tritium is a synthetic isotope that is unstable. This phenomenon is discussed later in this chapter.

Isotopes of elements are represented by showing the mass number in the upper left-hand corner of the symbol of the element. Often, the atomic number of the element is represented in the lower left-hand corner. This is strictly a convenience, however, since the atomic number determines the particular element. For example, if the atomic number is anything greater than 1, the element is not hydrogen. The three isotopes of neutral hydrogen atoms are represented in Figure 3-7.

Example 3-1

How many protons, neutrons, and electrons are present in $^{90}_{38}\text{Sr}$?

Atomic number = number of protons = $\underline{\underline{38}}$

Number of neutrons = mass number − number of protons

$90 - 38 = \underline{\underline{52}}$

In a neutral atom the number of protons and electrons is the same.

Number of electrons = $\underline{\underline{38}}$

As mentioned earlier, some atoms are present in compounds in the form of ions, which are charged atoms. An ion has a charge because the atom in question does not have an equal number of protons and electrons. *Anions have more electrons than the neutral atom (one extra electron for each negative charge) and cations have less electrons than the neutral atom (one less electron for each positive charge).* Therefore, the charge on an ion is equal to the sum of the charges of all of the electrons and the charges of all of the protons. The charge is indicated in the upper right-hand corner.

$^{1}_{1}\text{H}$

One proton
One electron

$^{2}_{1}\text{H}$ (deuterium)

One proton
One neutron
One electron

$^{3}_{1}\text{H}$ (tritium)

One proton
Two neutrons
One electron

Figure 3-7 THE ISOTOPES OF HYDROGEN. Isotopes of an element are characterized by different numbers of neutrons in the nucleus.

Example 3-2

How many protons, neutrons, and electrons are present in $^{79}_{34}\text{Se}^{2-}$?

$$\text{Number of protons} = \underline{\underline{34}}$$

$$\text{Number of neutrons} = 79 - 34 = \underline{\underline{45}}$$

The superscript 2− on the right indicates that this is an ion with a charge of 2−. This tells us that there are *two* more electrons than in a neutral atom. Therefore, there are

$$34 + 2 = \underline{\underline{36}} \text{ electrons}$$

See Problem 3-3

The mass number of an isotope is a convenient but imprecise measure of its comparative weight. A more precise measure of the weight of one isotope relative to another is known as the atomic weight. *The* **atomic weight** *of an isotope is determined by comparison to a standard. This standard is* ^{12}C*, which is* defined *as having an atomic weight of exactly 12 amu.* By comparison to ^{12}C, the atomic weight of ^{10}B is 10.013 amu and that of ^{11}B is 11.009 amu. Boron and all other naturally occurring elements are found as mixtures of isotopes. Thus the atomic weight of an element found in nature is the average of the atomic weights of all of the isotopes present. The average atomic weight is calculated like any other average weight, as illustrated by the following examples.

Example 3-3

A sack of coins is found to be 45.5% quarters, 31.2% dimes, and 23.4% nickels. If a quarter weighs 5.55 g, a dime 2.24 g, and a nickel 4.98 g, what is the average weight of a coin in the sack?

Procedure:

The contribution to the average weight of each type of coin is the percent in decimal form times the weight of the coin.

Quarter	0.455×5.55 g $=$	2.525 g
Dime	0.312×2.24 g $=$	0.699 g
Nickel	0.234×4.98 g $=$	1.165 g
		4.389 g $= \underline{\underline{4.39 \text{ g}}}$

Example 3-4

In nature boron is present as 19.9% $^{10}_{5}\text{B}$ and 80.1% $^{11}_{5}\text{B}$. If the atomic weight of $^{10}_{5}\text{B}$ is 10.013 amu and that of $^{11}_{5}\text{B}$ is 11.009 amu, what is the average atomic weight of boron?

$^{10}_{5}\text{B}$	0.199×10.013 amu $=$	1.99 amu
$^{11}_{5}\text{B}$	0.801×11.009 amu $=$	8.82 amu
		10.81 amu

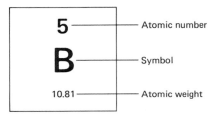

Figure 3-8 ATOMIC WEIGHT AND ATOMIC NUMBER FROM THE PERIODIC TABLE.

Thus we find the atomic weight of boron listed in the periodic table as 10.81 amu, which is the weighted average of the atomic weights of the two boron isotopes found in nature (see Figure 3-8).

The elements with whole-number atomic weights in parentheses are synthetic elements and do not occur in nature. These elements undergo nuclear disintegration as discussed in the next section. The atomic weight represents the most stable isotope of the element made.

See Problems 3-4, 3-5, 3-6, 3-7

(*Note:* The material in the remainder of this chapter concerns a more in-depth study of the nucleus of the atom. It is not directly relevant to the chapters that follow. It may be covered now as a logical extension of the nature of the nucleus, omitted completely, or covered later as desired.)

3-6 Unstable Nuclei

In 1896 a French scientist by the name of Henri Becquerel discovered (quite by accident) that one element can spontaneously change into another. This was the process that many an alchemist in the middle ages had hoped to promote. The actual process was far different than what the alchemist expected, however. It was eventually discovered that the major isotope of uranium, $^{238}_{92}U$, can spontaneously (without warning or stimulation) emit from its nucleus a helium nucleus. This occurrence can be symbolized by a nuclear equation. *A* **nuclear equation** *shows the change from the original nucleus or nuclei on the left of the arrow to the product nuclei on the right of the arrow.*

$$^{238}_{92}U \longrightarrow {}^{234}_{90}Th + {}^{4}_{2}He$$

Notice that the loss of four nucleons from the original (parent) nucleus leaves the remaining (daughter) nucleus with 234 nucleons ($238 - 4 = 234$); the loss of the two protons leaves 90 protons in the daughter nucleus ($92 - 2 = 90$). Since a nucleus with 90 protons is now an isotope of thorium, one element has indeed changed into another.

The helium nucleus, when emitted from another nucleus, is called an **alpha (α) particle** *(see Figure 3-9). The process of emitting energy or particles from a nucleus is known as* **radioactive decay** *or simply* **radioactivity**. *The particles or "rays" that are emitted are called* **radiation**. *The radioactive isotopes that undergo decay are sometimes referred to as* **radionuclides**.

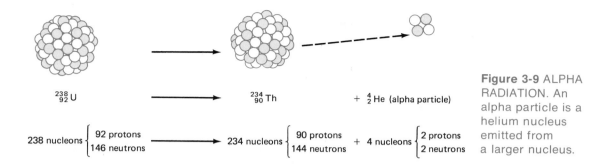

Figure 3-9 ALPHA RADIATION. An alpha particle is a helium nucleus emitted from a larger nucleus.

A second form of radioactive decay involves the emission of an electron from the nucleus. *An electron emitted* from a nucleus *is known as a* **beta (β) particle.** The following equation illustrates beta particle emission:

$$^{131}_{53}\text{I} \longrightarrow ^{131}_{54}\text{Xe} + ^{0}_{-1}e$$

(See Figure 3-10.) Notice that the loss of a beta particle from a nucleus has the effect of changing a neutron in the nucleus into a proton, which increases the atomic number by one but leaves the mass number (number of nucleons) unchanged.

A *third type of radioactive decay involves the emission of a high-energy form of light called a* **gamma (γ) ray.** Like all light, gamma rays travel at 3.0×10^{10} cm/sec (186,000 miles/sec). Light as a form of energy will be discussed in the next chapter. This type of radiation may occur alone or in combination with alpha or beta radiation as discussed above. The following equation illustrates gamma ray emission:

$$^{60}_{27}\text{Co}^\star \longrightarrow ^{60}_{27}\text{Co} + \gamma$$

(* means nucleus is in a high-energy state)

Notice that gamma radiation by itself does not result in a change in the nucleus. Excess energy contained in a nucleus has simply been released, much as burning coal releases chemical energy.

See Problems 3-8, 3-9

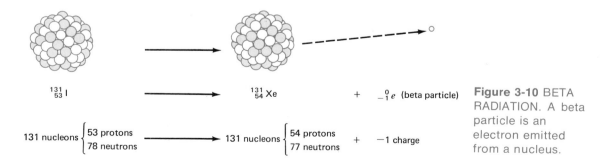

Figure 3-10 BETA RADIATION. A beta particle is an electron emitted from a nucleus.

Example 3-5
Complete the following nuclear equations:

(a) $^{218}_{84}\text{Po} \rightarrow$ _____ $+ ^{4}_{2}\text{He}$
(b) $^{210}_{81}\text{Tl} \rightarrow ^{210}_{82}\text{Pb} +$ _____

Procedure:
For the missing isotope or particle:

1 Find the number of nucleons; they are equal on both sides of the equation.

2 Find the charge or atomic number; this also is equal on both sides of the equation.

3 If what's missing is the isotope of an element, find the symbol of the element that matches the atomic number from the listing of the elements inside the front cover.

Solution:
(a) Nucleons: $218 = x + 4$, so $x = 214$.
     Atomic number: $84 = y + 2$, so $y = 82$.
From the inside cover, the element having an atomic number of 82 is Pb. The isotope is

$$^{214}_{82}\text{Pb}$$

(b) Nucleons: $210 = 210 + x$, so $x = 0$.
     Atomic number: $81 = 82 + y$, so $y = -1$.
An electron or beta particle has negligible mass (compared to a nucleon) and a charge of -1.

$$^{0}_{-1}e$$

3-7 Rates of Decay of Radioactive Isotopes

The isotopes of all elements heavier than $^{209}_{83}\text{Bi}$ are radioactive. If this is true, how can elements such as $^{238}_{92}\text{U}$ occur naturally in the earth? The answer is that many of these isotopes have very long "half-lives." *The* **half-life** *($t_{1/2}$) of a radioactive isotope is the time required for one-half of a given sample to decay.* Each radioactive isotope has a specific and constant half-life. For example, $^{238}_{92}\text{U}$ has a half-life of 4.5×10^{9} years. Since that is roughly the age of the earth, about one-half of the $^{238}_{92}\text{U}$ originally present when the earth formed from hot gases and dust is still present. Half of what is now present will be gone in another 4.5 billion years. The half-lives and mode of decay of some radioactive isotopes are listed in Table 3-2.

Notice in Table 3-2 that when $^{238}_{92}\text{U}$ decays it forms an isotope ($^{234}_{90}\text{Th}$) that also decays (very rapidly compared to $^{238}_{92}\text{U}$). The $^{234}_{91}\text{Pa}$ formed from the decay of $^{234}_{90}\text{Th}$ also decays, and so forth until finally the stable isotope $^{206}_{82}\text{Pb}$ is

Table 3-2 Half-lives

Isotope	$t_{1/2}$	Mode of Decay	Product
$^{238}_{92}U$	4.5×10^9 years	α, γ	$^{234}_{90}Th$
$^{234}_{90}Th$	24.1 days	β	$^{234}_{91}Pa$
$^{226}_{88}Ra$	1620 years	α	$^{222}_{86}Rn$
$^{14}_{6}C$	5760 years	β	$^{14}_{7}N$
$^{131}_{53}I$	8.0 days	β, γ	$^{131}_{54}Xe$
$^{218}_{85}At$	1.3 sec	α	$^{214}_{83}Bi$

formed. This is known as the $^{238}_{92}U$ radioactive decay series. *A* **radioactive decay series** *starts with a naturally occurring radioactive isotope with a half-life near the age of the earth* (if it was very much shorter, there wouldn't be any left). *The series ends with a stable isotope.* There are two other naturally occurring decay series: the $^{235}_{92}U$ series, which ends with $^{207}_{82}Pb$; and the $^{232}_{90}Th$ series, which ends with $^{208}_{82}Pb$. As a result of the $^{238}_{92}U$ decay series, where uranium is found in rocks we also find other radioactive isotopes as well as lead. In fact, by examining the ratio of $^{238}_{92}U/^{206}_{82}Pb$ in a rock, its age can be determined. For example, a rock from the moon showed about half of the original $^{238}_{92}U$ had decayed to $^{206}_{82}Pb$. This meant that the rock was about 4.5×10^9 years old.

Example 3-6

If we started with 4.0 mg of $^{14}_{6}C$, how long would it take until only 0.50 mg remained?

Solution:

$$\text{After 5,760 years,} \quad \frac{1}{2} \times 4.0 \text{ mg} = 2.0 \text{ mg remaining}$$

$$\text{After another 5,760 years,} \quad \frac{1}{2} \times 2.0 \text{ mg} = 1.0 \text{ mg remaining}$$

$$\text{After another } \frac{5,760 \text{ years}}{17,280 \text{ years}}, \frac{1}{2} \times 1.0 \text{ mg} = 0.50 \text{ mg remaining}$$

Therefore, in 17,280 years, 0.50 mg remains.

3-8 The Effects of Radiation

We mentioned in Chapter 1 that energy-rich compounds undergo chemical reactions to produce energy-poor compounds, releasing the chemical energy as heat. Radioactive isotopes can be considered in a similar manner. When the nucleus of an isotope is radioactive, it is considered to be an energy-rich nucleus. When it decays the radiation carries away this excess energy from the nucleus by means of a high-energy particle (alpha or beta) or high-energy light (gamma). This nuclear energy carried away by the particles or rays may also be converted to heat energy. In fact, much of the internal heat generated in the interior of the earth and other planets is a direct result of natural radiation from unstable elements. The presence of radioactive elements in wastes from nuclear reactors make these materials extremely hot so that they must be continuously cooled to avoid melting.

Radiation also affects matter when the high-energy particles or rays penetrate. *The radiation changes neutral molecular compounds into ions in a process known as* **ionization.** For example, if high-energy particles collide with an H_2O molecule, an electron may be removed from the molecule leaving an H_2O^+ ion behind. The properties of the ion are distinctly different than those of the neutral molecule. If the molecule in question is a large complex molecule that is part of a cell of a living system, the ionization causes damage or even eventual destruction of the cell. As shown in Figure 3-11, an alpha particle causes the most ionization and is the most destructive along its path. However, these particles do not penetrate matter to any extent and can be stopped even by a piece of paper. The danger of alpha emitters (such as uranium and plutonium) is that they can be ingested through food or inhaled into the lungs where these heavy elements tend to accumulate in bones. There, in intimate contact with the blood-producing cells of the bone marrow, they slowly do damage. Ultimately, the radiation can cause certain cells to change into abnormal cells that reproduce rapidly. These are the dreaded cancer cells such as found with leukemia.

Beta radiation is less ionizing than alpha radiation but is more penetrating. Still, a couple of inches of solid or liquid matter absorbs beta radiation. This type of radiation can cause damage to surface tissue such as skin and eyes but does not reach internal organs unless ingested. The most damaging radiation is gamma radiation. Although much less ionizing along its path than the others, it has tremendous penetrating power. Several feet of concrete or blocks of lead are needed for protection from gamma radiation.

The damage done to cells by gamma radiation is cumulative. That means small doses over a long period of time can be as harmful as one large dose at once. That is why the dosage of radiation absorbed by those who work

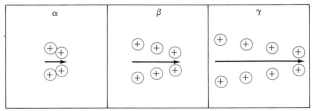

α	β	γ
High ionization but low penetration	Moderate ionization but low penetration	Low ionization but very high penetration

Figure 3-11 IONIZATION AND PENETRATION. Gamma radiation is the most damaging of the three types of radiation.

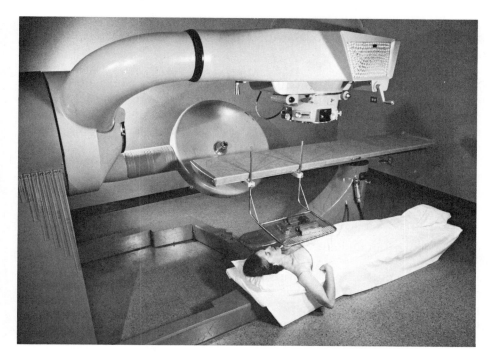

Figure 3-12 ^{60}Co RADIATION TREATMENT. A beam of gamma rays can be used to destroy cancerous tissue.

in radioactive environments such as a nuclear research laboratory must be carefully and continually monitored. If one receives too much radiation over a specified period, that person must be removed from additional exposure for a length of time.

Isotopes that emit gamma radiation (e.g., $^{60}_{27}Co$) can be useful as well as destructive because of their penetrating rays. The gamma radiation of $^{60}_{27}Co$ can be narrowly focused like a beam of light to destroy cancer cells in localized tumors located deep within the body (see Figure 3-12). Healthy cells are destroyed as well in the process, so this form of cancer treatment has many undesireable side effects. The purpose is to destroy or damage more cancer cells than normal cells. The normal cells can recover faster than the damaged abnormal cells.

See Problems 3-10, 3-11, 3-12, 3-13

3-9 Nuclear Reactions

Transmutation *of an element is the conversion of that element into another.* We have discussed how this occurs naturally by the spontaneous emission of an alpha or beta particle from a nucleus. Transmutation can also occur by artificial or synthetic means. The first example of transmutation was given by Ernest Rutherford in 1919. (Rutherford had earlier proposed the nuclear model of the atom.) Rutherford found that $^{14}_{7}N$ atoms could be bombarded with alpha particles, causing a nuclear reaction that produced $^{17}_8O$ and a proton. This nuclear reaction is illustrated by the nuclear equation

$$^{14}_7N + {}^4_2He \longrightarrow {}^{17}_8O + {}^1_1H$$

Notice that the total number of nucleons is conserved by the reaction [14 + 4 (on the left) = 17 + 1 (on the right)]. The total charge is also conserved during the reaction [7 + 2 (on the left) = 8 + 1 (on the right)]. Both the number of nucleons and the total charge must be balanced (the same on both sides of the equation) in a nuclear reaction.

Since the time of Rutherford, thousands of isotopes have been prepared by nuclear reactions. Two other examples are

$$^{209}_{83}Bi + {}^2_1H \longrightarrow {}^{210}_{84}Po + {}^1_0n$$

$$^{27}_{13}Al + {}^4_2He \longrightarrow {}^{30}_{15}P + {}^1_0n$$

(^2_1H is a deuterium nucleus or a deuteron, ^1_0n is a neutron).

Since the nucleus has a positive charge and many of the bombarding nuclei (except neutrons) also have a positive charge (e.g., ^4_2He, ^1_1H, ^2_1H), the particles must have a high energy (velocity) in order to overcome the natural repulsion between these two like charges. From the collision of the nucleus and the particle, a high-energy nucleus is formed. As a result, another particle (usually a neutron) is then emitted by the nucleus to carry away the excess energy from the collision. This is analogous to two cars in a head-on collision where fenders, doors, and bumpers come flying out after impact. This is illustrated in Figure 3-13.

The invention of particle accelerators, which accelerate particles to high velocities and energies, has opened up vast possibilities for artificial nuclear reactions. These accelerators can be large, expensive devices. For example, the linear accelerator at Stanford University is 2 miles long. It was completed in 1961 at a cost of $115 million. (See Figure 3-14.)

One interesting development from particle accelerators has been the

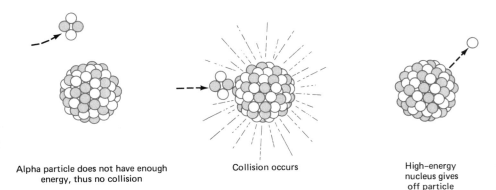

Figure 3-13 COLLISION OF AN α PARTICLE WITH A NUCLEUS. Alpha particles must have sufficient energy to overcome the repulsion of the target nucleus in order for a collision to take place.

Alpha particle does not have enough energy, thus no collision

Collision occurs

High–energy nucleus gives off particle

Figure 3-14 THE STANFORD LINEAR ACCELERATOR. In this device, charged particles are accelerated to extremely high velocities.

preparation of new, heavy elements. In fact, all elements between neptunium (atomic number 93) and the last element made (atomic number 106, which has not yet been named) are synthetic elements. The following nuclear reactions illustrate the formation of heavy elements:

$$^{238}_{92}U + ^{12}_{6}C \longrightarrow ^{244}_{98}Cf + 6^{1}_{0}n$$

$$^{238}_{92}U + ^{16}_{8}O \longrightarrow ^{250}_{100}Fm + 4^{1}_{0}n$$

Example 3-7

Complete the following nuclear reactions:

(a) $^{9}_{4}Be + ^{2}_{1}H \rightarrow$ _____ $+ ^{1}_{0}n$
(b) $^{252}_{98}Cf + ^{10}_{5}B \rightarrow$ _____ $+ 5^{1}_{0}n$

Procedure:

Same as Example 3-5.
(a) Nucleons: $9 + 2 = x + 1$, so $x = 10$.
  Atomic number: $4 + 1 = y + 0$, so $y = 5$.
From a listing of the elements, we find that the element is boron (atomic number 5). The isotope is

$$\underline{^{10}_{5}B}$$

(b) Nucleons: $252 + 10 = x + (5 \times 1)$, so $x = 257$.
  Atomic number: $98 + 5 = y + (5 \times 0)$, so $y = 103$.
From the listing of elements, the isotope is

$$\underline{^{257}_{103}Lr}$$

Many nuclear reactions form isotopes that are themselves radioactive. For example, ^{60}Co is prepared as follows:

$$^{59}_{27}\text{Co} + ^{1}_{0}n \longrightarrow ^{60}_{27}\text{Co}\star$$

This type of procedure (called neutron activation) is used to prepare many radioactive isotopes whose naturally occurring isotopes are stable. There has been a great deal of application of these synthetic radioactive isotopes in the field of medicine, mainly in the area of diagnosis of diseases. In fact, in 1976 there were 10 million diagnostic tests involving the use of radioactive isotopes.

3-10 Nuclear Fission

In 1938 two German scientists, Otto Hahn and Fritz Strassmann, first explained some strange and unexpected results of a nuclear reaction. When naturally occurring uranium (99.3% ^{238}U, 0.7% ^{235}U) was bombarded with neutrons, it was expected that elements with a higher atomic number than uranium could be synthesized as follows:

$$^{238}_{92}\text{U} + ^{1}_{0}n \longrightarrow ^{239}_{92}\text{U} \longrightarrow ^{239}_{93}\text{X} + ^{0}_{-1}e$$

Indeed, this reaction seemed to occur, but of more importance was the discovery by these scientists of isotopes present that were about half of the mass of the uranium atoms (e.g., $^{139}_{56}$Ba, $^{90}_{38}$Sr). The German scientists were able to show that the smaller atoms resulted from the reaction of ^{235}U and not the more abundant ^{238}U. The following represents a typical nuclear reaction that occurs when ^{235}U absorbs a neutron:

$$^{235}_{92}\text{U} + ^{1}_{0}n \longrightarrow ^{139}_{56}\text{Ba} + ^{94}_{36}\text{Kr} + 3^{1}_{0}n$$

This type of nuclear reaction is called fission. **Fission** *is the splitting of a large nucleus into two smaller nuclei of similar size* (see Figure 3-15).

There were two points about this reaction that had monumental consequences for the world. Scientists in Europe and America were quick to grasp the meaning in a world about to go to war.

1   Calculation of the weights of the product nuclei compared with the original nuclei indicated that a significant mass loss occurs in the reaction. According to Einstein's equation (see Section 1-12), this mass loss must be converted to a tremendous amount of energy. Fission of a few kilograms of ^{235}U could produce energy equivalent to tens of thousands of tons of the conventional explosive, TNT.

2   What made the rapid fission of a large sample of ^{235}U feasible was the potential for a chain reaction. *A **nuclear chain reaction** is a reaction that is self-sustaining. The reaction generates the means to trigger additional reactions.* Notice in Figure 3-15 that the reaction of one original neutron

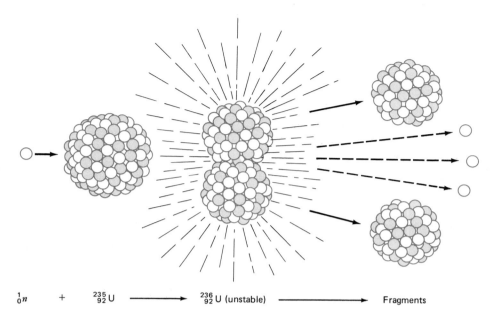

$$^{1}_{0}n \quad + \quad ^{235}_{92}U \quad \longrightarrow \quad ^{236}_{92}U \text{ (unstable)} \quad \longrightarrow \quad \text{Fragments}$$

Figure 3-15 FISSION. In fission, one neutron produces two fragments of the original atom and an average of three neutrons.

caused three to be released. These three neutrons cause the release of nine neutrons, and so forth as illustrated in Figure 3-16. If a certain densely packed "critical mass" of ^{235}U is present, the whole mass of uranium can undergo fission in an instant with a quick release of the energy in the form of radiation.

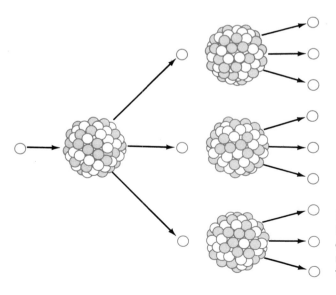

Figure 3-16 A CHAIN REACTION. The fission of one nucleus produces neutrons which induce fission in other nuclei.

Unfortunately, the world thus entered the nuclear age in pursuit of a bomb. After a massive but secret effort, the first nuclear bomb was exploded over Alamogardo Flats in New Mexico on July 16, 1945. This bomb and the bomb exploded over Nagasaki, Japan, was made of ^{239}Pu, which is a synthetic fissionable isotope. The bomb exploded over Hiroshima, Japan, was made of ^{235}U.

This enormously destructive device, however, can be tamed. The chain reaction can be controlled by absorbing excess neutrons with cadmium bars. A typical nuclear reactor is illustrated in Figure 3-17. In a reactor core the uranium in the form of pellets is encased in long rods called fuel elements. Cadmium bars are raised and lowered among the fuel elements to control the rate of fission by absorbing neutrons. If the cadmium bars are lowered all of the way the fission process can be halted all together. In normal operation, the bars are raised just enough that the fission reaction occurs at the desired level. Energy released by the fission and the decay of the radioactive products formed from the fission is used to heat water which circulates among the fuel elements. This water, which is called the primary coolant, becomes very hot (about 300°C) because it is under high pressure. The water heats a second source of water changing it to steam. The steam from the secondary coolant

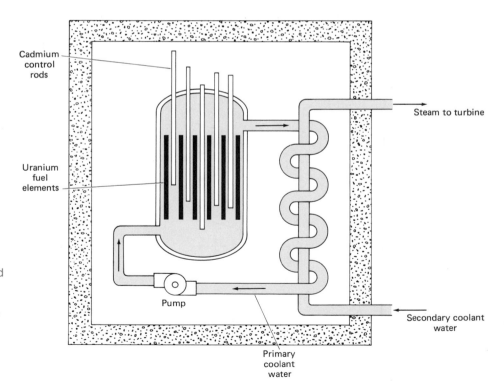

Figure 3-17 A NUCLEAR REACTOR. Nuclear energy is converted to heat energy, the heat energy to mechanical energy and the mechanical energy is converted to electrical energy.

is cycled outside of the containment building where it is used to turn a turbine which generates electricity.

Nuclear energy is a comparatively cheap source of power in the age of the "energy crisis." It does not deplete the limited supply of fossil fuels (natural gas, coal, and oil) and does not pollute the air. After years of use there have been few accidents and no loss of life from the use of commercial reactors. Proponents of nuclear power feel that adequate safeguards and backup systems are available to prevent a catastropic accident.

The disadvantages have recently been well publicized because of the accident at Three Mile Island in March 1979. A series of mechanical failures compounded by human error led to a loss of the primary coolant water. The loss of coolant water allowed the nuclear core to overheat, causing an emergency. Although there was some loss of airborne radiation, the main danger of a "melt down" was averted.

A "melt down" would occur if the temperature of the core exceeded 3000°C, the melting point of the uranium. Theoretically, the mass of molten uranium together with all of the highly radioactive decay products could accumulate on the floor of the reactor, melt through the many feet of protective concrete of the containment building, and eventually reach the ground water. If this happened, vast amounts of deadly radioactive wastes could be released into the environment.

Another disadvantage of nuclear reactors involves the used or spent fuel elements and the highly radioactive primary coolant water. Eventually, fuel elements must be replaced, as the buildup of radioactive isotopes from the fission of ^{235}U hinders the fission process. These fuel elements will remain highly radioactive for approximately 1000 years. The problem of processing, disposal, and transportation of wastes has not yet been solved to the extent necessary to handle the large number of nuclear power plants now in operation.

See Problems 3-14, 3-15, 3-16

3-11 Nuclear Fusion

Soon after the process of nuclear fission was demonstrated, an even more powerful type of nuclear reaction was shown to be theoretically possible. *This reaction involved the* **fusion** *or bringing together of two small nuclei.* An example of a fusion reaction is

$$^3_1H + {}^2_1H \longrightarrow {}^4_2He + {}^1_0n + \text{energy}$$

(^3_1H is called tritium; it is a radioactive isotope of hydrogen).

As in the fission process, a significant mass loss converts to energy. Fusion energy is the origin of almost all of our energy, since it powers the sun. Millions of tons of matter are converted to energy in the sun every second. Because of its large mass, however, the sun contains enough hydrogen to "burn" for additional billions of years.

The principle of fusion was first demonstrated on this planet with a tremendously destructive device called the hydrogen bomb. This bomb can be more than 1000 times more powerful than the atomic bomb, which uses the fission process.

Fission can be controlled, but what about fusion? This is a big and very important question at the present. Research into this area is currently of high priority in much of the industrial world, especially in the Soviet Union and the United States. Technically, controlling fusion for the generation of power is an extremely difficult problem. In order for fusion to take place, temperatures on the order of 100 million degrees Celsius are needed just to start the fusion process. This temperature is hotter than the interior of the sun. No known materials can withstand these temperatures, so alternative containment procedures are being examined. So far, the research efforts of several nations have not reached the "breakeven" point where as much energy is released from the fusion process as was put into it to get it started.

The advantages of controlled fusion power are impressive.

1   It would be clean. Few radioactive products are formed.

2   Fuel is inexhaustable. The oceans of the world contain enough deuterium, one of the reactants, to provide the world's energy needs for a trillion years. On the other hand, there is a very limited supply of fossil fuels and uranium.

3   There is no possibility of the reaction getting out of control and causing a melt down. Fusion will occur in power plants in short bursts of energy that can be easily stopped in case of mechanical problems.

The disadvantage is, of course, the expense of research and development. Billions of dollars have been spent and billions more will have to be spent before the technical problems are solved. Hopefully, before the end of this century there will be experimental plants in operation that will prove the feasibility of fusion as a source of power. Unfortunately, there is little choice in this matter as there are few alternatives for power in the next century.

Review Questions

1   What is an atom? What is a molecule?
2   How are the atoms of a molecule held together?
3   Describe what is meant by the formula of a compound.
4   What is an ion? What is a cation? What is an anion?
5   What holds an ionic compound together?

6   Where is the electron located in an atom? What is its charge?
7   Describe the nucleus of an atom.
8   Name the two nucleons. Give the mass (in amu) and charge of each.
9   What is atomic number? Mass number?
10   What determines the identity of the atoms of an element?

11 Define atomic weight.

12 What are isotopes of an element?

13 Describe how the atomic weight of an element as it occurs in nature is calculated from its isotopes.

Changes in the Nucleus

14 What is illustrated by a nuclear equation?

15 Define radiation. Radioactivity. Radioactive isotopes.

16 Describe the three main types of radiation.

17 What is meant by the half-life of an element?

18 How does radiation cause ionization? What type of radiation causes the most ionization along its path? Which is the most penetrating? Which type is the most destructive to living cells?

19 What is meant by the transmutation of an element?

20 Why is an accelerator needed for nuclear reactions involving charged particles?

21 Define nuclear fission. How does it differ from typical nuclear reactions?

22 What is a chain reaction?

23 How is fission controlled in a nuclear power plant?

24 Define nuclear fusion? Why is it harder to control than fission?

Problems

3-1 Determine the number of atoms of each element in the formula of each of the following compounds:

(a) $C_6H_4Cl_2$

(b) C_2H_5OH (ethyl alcohol)

(c) $CuSO_4 \cdot 9H_2O$ (H_2O's are parts of a single molecule)

■ (d) $C_9H_8O_4$ (aspirin)

■ (e) $Al_2(SO_4)_3$

■ (f) $(NH_4)_2CO_3$

3-2 Write the formula of each of the following compounds:

(a) Sulfur dioxide (one sulfur and two oxygen atoms)

■ (b) Carbon dioxide

(c) Sulfuric acid (two hydrogens, one sulfur, and four oxygens)

■ (d) Calcium perchlorate (one calcium ion and two ClO_4^- ions)

(e) Ammonium phosphate (three NH_4^+ ions and one PO_4^{3-} ion)

3-3 Give the number of protons, neutrons, and electrons in each of the following:

(a) $^{45}_{21}Sc$ (d) $^{90}_{38}Sr^{2+}$

(b) $^{235}_{92}U$ (e) $^{79}_{34}Se^{2-}$

■ (c) $^{223}_{87}Fr$ ■ (f) $^{35}_{17}Cl^-$

3-4 Using the periodic table, determine the atomic number and the atomic weight of each of the following elements:

(a) Re (b) Co (c) Br (d) Si

3-5 Silicon occurs in nature as a mixture of three isotopes: ^{28}Si (at. wt. 27.98 amu), ^{29}Si (at. wt. 28.98), and ^{30}Si (at. wt. 29.97). The mixture is 92.21% ^{28}Si, 4.70% ^{29}Si, and 3.09% ^{30}Si. Calculate the atomic weight of naturally occurring silicon.

★**3-6** Chlorine occurs in nature as a mixture of ^{35}Cl and ^{37}Cl. If the atomic weight of ^{35}Cl is approximately 35.0 amu and that of ^{37}Cl is 37.0 amu and the atomic weight of the mixture as it occurs in nature is 35.5 amu, what is the proportion of the two isotopes?

■ **3-7** If the atomic weight of ^{12}C was defined as exactly 8 instead of 12, what would be the atomic weight of the following elements to three significant figures? Assume that the elements have the same weights relative to each other as before. That is, a hydrogen still weighs about one-twelfth as much as carbon.

(a) H (b) N (c) Na (d) Ca

Changes in the Nucleus: Alpha and Beta Radiation

3-8 Complete the following nuclear equations involving alpha and beta particles:

(a) $^{214}_{83}Bi \rightarrow ^{214}_{84}Po + \underline{\hspace{1cm}}$

(b) $^{90}_{37}Rb \rightarrow \underline{\hspace{1cm}} + ^{0}_{-1}e$

(c) $^{235}_{92}U \rightarrow \underline{\hspace{1cm}} + ^{4}_{2}He$

(d) $\underline{\hspace{1cm}} \rightarrow ^{41}_{21}Sc + ^{0}_{-1}e$

(e) $\underline{\qquad} \rightarrow {}^{210}_{82}\text{Pb} + {}^{0}_{-1}e$

■ (f) ${}^{239}_{93}\text{Np} \rightarrow \underline{\qquad} + {}^{0}_{-1}e$

■ (g) ${}^{226}_{88}\text{Ra} \rightarrow {}^{222}_{86}\text{Rn} + \underline{\qquad}$

3-9 Write a nuclear equation that includes all isotopes and particles from the following information.

(a) ${}^{230}_{90}\text{Th}$ decays to form ${}^{226}_{88}\text{Ra}$

(b) ${}^{214}_{84}\text{Po}$ emits an alpha particle

(c) ${}^{210}_{84}\text{Po}$ emits a beta particle

■ (d) An isotope emits an alpha particle and forms ${}^{235}_{92}\text{U}$

■ (e) ${}^{14}_{6}\text{C}$ decays to form ${}^{14}_{7}\text{N}$

Half-life

3-10 The radioactive isotope ${}^{90}_{38}\text{Sr}$ can accumulate in bones, where it replaces calcium. It emits a high-energy beta particle, which eventually can cause cancer. (a) What is the product of the decay of ${}^{90}_{38}\text{Sr}$? The half-life of ${}^{90}_{38}\text{Sr}$ is 25 years. (b) How long would it take for a 0.10-mg sample of ${}^{90}_{38}\text{Sr}$ to decay to where only 2.5×10^{-2} mg was left?

■ **3-11** The radioactive isotope ${}^{131}_{53}\text{I}$ accumulates in the thyroid gland. On the one hand, this can be useful in detecting diseases of the thyroid and even in treating cancer at that location. On the other hand, exposure to excessive amounts of this isotope such as from a nuclear power plant can *cause* cancer of the thyroid. ${}^{131}_{53}\text{I}$ emits a beta particle with a half-life of 8.0 days. What is the product of the decay of ${}^{131}_{53}\text{I}$? If one started with 8.0×10^{-6} g of ${}^{131}_{53}\text{I}$, how much would be left after 32 days?

3-12 The isotope ${}^{14}_{6}\text{C}$ is used to date fossils of formerly living systems such as prehistoric animal bones. The radioactivity due to ${}^{14}_{6}\text{C}$ of a sample of the fossil is compared to the radioactivity of currently living systems.

The longer something has been dead, the lower will be the radioactivity due to ${}^{14}_{6}\text{C}$ of the sample. If the radioactivity of a sample of bone from a mammoth was one-fourth of the radioactivity of the current level, how old was the fossil? ($t_{1/2} = 5760$ years.)

★**3-13** The decay series of ${}^{238}_{92}\text{U}$ to ${}^{206}_{92}\text{Pb}$ involves alpha and beta emissions in the following sequence: $\alpha, \beta, \beta, \alpha, \alpha, \alpha, \alpha, \alpha, \beta, \alpha, \beta, \beta, \beta, \alpha$. Identify all isotopes formed in the series.

Nuclear Reactions

3-14 When ${}^{235}_{92}\text{U}$ and ${}^{239}_{94}\text{Pu}$ undergo fission, a variety of reactions actually take place. Complete the following:

(a) ${}^{235}_{92}\text{U} + {}^{1}_{0}n \rightarrow \underline{\qquad} + {}^{146}_{58}\text{Ce} + 3{}^{1}_{0}n$

(b) ${}^{235}_{92}\text{U} + {}^{1}_{0}n \rightarrow {}^{90}_{37}\text{Rb} + {}^{142}_{55}\text{Cs} + \underline{\qquad}$

■ (c) ${}^{239}_{94}\text{Pu} + {}^{1}_{0}n \rightarrow {}^{141}_{56}\text{Ba} + \underline{\qquad} + 2{}^{1}_{0}n$

3-15 Complete the following nuclear reactions:

(a) ${}^{35}_{17}\text{Cl} + {}^{1}_{0}n \rightarrow \underline{\qquad} + {}^{1}_{1}\text{H}$

(b) ${}^{27}_{13}\text{Al} + \underline{\qquad} \rightarrow {}^{25}_{12}\text{Mg} + {}^{4}_{2}\text{He}$

■ (c) ${}^{27}_{13}\text{Al} + {}^{4}_{2}\text{He} \rightarrow \underline{\qquad} + {}^{1}_{0}n$

★(d) ${}^{238}_{92}\text{U} + 15{}^{1}_{0}n \rightarrow {}^{253}_{100}\text{Es} + \underline{\qquad}$

(e) ${}^{244}_{96}\text{Cm} + {}^{12}_{6}\text{C} \rightarrow \underline{\qquad} + 2{}^{1}_{0}n$

■ (f) $\underline{\qquad} + {}^{2}_{1}\text{H} \rightarrow {}^{238}_{93}\text{Np} + {}^{1}_{0}n$

(g) ${}^{242}_{94}\text{Pu} + {}^{22}_{10}\text{Ne} \rightarrow {}^{260}_{104}\text{Ku} + \underline{\qquad}$

■ (h) ${}^{249}_{96}\text{Cm} + \underline{\qquad} \rightarrow {}^{260}_{103}\text{Lr} + 4{}^{1}_{0}n$

3-16 The fissionable isotope ${}^{239}_{94}\text{Pu}$ is made from the abundant isotope of uranium, ${}^{238}_{92}\text{U}$, in nuclear reactors. When ${}^{238}_{92}\text{U}$ absorbs a neutron from the fission process, eventually ${}^{239}_{94}\text{Pu}$ is formed. This is the principle of the "breeder reactor," although ${}^{239}\text{Pu}$ is formed in all reactors. Complete the following:

$${}^{238}_{92}\text{U} + {}^{1}_{0}n \longrightarrow \underline{\qquad} + {}^{0}_{-1}e$$

$$\longrightarrow {}^{239}_{94}\text{Pu} + \underline{\qquad}$$

The Nature of the Electrons in the Atom

The beautiful colors of a fireworks display originate from the presence of certain elements.

Water and table salt are two of the most common (and essential) compounds in our lives. But why are these particular compounds what they are? Why does the formula for water happen to be H_2O but not H_3O or HO_2? Why do crystals of salt consist of *ions* but crystals of ice consist of *molecules*? Why do the elements in both of these substances form compounds but some other elements have little or no tendency to form chemical bonds at all? These are fundamental questions concerned with the bonding of elements to form compounds. The answers to the questions are very logical if we first understand the basis for bond formation. In this chapter we will see that the basis for the formation of bonds lies in the arrangement of electrons in the atom.

Historically, to understand the nature of bonds a sophisticated model or theory of the atom was needed that emphasized the nature of electrons. This model was needed not only to explain bonding but to solve other puzzles seemingly related to electrons in atoms. One of these phenomena was the periodic similarity exhibited by the elements when arranged in order of increasing atomic weight in what is called the *periodic table*. The other was the discovery that under certain conditions elements emit light (called a spectrum) that is unique and characteristic of that element much as fingerprints are unique to each individual.

Rutherford's nuclear model of the atom, discussed in Chapter 3, was a start, but it did not go far enough. In 1913, one of Rutherford's students, Niels Bohr, from Denmark, expanded on the nuclear model and provided an ingenious description of the electrons in the atom. The immediate success of Bohr's model was assured when he was able to explain the known spectrum of hydrogen atoms, which at least partially explained one of the two puzzles. In addition to the hydrogen atom spectrum, Bohr's model laid the foundation for further developments that would eventually explain the other puzzle, the periodicity of the elements. In this chapter, one of our goals will be to understand the reason for the periodicity of the elements. This goal will serve us well in our ultimate purpose in future chapters, which is an understanding of bonding.

Before Bohr's model of the atom and subsequent modifications are discussed, we will set the stage with some background on the two puzzles we have mentioned: the periodicity of the elements and their emission spectra.

4-1 The Periodic Table of the Elements

Before Bohr's model of the atom, scientists were well aware that certain elements display close chemical similarities. This was demonstrated by an arrangement of the elements called the **periodic table.** In this arrangement, as shown inside the back cover, *elements with close chemical similarities are found in vertical columns called* **groups** or **families.**

An example of this vertical relationship, which you will recognize, is found in group IB: copper, silver and gold. These familiar metals are used for coins and jewelry because of their low chemical reactivity as well as their lustrous beauty and comparative rarity (see Figure 4-1). These elements can be used for this purpose because they do not rust the way iron does or become covered with a dull coating the way aluminum does. On the other hand, lithium, sodium, and potassium are also metals, but they are very reactive (see Figure 4-1). They easily form compounds with the oxygen in the air and dissolve explosively in contact with water. Notice that these closely related ele-

Figure 4-1 METALS IN WATER. Sodium is a metal that reacts violently with water. The copper, silver, and gold in coins, however, are completely unreactive in water. Here, sodium (on the left) explodes on contact with water, while the coins are unaffected.

ments are all members of group IA. The chemical relationships among the elements have been known and studied for more than 150 years, and the periodic table (with, of course, fewer elements) has been displayed for a little more than a century.

As you look at the periodic table, you may notice that the elements appear in the vertical columns in a somewhat repeating or "periodic" relationship. For example, the number of elements that come *between* the members of group IA are 2, 8, 8, 18, 18, and 32. Scientists felt that there must be some reason why certain elements have similarities and the groups are related in a regular, periodic manner.

This table is a work of art to the chemist, and is well worth the time to discuss it in some detail. We will do this in Chapter 5 after we have answered the question, "Why is it periodic?" in this chapter.

4-2 The Emission Spectra of the Elements

When any substance is heated to a high enough temperature, it glows. The tungsten filament in an incandescent lightbulb glows as a result of the heat generated by the flow of electricity through the filament. Let's take a close look at the white light that comes from the bulb. *When a beam of the light is allowed to pass through a glass prism, the light breaks down into a rainbow of colors called its* **spectrum.** *Since one color blends gradually into another, this is known as a* **continuous spectrum** (see Figure 4-2). Rain drops in the air also serve as prisms after a rain storm, breaking white sunlight into the rainbow.

Light is a form of energy, like electricity and heat. Since light travels

Figure 4-2 THE SPECTRUM OF INCANDESCENT LIGHT. Incandescent light is composed of all the colors of the rainbow.

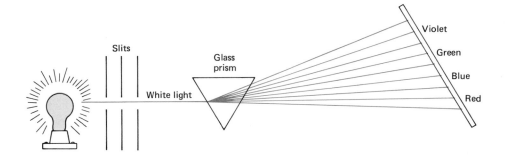

like waves on the ocean, dissipating in all directions (except that light travels at a speed of about 186,000 miles/sec), it has what is known as a wave nature. *A property of a wave, called its* **wavelength,** *is the distance between two equivalent points,* as shown below. Wavelength is given by the Greek symbol λ (lambda).

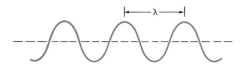

Each specific color of light in the rainbow has a specific λ. In the rainbow, red has the longest λ and violet has the shortest. When light has a somewhat longer λ than red it is invisible to the eye and is called **infrared** (beneath red) light. When light has a somewhat shorter λ than violet it is also invisible and is called **ultraviolet** (beyond violet) light.

The energy of light is inversely proportional to its wavelength (see Appendix B for a discussion of proportionalities). That is,

$$E \propto \frac{1}{\lambda} \qquad E = \frac{hc}{\lambda}$$

Review Appendix B-3 on proportionalities.

where h is a constant of proportionality and is known as Planck's constant, and c is also a constant and is equal to the speed of light.

Notice that the larger the value of λ, the smaller the energy and vice versa. Therefore, red light, with the longer λ, has less energy than violet light. Indeed, ultraviolet light has enough energy to damage living organisms. X-rays and gamma rays have extremely small values of λ and are accordingly extremely powerful. (Medical X-rays, however, are given in small doses in order to minimize the damage.) In Figure 4-3 these and other forms of light are shown in terms of relative wavelengths.

As shown in Figure 4-2, white light from a glowing tungsten filament is a combination of all wavelengths and energies of light in the visible range.

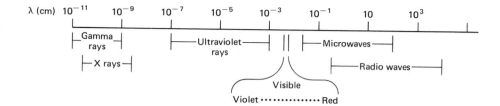

Figure 4-3 LIGHT. Light is classified according to its wavelength.

When an element is heated to such an extent that it breaks into individual gaseous atoms, the spectrum produced is different from the continuous spectrum of the rainbow. In this case, only definite or discrete colors are produced. Recall that if the light produced has a specific color (λ), it has a specific energy (E). The spectrum of a gaseous atomic element (known as its *atomic emission spectrum*) is particular and characteristic of that element. You may have observed this phenomenon in the chemicals that create colors when added to the logs in a fireplace or to the powder in fireworks. Lithium compounds create a red color, boron compounds create a green color, and sodium compounds create a yellow color. *Since only certain colors are produced by the atoms of hot, gaseous elements, their atomic spectra are referred to as* **discrete** *or* **line spectra.**

Hydrogen, the simplest of the elements, has the most orderly atomic spectrum, as shown in Figure 4-4. In a sense, this made the hydrogen spectrum the most puzzling to the scientists.

The fact that light, as a form of energy, is emitted by atomic hydrogen in a discrete and yet orderly manner was also a matter that required an explanation.

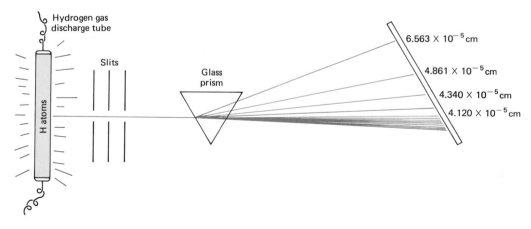

Figure 4-4 THE ATOMIC SPECTRUM OF HYDROGEN IN THE VISIBLE RANGE. The atomic spectrum of an element is composed of discrete colors.

4-3 A Model for the Electrons in the Atom

By 1913 the nuclear model of the atom proposed by Rutherford was generally accepted by scientists. In that year Niels Bohr set out to explain the origin of the atomic spectrum of hydrogen by expanding on this model. Bohr's contribution would provide not only a reasonable picture of what forces hold the atom together but also the mathematical relations corresponding to the known experimental facts about the hydrogen atom.

First, Bohr proposed that the electrons revolve around the nucleus in stable, circular orbits (see Figure 4-5). The attractive forces between the negative electron and the positive nucleus would be exactly balanced by the centrifugal force* of the orbiting electron. This was a controversial and bold statement by Bohr, because classical physics predicted that this would not be possible—the electron would collapse into the nucleus. Bohr neatly sidestepped this problem by postulating (suggesting without proof as a necessary condition) that classical physics does not apply in the small dimensions of the atom.

Second, Bohr postulated that electrons are confined to definite or stationary orbits around the nucleus. Electrons in a stationary orbit would have a discrete energy and be at a fixed distance from the nucleus. *Since only discrete energy states are available to the electron in an atom, the electronic energy is said to be* **quantized** *according to Bohr's theory.*

The discrete energy levels in the atom can be compared to the stairway in your home. In Figure 4-6 you can see that between floor A and floor B there are five steps which are available energy levels. Since you cannot stand

* The force on a rotating object that tends to move it radially outward. For example, when a discus thrower lets go of the discus, it goes straight out from the point of release.

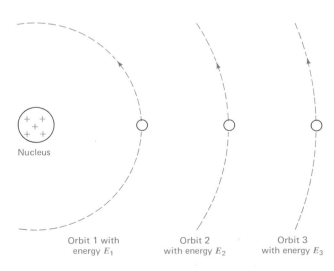

Nucleus

Orbit 1 with energy E_1 Orbit 2 with energy E_2 Orbit 3 with energy E_3

Figure 4-5 BOHR'S MODEL OF THE HYDROGEN ATOM. In this model the electron can exist only in definite or discrete energy states.

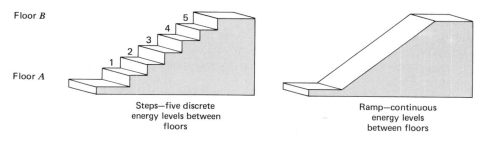

Figure 4-6 DIS-CRETE VERSUS CONTINUOUS ENERGY LEVELS. The steps represent discrete energy levels, the ramp continuous energy levels.

(with both feet together) between steps, you can see that only these five energy states are possible between floors. *A stairway is thus* **quantized.** On the other hand, a ramp would be continuous, as all positions or energy levels between floors are possible.

When an electron resides in the available *orbit that is lowest in energy* (*which is also the closest to the nucleus*), *the electron is said to be in the* **ground state.** The electron can jump from the ground state to a higher energy level farther from the nucleus if it is supplied exactly the right amount of energy to make the jump. (In the analogy to the stairway, it takes a definite amount of energy for a person to climb one step.) The electron then has potential energy and is said to be in an **excited state.** When an electron is in an excited state, it can fall back down to a lower level; but the electron must then give up the same amount of energy that it previously absorbed. The energy that is emitted by this process comes off in the form of light. Since the orbits have discrete energy, the difference in energy between the levels is fixed. This in turn means that the energy of the light is discrete, with a discrete wavelength and hence a specific color. (See Figure 4-7.)

Bohr did significantly more than just present a model to explain discrete

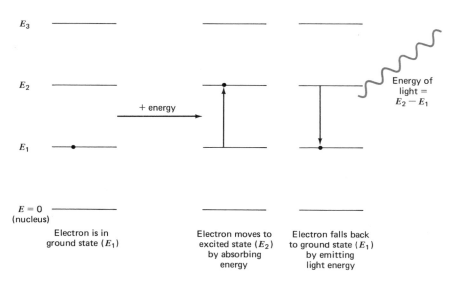

Figure 4-7 LIGHT FROM THE HYDROGEN ATOM. An electron in an excited state emits energy in the form of light when it drops to a lower energy level.

spectra. From the mathematical relationships suggested by his model, he was able to calculate the wavelengths of the light emitted by the hydrogen atom. The value of his theory was proven when the calculated values corresponded with the values measured from actual experiments (one series of lines is shown in Figure 4-4).

Because of the beautiful simplicity of Bohr's model of the atom, modern scientists still use his picture in certain situations. However, our theories have now gone well beyond this model. For one thing, Bohr's model did not work for atoms with more than one electron. That essentially limits the theory to hydrogen. Also, new experiments and new knowledge of the nature of matter have forced us away from viewing electrons as particles orbiting at definite distances from the nucleus with definite speeds and energies. Now we have a mathematical model that more accurately describes the energies and locations of the electrons but leaves no convenient picture for us to visualize. This mathematical description is a branch of physics and chemistry known as **wave mechanics.** Despite the complexity of wave mechanics, the results can be easily grasped and put to use in the effort to explain the chemical nature of the elements. What follows then, are results of calculations of wave mechanics.

4-4 Shells

Both Bohr's theory and modern wave mechanics agree that electrons reside in regions of space called **Shells.*** Both theories also agree that the shell closest to the nucleus is lowest in energy and that it takes a definite amount of energy for an electron to move from an inner shell to an outer shell. In this respect, shells in an atom are analogous to dormitories near a campus. A dormitory is a defined region of space where students reside. In a series of dorms

---

* The theories differ as to whether the shell is located at a precise distance from the nucleus (Bohr) or has only a certain *probability* of being at a certain distance.

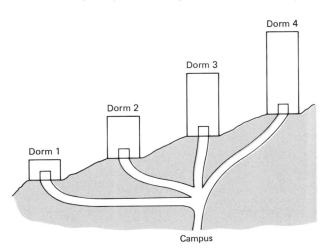

Figure 4-8 ENERGY LEVELS OF DORMS. The dorms on the hill represent different energy levels.

Table 4-1   Shells

Letter designation	K	L	M	N	O
Quantum number designation	1	2	3	4	5
Capacity ($2n^2$)	2	8	18	32	50

on a hill, obviously the one lowest on the hill requires the least amount of effort to reach and is thus considered the *lowest* in energy (see Figure 4-8). Also analogous to electrons in shells, a definite amount of energy is needed for a student to move uphill from dorm 1 to dorm 2.

The shells in an atom are labeled in two ways (see Table 4-1). Historically, the letters K, L, M, N, and so on, were used, with the K shell being closest to the nucleus (and the lowest in energy), the L shell next, and so on. The modern way is to label the shells by means of a number (n). *These numbers are called the* **principal quantum numbers** *and describe the energy of the shell.* The $n = 1$ shell corresponds to the K shell, $n = 2$ to the L shell, and so on. Another result of Bohr's theory that is still valid is that each shell holds $2n^2$ electrons. You can see that the shells get larger the farther they are from the nucleus. This result is reasonable, since the farther a shell is from the nucleus the greater the volume of space that is available to the electrons.

4-5 Subshells

Let's again look at the analogy of the dorms. If the careful student wants to live in the dorm of lowest energy, he or she would naturally look to the one lowest on the hill with space available. But another point should be raised as to the energy required to get to a certain room. That question is, "On what floor is the available room?" Since you can count on the elevators being full, on the top floor, or out of order, it is not far-fetched to assume that the student will have to use the stairs. In our example, the first dorm has only one floor, the second has two, the third has three, and so on. (This unusual situation is for the good of the analogy.) Thus, not only do the students need to know how far up the hill they must walk but how many flights of stairs must be climbed once they are in the dorm. Each floor within a dorm is at a different energy level (see Figure 4-9).

Just as a dorm has floors, *a shell has subshells. Subshells within a shell are also at different energy levels.* Subshells are labeled historically in order of increasing energy by the letters s, p, d, and f. *They are also designated by the quantum number l.* The quantum number l relates to n as shown below:

$$l = 0, 1, 2, \ldots, n - 1$$

For example, when $n = 4$, l can have four values, 0, 1, 2, and 3. Thus there are four subshells in the fourth shell. The $l = 0$ subshell corresponds to the s subshell, the $l = 1$ subshell corresponds to the p subshell, and so on. The subshells with both their letter and quantum number designations are shown in Table 4-2.

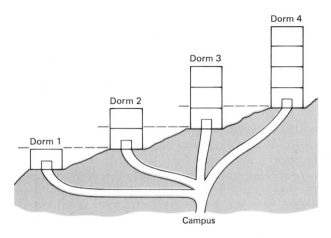

Figure 4-9 ENERGY LEVELS WITHIN A DORM. The floors in the dorms represent different energy levels within each dorm.

Analogous to the dorms in Figure 4-9, each successive shell has one additional subshell. Thus the $n = 1$ shell has only the s subshell; the $n = 2$ has two, the s and the $p;$ and so on.

The information can now be put together into Table 4-3 by expanding Tables 4-1 and 4-2. *We will adopt the accepted convention of indicating a shell by the quantum number and the subshell by the letter.* Notice how the total capacity of a shell is distributed among the subshells. The information in Table 4-3 should be committed to memory before proceeding.

See Problems 4-1, 4-2, 4-3, 4-4

In Figure 4-10, the first four shells and their subshells are shown in order of increasing energy analogous to the dorms shown in Figure 4-9.

4-6 Electron Notation of the Elements

We are now ready to examine the electron arrangement or notation of all the electrons in an atom. The logical place to start is with the one-electron atom, H, and work on from there. Two principles will guide us:

1 A shell will be designated by a quantum number n, a subshell by a letter corresponding to the quantum number l and a superscript number indicating the number of electrons in that particular subshell.

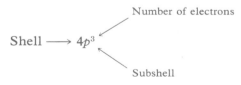

Table 4-2 Subshells

Letter designation	s	p	d	f
Quantum number designation	0	1	2	3
Capacity	2	6	10	14

Table 4-3 Shells and Subshells

Shell	1	2		3			4			
Subshell	s	s	p	s	p	d	s	p	d	f
Subshell capacity	2	2	6	2	6	10	2	6	10	14
Shell capacity	2	8		18			32			

2 *Electrons fill the lowest available energy levels (shell and subshell)* first. This is known as the **Aufbau principle.** It is analogous to the student who wants the available room of lowest energy. Obviously, the most desirable room is on the lowest available floor in the lowest available dorm on the hill.

The first element, H, has one electron, which goes into the first shell which has only the *s* subshell. The electron notation for the one electron is

$$1s^1$$

The next element, He, has two electrons. Since the first shell has room for two electrons, the electron notation is

$$1s^2$$

Next comes Li with three electrons. Two fill the first shell, but the third must move on to the second shell. The second shell has two subshells, the *s* and the *p*, but as you know the *s* is the lower in energy. The electron notation for Li is

$$1s^2 2s^1$$

The next seven elements complete the filling of the second shell:

Be	$1s^2 2s^2$	C	$1s^2 2s^2 2p^2$	O	$1s^2 2s^2 2p^4$
B	$1s^2 2s^2 2p^1$	N	$1s^2 2s^2 2p^3$	F	$1s^2 2s^2 2p^5$
				Ne	$1s^2 2s^2 2p^6$

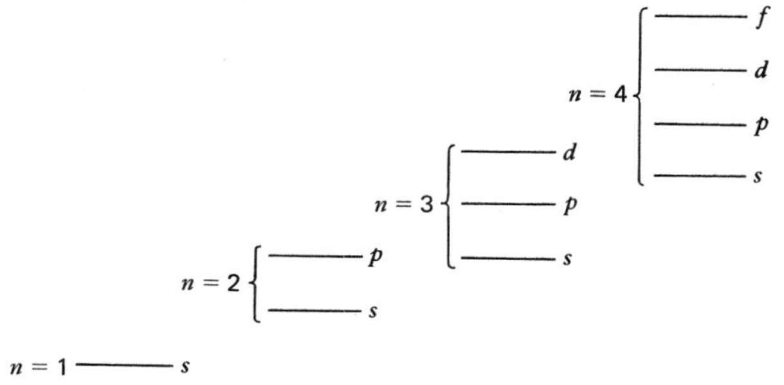

Figure 4-10 ENERGY LEVELS IN THE ATOM. Each successive shell contains one additional subshell.

If you will remember, one of our goals was to explain the *reason* for the periodic table. An early observation may lead to this end. He and Ne are in a vertical column, indicating that they are members of the same group. This group is known as the **noble gases.** Except for Xe and Kr they have no tendency to enter into chemical reactions and form chemical bonds (they are inert). Notice that the electron notation for He and Ne indicate that they do have something in common. That is, they both have "filled shells." Perhaps that is what makes for noble gas behavior. This premise can be tested later. Let us now continue with the electron notation for still heavier elements. At this point we may adopt a shorthand way of writing that saves us from repeating all of the filled inner subshells. A noble gas in brackets will be equivalent to all of the electrons of that element (e.g., $[Ne] = 1s^2 2s^2 2p^6$).

Na	$[Ne]3s^1$	P	$[Ne]3s^2 3p^3$
Mg	$[Ne]3s^2$	S	$[Ne]3s^2 3p^4$
Al	$[Ne]3s^2 3p^1$	Cl	$[Ne]3s^2 3p^5$
Si	$[Ne]3s^2 3p^2$	Ar	$[Ne]3s^2 3p^6$

Let's make some more observations. Note that although Ar is in the noble gas group, the element does not have a filled shell. The first premise must be modified or discarded. However, we might suggest that perhaps filled outer s and p subshells is what is characteristic of a noble gas (except for He, which doesn't have a $1p$ subshell). Notice also that Li and Na have something in common and that is one electron beyond a noble gas. Since Li and Na are also in the same group, it appears at this point that similar electron notations relate to the periodic relationship of the elements.

The next element after Ar presents a problem. Normally we might expect K, element 19, to have one electron in a $3d$ subshell. In fact it has its 19th electron in the $4s$ subshell. How could this be possible? For the answer let's go back to the analogy shown in Figure 4-9. Notice that although dorm 4 is higher up the hill than dorm 3, all of the floors in dorm 4 are not necessarily higher than those in dorm 3. In fact, the top floor of dorm 3 is higher than the ground floor of dorm 4. The sharp student would thus choose the ground floor of dorm 4 in preference to the top floor of dorm 3. In the case of elements a similar phenomenon is true. For the neutral atom the $4s$ subshell is lower in energy than the $3d$ subshell and fills first. After the $4s$ subshell, the $3d$ fills, followed by the $4p$. In Figure 4-10, notice that the $4s$ subshell is lower in energy than the $3d$ subshell.

K	$[Ar]4s^1$
Ca	$[Ar]4s^2$
Sc	$[Ar]4s^2 3d^1$

.
.
.

$$\text{Zn} \quad [\text{Ar}]4s^2 3d^{10}$$

$$\text{Ga} \quad [\text{Ar}]4s^2 3d^{10} 4p^1$$

.
.
.

$$\text{Kr} \quad [\text{Ar}]4s^2 3d^{10} 4p^6$$

Notice that our observations on periodicity seem valid. Li, Na and K, which are all in the same group, have an ns^1 electron notation. Kr, a noble gas, has filled outer $4s$ and $4p$ subshells, which was suggested earlier to be characteristic of this group. *Apparently, vertical columns of elements have the same subshell electron notation but different shells.*

At this point, a scheme may be helpful to remember the order of filling of the subshells. As you go to shells of higher energy, the shells get larger and much more mixing of subshells occurs. A way to remember is illustrated in Figure 4-11. Write in horizontal columns all of the subshells that exist starting with the $1s$. Under $1s$ put $2s$, followed by $2p$, and so on. Draw a stairstep down on the right. Now draw a diagonal arrow through *each* corner of the stairstep. The top arrow points to the first subshell filled. The second arrow points to the second subshell, the third arrow points to the $2p$ followed by the $3s$, and so on.

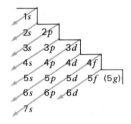

Figure 4-11 THE ORDER OF FILLING OF SUBSHELLS. The arrows indicate the order of filling.

Example 4-1
Write the electron notation for iron.

Solution:
From inside the front cover, we find that iron is element number 26, which means that it has 26 electrons. Write the subshells in the order of filling plus a running summation of the total number of electrons involved.

Subshell	Number of Electrons in Subshell	Total Number of Electrons
$1s$	2	2
$2s$	2	4
$2p$	6	10
$3s$	2	12
$3p$	6	18
$4s$	2	20
$3d$	10	30 (over 26)

Iron therefore has six electrons past the filled $4s$ subshell, but it does not completely fill the $3d$ subshell. The electron notation for iron is written out as

$$1s^2 2s^2 2p^6 3s^2 3p^6 4s^2 3d^6$$

or, using noble gas shorthand,

$$[\text{Ar}]4s^2 3d^6$$

The easiest way to predict the electron notation of the elements is by use of the periodic table. In this endeavor the fact that vertical columns have the same electron subshell notation can be used to advantage. In Figure 4-12 the subshell notation is listed above the group and the shell (n) is listed in the box for the element. With practice, one can quickly give the electron notation of any element with only a periodic table for reference. It should be pointed out that certain elements do deviate from this scheme, especially those elements where d and f subshells are partially filled. For example, Cr (element number 24) is actually $4s^1 3d^5$ rather than $4s^2 3d^4$. Another example of an exception is the element La (element number 57). Figure 4-12 indicates that La has the outer shell notation $6s^2 5d^1$ rather than $6s^2 4f^1$ as predicted by Figure 4-11. Figure 4-12 is correct. Although these exceptions are important in the discussion of the chemistry of these elements, it is not important at this point to know all the exceptions to the rules.

Example 4-2
What is the electron notation for (a) vanadium and (b) lead?

Solution:
(a) In the front cover of the book, note that vanadium (V) is element number 23. Locate V in the periodic table shown in Figure 4-12. V has the electron notation

$$1s^2 2s^2 2p^6 3s^2 3p^6 4s^2 3d^3$$

Using the noble gas notation shorthand, the notation is

$$[\text{Ar}]4s^2 3d^3$$

(b) Lead (Pb) is element number 82. Locate Pb in the periodic table shown in Figure 4-12. Using the noble gas shorthand, notice that the following subshells come after [Xe] in lead:

$$[\text{Xe}]6s^2 5d^{10} 4f^{14} 6p^2$$

Example 4-3
What element has the following electron notation:

$$[\text{Kr}]5s^2 4d^{10} 5p^5$$

Figure 4-12 ELECTRON NOTATION AND THE PERIODIC TABLE. The electron notation of an element can be determined from its position in the periodic table.

s^2
1

s^1
1

s^1	s^2
2	2
3	3
4	4
5	5
6	6
7	7
8	8

d^1	d^2	d^3	d^4	d^5	d^6	d^7	d^8	d^9	d^{10}
3	3	3	3	3	3	3	3	3	3
4	4	4	4	4	4	4	4	4	4
5	5	5	5	5	5	5	5	5	5
6	6	6							

s^2p^1	s^2p^2	s^2p^3	s^2p^4	s^2p^5	s^2p^6
2	2	2	2	2	2
3	3	3	3	3	3
4	4	4	4	4	4
5	5	5	5	5	5
6	6	6	6	6	6
7	7	7	7	7	7

f^1	f^2	f^3	f^4	f^5	f^6	f^7	f^8	f^9	f^{10}	f^{11}	f^{12}	f^{13}	f^{14}
4	4	4	4	4	4	4	4	4	4	4	4	4	4
5	5	5	5	5	5	5	5	5	5	5	5	5	5

See Problems 4-5
through 4-11

Solution:
In Figure 4-12, locate s^2p^5 in the upper right. Element number 53 will have the given electron notation. Element number 53 is

<u>iodine</u>

(*Note:* The topics in the remainder of this chapter continue the discussion of the nature of the electrons in the atom. However, this material may be omitted, since it is not referred to in the following chapters except in several footnotes.)

4-7 Orbitals

If one were to attempt to locate a student in a dorm, more information would be needed than just the particular dorm and the floor. One would have to know the student's room number. Just as floors are divided into individual rooms, subshells are divided into individual orbitals. **Orbitals** *are the regional divisions of a subshell in which the electron resides.* Like a two-person dorm room, orbitals can hold two electrons. An *s* subshell, however, is like a floor that is all one room, so it is not divided and is thus all one orbital. A *p* subshell, on the other hand, has three separate orbitals, each having a capacity of two electrons for a total of six (see Figure 4-13).

The orbitals of a subshell are designated by a third quantum number, m_l. The values of m_l depend on the value of the second quantum number l. The relationship between l and m_l is

$$m_l = 0, \pm 1, \pm 2, \ldots, \pm l$$

For example, when $l = 2$ (a *d* subshell), $m_l = 2, 1, 0, -1, -2$, or five values. Each value for m_l represents a different *d* orbital; thus there are $2l + 1$ orbitals per subshell. Only the shell and the subshell determine the energy of an electron in an atom. Under normal conditions each orbital within a particular subshell has the same energy. In the dorm analogy, once you have reached the floor, it doesn't matter whether you go to the room on the right, left, or straight ahead. The same amount of energy will be required.

See Problems 4-12,
4-13, 4-14

Example 4-4
What are the values of m_l for a *p* subshell? How many different orbitals does this represent?

Solution:
In Table 4-2, notice that $l = 1$ for a *p* subshell. Therefore, m_l has the values $+1, 0$, and -1. Thus there are three different orbitals in a *p* subshell.

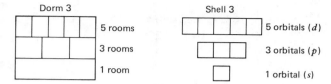

Figure 4-13 DORM ROOMS AND ORBITALS. The rooms on each floor of a dorm represent different regions of space at the same energy level, which is analogous to the orbitals of a subshell.

4-8 Electron Spin

Finally, there is a fourth quantum number that differentiates between the two electrons in a particular orbital. This is called the **spin quantum number** m_s. It relates to the fact that an electron has properties of a spinning charge. Since it can spin in only one of two directions, clockwise or counterclockwise, there are only two possible values for m_s. These values are $m_s = +\frac{1}{2}$ or $m_s = -\frac{1}{2}$.

If two electrons occupy the same orbital, they must have different spin quantum numbers. In that case the spins are said to be "paired." This is a way of stating the **Pauli exclusion principle** that reads: *No two electrons in an atom can have the same four quantum numbers.* In the dorm analogy the reader is invited to proceed on his own. It could simply mean that everyone gets his or her own bed.

We can now summarize the discussion of quantum numbers. An electron in an atom is completely defined by a set of four quantum numbers. The restrictions as to the values of these numbers are as follows:

Shell: $n = 1, 2, 3,$ etc.

Subshell: $l = 0, 1, 2, \ldots, n-1$

Orbital: $m_l = 0, \pm 1, \pm 2, \ldots, \pm l$

Spin: $m_s = +\frac{1}{2}$ or $-\frac{1}{2}$

The first two quantum numbers (shell and subshell) define the energy of the electron. The third defines a particular orbital within the subshell; and finally, the fourth defines the spin of the electron.

See Problems 4-15, 4-16

4-9 Paramagnetism and Diamagnetism

When an atom has an electron structure with one or more unpaired electrons in orbitals it is **paramagnetic.** Paramagnetic matter is slightly attracted to a magnetic field. *If all electrons are paired, the atom is* **diamagnetic.** Diamagnetic matter is slightly repelled by a field.

Figure 4-14 MAGNETISM OF GROUP IA AND IIA. IA elements are paramagnetic and IIA elements are diamagnetic.

In the following scheme, orbitals are represented by a box □ and an electron by an arrow ↑. To represent the two possible spin quantum numbers of an electron, the arrow can point either up or down.

Figure 4-14 indicates that all atoms in group IA must be paramagnetic since they all have one unpaired electron in an *s* orbital. At this point it doesn't matter whether the arrow points up or down. Both possibilities have the same energy. Also, the spins of electrons on different atoms do not cancel for free atoms. Group IIA atoms are all diamagnetic, with the electrons in the *s* orbital paired.

Group IIIA atoms are again paramagnetic, with an unpaired electron in one of the three *p* orbitals. Group IVA atoms have three possibilities for the two electrons in the *p* orbitals:

Hund's rule of maximum multiplicity *states that electrons occupy separate orbitals of the same energy with their spins parallel.* Thus group IVA atoms are found to be paramagnetic, as predicted by possibility 3. This result is understandable, since electrons, having the same negative charge, would prefer to get as far away from each other as possible in the same energy level. In the dorm analogy, a similar situation develops between two students of the same charge (sex). Given two unoccupied rooms in the same building on the same floor, the two students will probably prefer private rooms. Pairing occurs when no more empty rooms are available. The arrangement of electrons in the *p* orbitals is represented in Figure 4-15. All are paramagnetic except the noble gases.

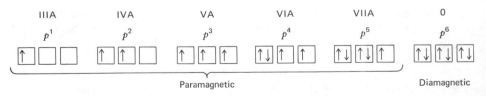

Figure 4-15 MAGNETISM OF GROUP IIIA TO GROUP O. In these groups all elements are paramagnetic except the noble gases, (Group O).

This same phenomenon occurs when the d subshell is filling. For example, Fe (element number 26) has the electron notation $4s^23d^5$, with five unpaired electrons, one in each of the five different $3d$ orbitals.

Magnetism of individual atoms will be affected when they combine to form molecules.

See Problems 4-17, 4-18, 4-19

4-10 The Shapes of Orbitals

An important topic of chemistry concerns the shape or geometry of molecules. The geometry of molecules is related to the shape of the orbitals of the electrons involved in bonding.

An s orbital has a perfectly spherical shape. Electrons in s orbitals are free to occupy any position within the sphere (see Figure 4-16).

The three different p orbitals, on the other hand, are roughly shaped like a weird baseball bat with two fat ends called "lobes." If the axis through one of the p orbitals is defined as the x axis, the orbital is designated as the p_x orbital. The other two p orbitals are at 90° angles to the p_x orbital and to each other. They are referred to as the p_y and p_z orbitals (see Figure 4-17).

The d orbitals are more complex. Four of the orbitals consist of four lobes each. The fifth consists of two lobes along the z axis and a torus (a doughnut shape) in the xy plane (see Figure 4-18). No matter how many lobes an orbital has, however, it still holds no more than two electrons.

The f orbitals are even more complex. However, since they are rarely involved in bonding, their shapes are not usually important to the chemist.

The shapes of the orbitals that have been shown represent the regions of space where the probability of finding the electron in a particular orbital is highest. In this respect, it is like locating students in their rooms. The highest probability is at the desk or in bed, but there is a probability that the students won't be there at all. Likewise, there is a very small probability that the electron in an orbital is not in the region indicated by the shape of the orbital. In summary, these pictures of orbitals represent a "cloud" of maximum

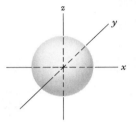

Figure 4-16 AN s ORBITAL. An s orbital has a spherical shape.

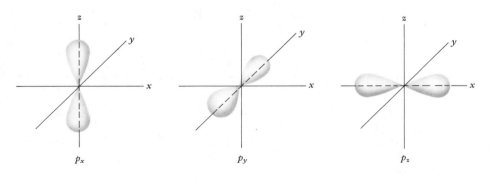

p_x p_y p_z

Figure 4-17 THE p ORBITALS. Each p orbital has two lobes.

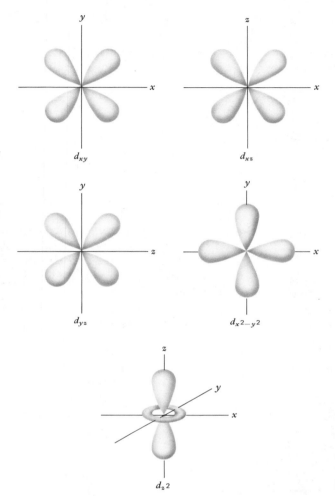

Figure 4-18 THE *d* ORBITALS. Except for the *d*$_{z^2}$ the *d* orbitals have four lobes.

probability. It is *not* proper in the modern sense to think of an electron orbiting the nucleus in some path within the orbital.

See Problem 4-20

Review Questions

1   Define a group in the periodic table.
2   What is meant by a continuous spectrum?
3   Under what conditions does tungsten give off a continuous spectrum?
4   What is meant by the wavelength of light?

5   Describe the relationship between the wavelength of light and energy.
6   What is meant by a discrete spectrum? Under what conditions would tungsten give off a discrete spectrum?

7 Explain the meaning of a stationary orbit.
8 Describe the ground state of an atom; an excited state.
9 How can an electron move to an excited state?
10 What happens when an electron in an outer orbit returns to an inner orbit?
11 What is a shell? How are shells labeled? How many electrons can go into each shell? What are the possible values of the n quantum number?
12 Describe a subshell. How are subshells labeled? How many electrons can go into each subshell? What are the possible values of the l quantum number?
13 What is the Aufbau principle?
14 What is meant by electron notation for an element?

15 How does electron notation of the elements relate to the periodic table?
16 Describe an orbital. How does it differ from an orbit? How many electrons can go into each orbital?
17 What is the relationship between the l quantum number and the m_l quantum number? How many orbitals are in each type of subshell?
18 What is meant by the Pauli exclusion principle?
19 When is an atom paramagnetic? Diamagnetic?
20 What is Hund's rule of maximum multiplicity? What does it mean?
21 Describe the shapes of the s, p, and d orbitals.

Problems

Shells and Subshells

4-1 When $n = 4$, what values can l have? What subshell (letter) does each value of l designate?

4-2 How many subshells are present in the fifth shell? What is the electron capacity of the fifth subshell (called the g subshell)?

■ **4-3** What is the l quantum number for a d subshell?

4-4 Which of the following subshells do not exist: $6s$, $3p$, $2d$, $5f$, $3f$, $1p$? [Hint: Look at the values of l for the shell in question.]

Electron Notation

4-5 Using only the periodic table, determine the electron notation for S, Zn, Y, ■ Tl, and ■ Pr.

4-6 What elements have the following electron notations? Use only the periodic table.
(a) $1s^2 2s^2 2p^5$
(b) $[Ar]4s^2 3d^{10} 4p^1$
(c) $[Xe]6s^2$
(d) $[Xe]6s^2 5d^1 4f^7$
■ (e) $[Ar]4s^1 3d^{10}$ (exception to rules)
■ (f) $[Kr]5s^2 4d^7$

4-7 What would be the atomic number of the element with one electron in the $5g$ subshell?

4-8 Element number 114 has not yet been made, but scientists believe that it will form a comparatively stable isotope. What would be its electron notation and in what group would it be?

■ **4-9** What electron notation is common for all elements in group VIIA? For group VIIB?

4-10 Using only the periodic table, determine which subshell fills first:
(a) $5s$ or $5p$ (d) $5s$ or $4d$
(b) $6s$ or $5p$ ■ (e) $4d$ or $5p$
(c) $6s$ or $4f$ ■ (f) $4f$ or $6p$

4-11 Which of the following elements fits the electron notation $ns^2(n-1)d^{10}np^4$?
(a) Cr (b) Te (c) S (d) O (e) Si

Orbitals

4-12 What are the values of m_l for an f subshell? How many orbitals are in the f subshell?

■ **4-13** When $l = 4$, how many values can m_l have? How many orbitals does this represent?

4-14 How many orbitals total are in the $n = 3$ shell?

Quantum Numbers

4-15 The following sets of four quantum numbers each describe an electron in an atom. Give the shell and subshell designated by each set and arrange them in order of increasing energy (n, l, m_l, m_s).

(a) 2, 0, 0, $+\frac{1}{2}$ (d) 4, 2, 2, $+\frac{1}{2}$
(b) 3, 2, 1, $-\frac{1}{2}$ (e) 5, 3, -3, $-\frac{1}{2}$
(c) 5, 1, -1, $+\frac{1}{2}$ (f) 4, 2, -1, $-\frac{1}{2}$

4-16 Which of the following sets of four quantum numbers is impossible?

(a) 2, 1, 0, $+\frac{1}{2}$ (c) 3, 3, -2, $-\frac{1}{2}$

(b) 3, 2, -3, $+\frac{1}{2}$ (d) 6, 3, 0, $+\frac{1}{2}$

■ 4-17 The atoms of the elements of three groups in the periodic table are diamagnetic. What are the groups?

4-18 On page 86, why couldn't group IVA elements be paramagnetic with both electrons in the same orbital (i.e., $\boxed{\uparrow\ \uparrow}$)?

4-19 How many unpaired electrons are in groups IIB, VB, VIA, VIB $[ns^1(n-1)d^5]$, and Pm (element number 61)?

Shapes of Orbitals

4-20 Describe the shape of the orbitals given by the following set of four quantum numbers: 4, 1, 0, $-\frac{1}{2}$

The Periodic Nature of the Elements

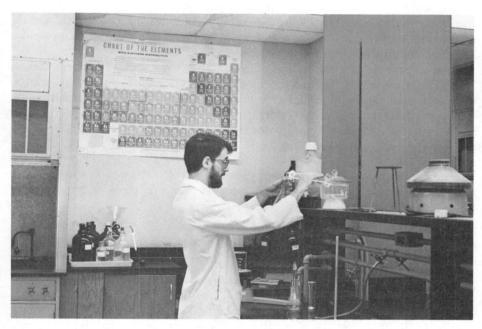

A periodic table is usually in view of the laboratory chemist.

It would appear that the periodic table adorns more walls in the chemistry department than Lenin's picture in the Kremlin. Why is this one table considered so beautiful or else so important that it graces so many classrooms, laboratories, and offices? The answer lies in the vast amount of information displayed by this instrument. In fact, this one-page table summarizes volumes of data about the elements and makes the table to a chemist what a brush is to a painter. Although it may be an exaggeration to suggest that a chemist will break out in a cold sweat if a periodic table isn't within easy grasp or sight, the table certainly is a faithful companion. The purpose of this chapter is to examine the trends in properties that allow us to utilize this one table for so much information.

The table in roughly its present form was introduced in 1869 by Dimitri Mendeleev of Russia and independently in 1870 by Lothar Meyer of Germany. At that time, elements were arranged by increasing atomic weights since the concept of atomic number was still unknown. This presented some problems, since several elements would be in reverse order if this rule were strictly followed (e.g., Co and Ni, I and Te). However, since many atomic weights were not accurately known at the time, it was assumed that the atomic weights of some of the elements that would otherwise be out of order were among those in error. We now know that the elements are arranged in increasing atomic number with vertical columns having the same notation for electrons in their outermost subshells.

An essential part of any theory is its ability to predict new phenomena. The gentlemen who proposed the modern periodic table were bold enough to leave vacant spaces in the table for undiscovered elements. Eventually, Ga (element number 31) and Ge (element number 32) were discovered and fit nicely into the spaces left vacant for them. As predicted by Mendeleev, the properties of these elements turned out to be similar to their nearest vertical neighbors in the table.

As background for this chapter, you should be familiar with the electron notation of the elements, especially for any electrons beyond the previous noble gas. This has all been discussed in Chapter 4.

5-1 The Physical Properties of the Elements

The 106 elements come in all physical states and a variety of colors. Chlorine is a light green gas, bromine is a red-brown liquid, and carbon is a black solid (graphite) or a colorless crystal (diamond).* *Perhaps the most basic division of the elements is between* **metals** *and* **nonmetals.** You may be familiar with certain metallic properties, such as the ability to conduct electricity and heat, but other properties such as hardness and luster are not solely metallic properties. Of more importance to the chemist are the differences in chemical properties between metals and nonmetals. The basis for this fundamental difference will be evident before this chapter is completed.

In the periodic table shown in Figure 5-1, the stairstep line separates the metals (M) on the left from the nonmetals (N) on the right. Elements on either side of the borderline have some metallic and some nonmetallic properties. Most of these elements are sometimes referred to as **metalloids** (MN). You will notice that almost 80% of the elements are classified as metals.

Figure 5-1 also indicates the physical state of the elements at room conditions (25°C). The temperature must be specified, since at very high temperatures all elements are gaseous atoms and at very low temperatures all elements are solids. About 85% of the elements are solids at room temperature,

---

* Most naturally occurring solids exist as crystals that have a defined and regular three-dimensional geometrical shape such as a cube (table salt and table sugar) or a tetrahedron (naturally occurring diamond).

Table 5-1   Periods

Periods	Number of Elements	Subshells	Noble Gas at End of Period
1	2	$1s$	He
2	8	$2s, 2p$	Ne
3	8	$3s, 3p$	Ar
4	18	$4s, 3d, 4p$	Kr
5	18	$5s, 4d, 5p$	Xe
6	32	$6s, 5d, 4f, 6p$	Rn
7	20	$7s, 6d, 5f$	?

with only two being liquids* (the metal Hg and the nonmetal Br) and the rest gases. (The numbers at the bottom of each box in Figure 5-1 refer to the radius of the atom and will be referred to later in this chapter.)

See Problems 5-1, 5-2

5-2 Periods

Periods *are horizontal rows of elements in the periodic table. Each period is composed of all of the elements between two noble gases.* The periods and the subshells that comprise each are listed in Table 5-1.

See Problem 5-3

5-3 Groups

Groups *of elements are vertical columns in the periodic table.* Elements in groups have the same number and arrangement of electrons in their outermost subshell as was shown in Figure 4-12. *The groups of elements can be classified into four general categories called* **Bohr Groups.**

1   The **representative elements** are metals and nonmetals where s and p subshells are being filled. They are shown as the A Group elements.

2   The **noble** or **rare gases** (Group 0) are nonmetals at the end of each period. Except for He all have filled outer s and p subshells.

3   The **transition elements** are metals where the d subshell is being filled. They are shown as the B Group elements.

4   The **inner transition elements** are metals where the f subshell is being filled. The elements where the $4f$ subshell is being filled are known as the **lanthanides** or **rare earths** (no. 58 to no. 71) and the elements where the $5f$ subshell is being filled are known as the **actinides** (no. 90 to no. 103).

* The metals cesium and gallium both melt just above room temperature (e.g., at about 29°C).

Figure 5-1 PHYSICAL STATES OF THE ELEMENTS.

M—metals
N—nonmetals
MN—metalloids

N 1								
H 0.37								
M 3 Li 1.33	M 4 Be 0.89							
M 11 Na 1.57	M 12 Mg 1.36							
M 19 K 2.03	M 20 Ca 1.74	M 21 Sc 1.44	M 22 Ti 1.32	M 23 V 1.22	M 24 Cr 1.17	M 25 Mn 1.17	M 26 Fe 1.14	M 27 Co 1.15
M 37 Rb 2.16	M 38 Sr 1.91	M 39 Y 1.62	M 40 Zr 1.45	M 41 Nb 1.34	M 42 Mo 1.29	M 43 Tc 1.27	M 44 Ru 1.24	M 45 Rh 1.25
M 55 Cs 2.35	M 56 Ba 1.97	M 57 La 1.69	M 72 Hf 1.44	M 73 Ta 1.34	M 74 W 1.30	M 75 Re 1.28	M 76 Os 1.26	M 27 Ir 1.26
M 87 Fr —	M 88 Ra —	M 89 Ac —	M 104 Ku —	M 105 Ha —	M 106 —			

M 58 Ce 1.65	M 59 Pr 1.65	M 60 Nd 1.64	M 61 Pm —	M 62 Sm 1.66
M 90 Th 1.65	M 91 Pa —	M 92 U 1.42	M 93 Np —	M 94 Pu —

There is very little vertical chemical similarity among these elements but there is a strong horizontal similarity especially among the lanthanides.

Since much of the chemistry that will be discussed in this book involves the representative elements, they will be singled out for a formal introduction. Before long most of these elements will seem like old friends, so we should get to know them better.

See Problem 5-4

Solids—clear
Liquids—⊠
Gases— ▨

									N 2 He 0.93
			MN 5 B 0.80	N 6 C 0.77	N 7 N 0.74	N 8 O 0.74	N 9 F 0.72	N 10 Ne 1.12	
			M 13 Al 1.25	MN 14 Si 1.17	N 15 P 1.10	N 16 S 1.04	N 17 Cl 1.00	N 18 Ar 1.54	
M 28 Ni 1.15	M 29 Cu 1.17	M 30 Zn 1.25	M 31 Ga 1.24	MN 32 Ge 1.21	MN 33 As 1.21	N 34 Se 1.14	35 Br 1.14	N 36 Kr 1.69	
M 46 Pd 1.28	M 47 Ag 1.44	M 48 Cd 1.44	M 49 In 1.44	M 50 Sn 1.41	MN 51 Sb 1.41	MN 52 Te 1.37	N 53 I 1.35	N 54 Xe 1.90	
M 78 Pt 1.29	M 79 Au 1.44	M 80 Hg 1.44	M 81 Tl 1.80	M 82 Pb 1.54	M 83 Bi 1.52	MN 84 Po 1.64	MN 85 At —	N 86 Rn 2.20	

M 63 Eu 1.85	M 64 Gd 1.61	M 65 Tb 1.59	M 66 Dy 1.59	M 67 Ho 1.58	M 68 Er 1.57	M 69 Tm 1.56	M 70 Yb 1.70	M 71 Lu 1.56
M 95 Am —	M 96 Cm —	M 97 Bk —	M 98 Cf —	M 99 Es —	M 100 Fm —	M 101 Md —	M 102 No —	M 103 Lr —

Group IA ([Noble Gas]ns^1) These metals follow the noble gases and are known as the **alkali metals.** They are all soft, shiny metals but are very reactive with air and water. All of these elements react with Group VIIA elements to form ionic compounds with the general formula MF, MCl, MBr, and MI. The properties of this group vary smoothly from top to bottom.

Hydrogen is often shown as a IA element because it has a $1s^1$ electron notation. However, hydrogen is not an alkali metal and this nonmetal actu-

Figure 5-2 GROUP IA, THE ALKALI METALS. Sodium is a typical alkali metal which must be stored under mineral oil because of its chemical reactivity.

ally has little in common with these metals other than the formulas of some of its compounds (i.e., HCl and LiCl). Hydrogen is a unique element with little periodic relationship to any other element. It should (and sometimes is) given a position in the table all to itself.

Group IIA ([Noble Gas]ns^2) The group IIA elements are known as the **alkaline earth metals.** They are not as reactive as the alkali metals, since most of the metals can exist in air (see Figure 5-3). The elements form compounds with oxygen with the general formula MO. With group VIIA elements they have the general formula MCl_2, MBr_2, and so on. Beryllium chemistry is somewhat unique compared to the chemistry of the rest of the group. Magnesium forms a strong, lightweight **alloy** (*a homogeneous mixture of metals*). It is also used in flashbulbs because of the bright light it gives off when ignited.

Group IIIA ([Noble Gas]$ns^2(n-1)d^{10}np^1$)* In group IIIA there is a significant difference between the nonmetal at the top of the group (boron) and the next element, aluminum. Because of its unique chemistry, boron chemistry is usually described separately from that of the other members of the group. Aluminum is an important metal in this group because of its combination

* As shown in Table 5-1, a filled *d* subshell is not present until the fourth period (i.e., Ga).

Figure 5-3 GROUP IIA, THE ALKALINE EARTH METALS. Magnesium is a typical alkaline earth metal. It is not as reactive as the alkali metals and can be stored in the open air. Shown here is a piece of magnesium ribbon.

of low density and high strength. Aluminum is also the most common metal in the earth's crust (the outer layers including the atmosphere). Compounds of aluminum are components of most clays (see Figure 5-4).

Group IVA ([Noble Gas]$ns^2(n - 1)d^{10} np^2$) Organic chemistry is a major branch of chemistry that focuses mainly on the millions of compounds formed by carbon, the lightest member of group IVA. The complex organic compounds involved in life processes are studied in another but closely allied branch of chemistry called **biochemistry.** Pure carbon occurs in two crystalline forms as an element, graphite and diamond. *Different forms of a pure element are called* **allotropes** (see Figure 5-5).

Silicon and germanium, known as semiconductors, are metalloids and find use in transistors because of their ability to conduct a limited amount of electricity. Silicon, a major component of sand, is also a plentiful element in the earth's crust. Tin and lead have properties typical of other metals.

Figure 5-4 GROUP IIIA. In this group boron (left) is a nonmetal but aluminum (right) and other members of the group are metals.

Figure 5-5 GROUP IVA. Carbon is found in two crystalline allotropes, graphite and diamond (left and center). Tin is a typical metal (right).

Group VA ([Noble Gas]$ns^2(n - 1)d^{10}np^3$) Group VA shows the most startling change in properties from top to bottom of any group. At the top nitrogen exists as a diatomic gas, N_2. It is important as the major component (about 80%) of our atmosphere as well as a component of living organisms. Phosphorus is a nonmetal solid that exists as P_4 molecules in its most common allotrope, called white phosphorus. It is very reactive in air, burning violently with the evolution of large amounts of heat. Arsenic and antimony are metalloids, but bismuth is definitely metallic. (See Figure 5-6.)

Group VIA ([Noble Gas]$ns^2(n - 1)d^{10}np^4$) Group VIA (see Figure 5-7) is sometimes (actually very rarely) referred to as the **chalcogens.** As in group

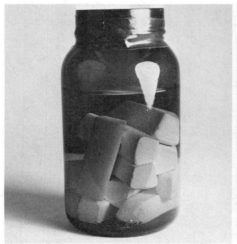

Figure 5-6 Phosphorus (left) is a reactive nonmetal and must be stored under water. Bismuth (right) is a typical metal.

Figure 5-7 GROUP VIA. The two famous members of this group are oxygen, which is a gas, and sulfur, which is a nonmetal solid.

VA, there is a large variation in properties from top to bottom. At the top, oxygen has a unique chemistry and for this reason is usually discussed separately from the other elements in the group (like other elements in the second period). Oxygen is the most abundant element in the earth's crust and makes up about 20% of the atmosphere. It exists mainly as O_2 in the air, but there is another important allotrope of oxygen, O_3, which is known as **ozone.** Ozone in the upper atmosphere is essential to life on earth, since it absorbs damaging ultraviolet light from the sun. In the lower atmosphere, however, it is a toxic, pungent-smelling pollutant. Oxygen forms compounds either directly or indirectly with all other elements except helium, neon, and argon. Sulfur and selenium are also nonmetals and form many important compounds. Tellurium and polonium are both metalloids, although the chemistry of polonium has not been studied extensively. It is a highly radioactive and unstable element that occurs in nature only in trace amounts, making it difficult to study.

Group VIIA ([Noble Gas]$ns^2(n-1)d^{10}np^5$) Like the alkali metals, group VIIA elements show a strong chemical similarity from top to bottom. This group, known as the **halogens,** are all nonmetals and exist as diatomic molecules in the elemental state. Astatine is classified as a metalloid, but because of its radioactivity it is also difficult to study. All of the halogens are very reactive elements. Fluorine is especially reactive and, like oxygen, forms compounds with all other elements except helium, neon, and argon. At room temperature, fluorine and chlorine are gases, bromine is a liquid, and iodine is a solid (see Figure 5-8).

Figure 5-8 GROUP VIIA, THE HALOGENS. At room temperature chlorine (left) is a gas, bromine (center) is a liquid,
and iodine (right) is a solid. The elements of this group are
very reactive chemically. (Left, Time/Life Books, Inc.)

Sometimes hydrogen is also shown in group VIIA as well as IA. However, hydrogen is not a halogen although it does have some chemical resem-
blance to the group VIIA elements. Hydrogen forms a negative ion and exists
as a diatomic gas, H_2. Hydrogen is not nearly as reactive as the halogens,
however. It is sometimes placed in group VIIA because, like the halogens, it
See Problem 5-5   is the element before a noble gas.

5-4 Periodic Trends: Atomic Radius

The **radius** *of an atom is the distance from the nucleus to the outermost electrons.* This distance is in fact neither clearly defined nor easily measured. As
mentioned in Chapter 4, modern theory of the atom discourages a model of
the atom where an electron orbits the nucleus at a fixed distance. Instead, the
electrons have a certain *probability* of being at a range of distances from the
nucleus. Therefore, the radius does not actually have a fixed value. Besides
the theoretical problem, there is a more practical problem. It is possible to determine the distance between atoms in a solid, but it is difficult to decide
where one atom starts and the other leaves off. Nevertheless, consistent values for the radii of neutral atoms have been compiled and are listed under the
element in Figure 5-1. (The values are in angstrom units. One angstrom (Å)
equals 10^{-8} cm.)

There are two trends that are apparent:

1   Horizontally, from left to right across a period, the atoms generally get
    smaller.

2   Vertically, down a group, the atoms generally get larger.

Notice that these are simply trends and are certainly not true in all cases. However, since the radius of an atom has an important effect on other properties, it is worth taking the time to explore the reasons for the trends noted.

As mentioned, the radius of an atom is determined by the distance from the nucleus to the *outermost* electrons. To a large extent, that distance depends on the forces of attraction between the outermost electrons and the nucleus. These forces of attraction depend on (1) the charge on the nucleus, (2) the shell or distance from the nucleus of the outermost electron and, (3) the location of electrons in an atom relative to each other (known as shielding effects).

The force of attraction (F) between the positively charged nucleus (represented as q^+) and the negatively charged electron (represented as q^-) is determined by **Coulomb's law,** which states:

$$F = \frac{q^+ q^-}{r^2}$$

Coulomb's law tells us that:

1 There is an inverse square relationship between the attraction (F) and the distance (r) between the nucleus and the electron. The *smaller* the value of r, the *greater* is the attraction.

2 There is a direct relationship between F and the magnitude of the nuclear charge (q^+). The *greater* the charge, the *greater* is the attraction.

See Problem 5-6

In an atom, F and r interact with each other to some extent. If the attraction (F) is large between an electron and a nucleus, the nucleus pulls the electron in closer, thus shrinking the atom. (You may have noticed how a large force of attraction tends to draw charged species together in dealings with the opposite sex.)

The amount of nuclear charge (q^+) an outer electron feels will depend to an extent on the location of the electrons in an atom *relative to each other*. To illustrate this, let's take an atom with a nucleus ⊕ and two electrons that we are free to move around as we please. In this endeavor, Bohr's model with electrons as particles in specific orbits will serve us well. In Figure 5-9, case A represents a situation in which electron 1 is in between electron 2 and the nucleus. The radius of the atom is determined by electron 2, which lies in an outer subshell compared to electron 1. In this case electron 1 "eclipses" electron 2, thus *shielding* some of the nuclear charge from the outer electron. This is similar to what happens during a solar eclipse when the moon "shields" the earth from some of the light from the sun.

Now we will move electron 1 out to the same subshell as electron 2, as in case B. Notice that electron 1 no longer eclipses electron 2, so that neither electron is shielded from the nuclear charge. Since the electron feels more of the nuclear charge, it feels a stronger attraction according to Coulomb's law. As mentioned before, the stronger the attraction, the more the nucleus and

Figure 5-9 SHIELD-ING. An electron shields the nuclear charge from another electron more effectively if the former electron lies in an inner subshell.

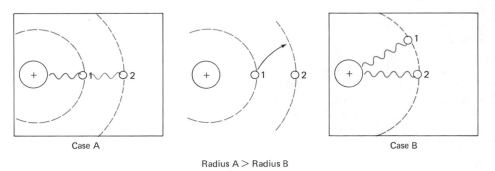

Case A

Case B

Radius A > Radius B

the electrons draw together. The result is that the atom in case B is smaller than that in case A.

This discussion can now be used to rationalize why successive elements have smaller radii across a period when a subshell is being filled. The following facts can be listed. Notice that each fact follows directly as a result of the previous fact. From element to element when a subshell is being filled:

1 Each additional electron does not effectively shield the additional nuclear charge from the other electrons in the same subshell, since it does not come between the nucleus and the other outer electrons.

2 All of the electrons in the outer subshell will feel a greater nuclear charge.

3 The electrons are more firmly attracted to the nucleus.

4 The radius of the atom is smaller.

Figure 5-10 illustrates how the radius of a carbon atom compares to that of a boron atom. It should be noted that this is a simplification of the situation, as electrons in the same subshell do shield each other to a small extent. More accurate calculations of shielding and the actual nuclear charge that an outer electron "feels" is possible if all factors are taken into account. The trend discussed also has many exceptions, as you will notice in Figure 5-1.

Now let's go down a group (see Figure 5-11). Each successive element has a greater nuclear charge, which by itself would mean smaller atoms. In this case, however, the additional electrons lie between the outer electrons and the nucleus, which means that the additional nuclear charge is effectively shielded by the additional electrons. The result is a stand-off. One effect (higher nuclear charge) is canceled by the other (shielding). The outer electrons therefore feel about the same nuclear charge in each case. The atoms have a larger radius for each successive element because the outer electrons are in a higher-energy shell which lies farther from the nucleus.

An understanding of the trends in the size of the atoms is important, since the trend is directly linked to ionization energy, which is discussed

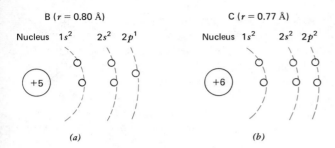

Figure 5-10 THE RADII OF BORON AND CARBON.
(a) One outer $2p$ electron is shielded from $+5$ charge
by four inner electrons. The outer electron "feels"
about $+5 - 4 = +1$ nuclear charge. (b) Two
outer $2p$ electrons are shielded from $+6$ charge by
four inner electrons. The outer electron "feels"
about $+6 - 4 = +2$ nuclear charge.

next. Understanding ionization energy is basic to understanding the bonding
properties of the elements.

**See Problems 5-7,
5-8, 5-9**

5-5 Periodic Trends: Ionization Energy

Ionization energy (*I.E.*) *is defined as the energy required to remove an electron
from a gaseous atom to form a gaseous ion.* Since the outermost electron is
generally the least firmly attached, it is the first to go.

$$M(g) \longrightarrow M^+(g) + e^-$$

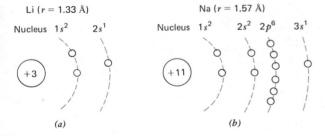

Figure 5-11 THE RADII OF SODIUM AND LITHIUM.
(a) One outer $2s$ electron is shielded from $+3$
charge by two inner electrons. The outer electron
"feels" about $+3 - 2 = +1$ nuclear charge.
(b) One outer $3s$ electron is shielded from
$+11$ charge by 10 inner electrons. The outer electron
"feels" about $+11 - 10 = +1$ nuclear charge
but it is in a shell further from the nucleus.

This is always an endothermic process (requires energy), since an electron is held in the atom by its attraction to the nucleus. This attractive force must be overcome to separate the electron from the atom, leaving the rest of the atom as a positive ion. *The amount of energy that this process requires depends on how strongly the outermost electron is attracted to the nucleus.* Fortunately, the same reasoning that explains the trends in atomic radii also explains the trends in ionization energy. The ionization energies for the second and third period are shown in Table 5-2. The energy unit abbreviated kJ/mol stands for kilojoules per mole, which is energy per a certain defined quantity of atoms.

Across a period the I.E. generally gets larger, which means increased difficulty in removing an electron. This corresponds to what was discussed in Section 5-3. Since the outer electrons are more strongly attracted to the nucleus, the atoms of each successive element are smaller and the outermost electron harder to remove.

Down a group the I.E. gets smaller for removal of the first electron. As the atoms become larger, the outer electron is less firmly attracted to the nucleus because of the greater distance. Thus the outer electron is easier to remove for each successive element.

See Problem 5-10

The ease with which an atom forms a positive ion by losing an electron is a fundamental difference between a metal and a nonmetal. Notice that metallic properties increase to the left and down in the periodic table. This also corresponds to the trend toward lower I.E. Therefore, metals form cations more easily than nonmetals. Nonmetal behavior, to the right and up in the periodic table, is indicated by a comparative difficulty in forming cations. This fundamental trend affects the difference in bonding behavior in the metals and nonmetals, which will be discussed in the next chapter.

The second I.E. for an ion involves the removal of an electron from a +1 ion to form a +2 ion:

$$M^+(g) \longrightarrow M^{2+}(g) + e^-$$

In a similar manner, the third I.E. forms a +3 ion and so forth. In all cases it becomes increasingly difficult to remove each succeeding electron. When an

Table 5-2 Ionization Energy of Some Elements

Element	I.E. (kJ/mol)	Element	I.E. (kJ/mol)
Li	520	Na	496
Be	900	Mg	738
B	801	Al	578
C	1086	Si	786
N	1402	P	1102
O	1314	S	1000
F	1681	Cl	1251
Ne	2081	Ar	1520

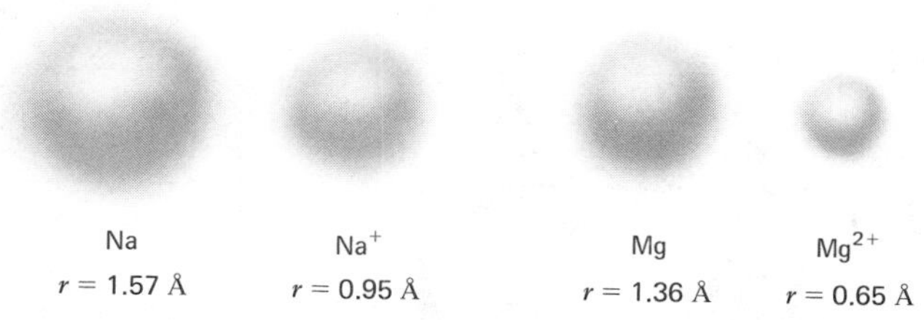

Figure 5-12 IONIC RADII. A positive ion is always smaller than the neutral atom.

Na
$r = 1.57$ Å

Na$^+$
$r = 0.95$ Å

Mg
$r = 1.36$ Å

Mg^{2+}
$r = 0.65$ Å

electron is removed from an atom or ion, the ion remaining shrinks as a result of the increased attraction that the nucleus will have for the remaining electrons (see Figure 5-12). Since the remaining electrons are held more firmly, the next electron is more difficult to remove. The trends in ionization energies for Na through Al are shown in Table 5-3.

Notice that the second I.E. for Na, the third I.E. for Mg, and the fourth I.E. for Al are all very large compared to the preceeding number. For example, it takes 2188 kJ (1450 + 738) to remove the first two electrons from Mg but about three times as much energy to remove the third electron (7732 kJ). In the case of Al, the first three electrons can be removed (to form Al^{3+}) at a cost of 5137 kJ, but the fourth would require an input of energy of 11,580 kJ (to form Al^{4+}).

For Na, the first electron removed comes from the outer $3s$ subshell but the second must be removed from the inner $2p$ subshell. Removal of the first two electrons from Mg to form a Mg^{2+} ion is comparatively easy compared to formation of a Mg^{3+} ion, which also would require removal of one electron from the inner $2p$ subshell. *Apparently, formation of a cation by removal of electrons from filled noble gas electron structures is energetically a very expensive process.* When electrons are actually removed from an atom the energy must be supplied from somewhere. The source of this energy is the chemical reaction. There are limits to this energy, however, so there are limits to the formation of positive ions. We will pursue this in the next chapter.

See Problems 5-11, 5-12

5-6 Periodic Trends: Electron Affinity

Electron affinity (E.A.) *is defined as the energy released when a gaseous atom adds an electron to form a gaseous ion.*

$$X(g) + e^- \longrightarrow X^-(g)$$

Table 5-3   Ionization Energies[a]

Element	First I.E.	Second I.E.	Third I.E.	Fourth I.E.
Na	496	4565	6912	9540
Mg	738	1450	7732	10,550
Al	577	1816	2744	11,580

[a] All values are in kJ/mol.

Table 5-4   Electron Affinity[a]

Element	E.A. (kJ/mol)	Element	E.A. (kJ/mol)	Element	E.A. (kJ/mol)
B	-30				
C	-121	Si	-140		
N	$+9$	P	-75		
O	-140	S	-200	Se	-160
F	-334	Cl	-349	Br	-325

See Problems 5-13, 5-14

[a] A negative value means that the process is exothermic.

The electron affinities of some nonmetals are given in Table 5-4. The trends are somewhat uneven but generally the formation of a negative ion becomes more favorable to the right in the periodic table. The *general* trend is to lower electron affinities down a group although we do find that the third period (Si − Cl) is actually higher than the second. Thus a high electron affinity (ease in formation of an anion) is generally a nonmetal property just as a low ionization energy (ease in formation of a cation) is a metallic property. Although it is energetically unlikely to form a nonmetal cation in chemical reactions formation of a nonmetal anion is favorable.

Review Questions

1 What is a metalloid? Where are they located in the periodic table?

2 Name the elements that are gases at room temperature.

3 What is a period in the periodic table?

4 Name the Bohr groups. What are lanthanides?

5 What is the common name for the group IA elements? Group IIA? Group VIIA?

6 Define an allotrope.

7 Identify the two most reactive elements.

8 Identify the group among the "A" elements that has the largest variation of properties from top to bottom. Which groups have the least?

9 What is meant by shielding of the nucleus?

10 Explain why a carbon atom is smaller than a boron atom.

11 Explain why a phosphorus atom is larger than a nitrogen atom.

12 Of the representative elements, which has the smallest radius? Which has the largest radius?

13 What is meant by ionization energy (I.E.)? How do trends in atomic radii correlate with trends in I.E.?

14 Which representative element has the highest I.E.? Which has the smallest?

15 How do the trends in I.E. relate to metallic properties?

16 What is the second ionization energy for an element?

17 Define electron affinity.

18 How do trends in electron affinity relate to nonmetallic properties?

Problems

The Periodic Table

5-1 From the periodic table, identify each of the following elements as a metal or a non-metal:
(a) Ru (b) Sn (c) Hf (d) Te (e) Ar ■ (f) B ■ (g) Se ■ (h) W

5-2 Which, if any, of the elements in Problem 5-1 are classified as metalloids?

5-3 How many elements would be in the seventh period if it were complete? How many elements would be in a full eighth period?

5-4 Classify the following elements into one of the four Bohr Groups:
(a) Fe (b) Te (c) Pm (d) La ■ (e) Xe ■ (f) H ■ (g) In

5-5 Why is hydrogen sometimes classified with the IA elements? How is H different from IA elements? Why is hydrogen sometimes classified with the VIIA elements?

Atomic Radius

5-6 Using Coulomb's law, place the following interactions between an electron (-1 charge) and a nucleus in order of increasing attraction (assume no shielding).
(a) A $+1$ nucleus at a distance of 1 Å
(b) A $+2$ nucleus at a distance of 2 Å
(c) A $+4$ nucleus at a distance of 4 Å
(d) A $+4$ nucleus at a distance of 5 Å
(e) A $+2$ nucleus at a distance of 4 Å
(f) A $+6$ nucleus at a distance of 5 Å

*5-7 Using Coulomb's law, place the following interactions between an electron and a nucleus shielded by inner electrons in order of increasing attraction.
(a) A $+11$ nucleus shielded by 10 inner electrons at a distance of 3 Å
(b) A $+12$ nucleus shielded by 10 inner electrons at a distance of 3 Å
(c) A $+6$ nucleus shielded by 4 inner electrons at a distance of 2 Å
(d) A $+8$ nucleus shielded by 7 inner electrons at a distance of 2 Å

5-8 Which of the following atoms has the largest radius?

(a) As or Se (b) Ru or Rh (c) Sr or Ba ■ (d) F or I ■ (e) Sc or Y

*5-9 Zirconium and hafnium are in the same group and have almost the same radius despite the general trends. As a result the two elements have almost identical chemical and physical properties. The fact that these elements have almost the same radius is due to the *lanthanide contraction*. With the knowledge that atoms get progressively smaller as a subshell is being filled, can you explain this phenomenon? (Hint: Follow all of the expected trends between the two elements.)

Ionization Energy

5-10 Which of the following elements has the higher ionization energy?
(a) Ti or V (b) P or Cl (c) Mg or Sr
(d) Fe or Os (e) B or Br ■ (f) B or Al ■ (g) Ba or Cs ■ (h) Ga or S

5-11 The first four ionization energies for Ga are 578.8 kJ (per mol), 1979 kJ, 2963 kJ, and 6200 kJ. How much energy is required to form each of the following ions: Ga^+, Ga^{2+}, Ga^{3+}, and Ga^{4+}? Why does the formation of Ga^{4+} require a comparatively large amount of energy?

■ **5-12** The first five ionization energies for carbon are 1086 kJ (per mol), 2353 kJ, 4620 kJ, 6223 kJ, and 37,830 kJ. How much energy is required to form the following ions: C^+, C^{2+}, C^{3+}, C^{4+}, and C^{5+}. In actual fact even C^+ does not form in compounds. Compare the energy required to form this ion with that needed to form some metal ions and explain.

5-13 From an energy standpoint, predict which of the following ions might be expected to exist. If the answer is no, explain.
(a) Cs^+ (b) Rb^{2+} (c) Ne^+ (d) Tl^{3+} (e) S^+
(f) Pb^{4+} (g) Ba^{2+} ■ (h) Ca^{3+} ■ (i) F^{2-} ■ (j) Na^{3+} ■ (k) Te^{2+} ■ (l) Sc^{3+}

★5-14 On the planet Zerk, the periodic table of elements is slightly different from ours. On Zerk, there are only two p orbitals, so a p subshell holds only four electrons. There are only four d orbitals, so a d subshell holds only eight electrons. Everything else is the same as on earth, such as the order of filling ($1s$, $2s$, etc.) and what is characteristic of noble gases, metals, and nonmetals. Construct a Zerkian periodic table using numbers for elements up to element number 50. Then answer these questions.

(a) How many elements are in the second period? In the fourth period?

(b) What are the atomic numbers of the noble gases at the ends of the third and fourth periods?

(c) What is the atomic number of the first inner transition element?

(d) Which element is most likely to be a metal: element number 5 or element number 11, element number 17 or element number 27?

(e) Which element has the largest radius: element number 12 or element number 13; element number 6 or element number 12?

(f) Which atom has a higher ionization energy: (1) element number 7 or element number 13; (2) element number 7 or element number 5; (3) element number 7 or element number 9?

(g) Which ions are reasonable? (1) 16^+ (2) 9^{2+} (3) 7^+ (4) 13^- (5) 17^{4+} (6) 15^+ (7) 1^-

The Nature of Bonding

Atoms and people have much in common. They both have a tendency to form bonds with other atoms or people that are mutually beneficial.

At last we have the tools necessary to seek answers to the fundamental questions raised in the introduction to Chapter 4. These questions concerned the nature of the chemical bond and why bonds between elements differ, yet in a periodic manner. In Chapter 4, the basis for this periodic relationship was found to lie in the electronic structure of the elements. The periodic nature of electron structure implied trends in ionization energy and electron affinity which were examined in Chapter 5. This all leads us to this chapter, in which another fundamental periodic property will be developed: the chemical bond itself. This topic is of fundamental importance in the study of chemistry.

The background material necessary for this chapter includes:

1 The nature of ionic and molecular compounds (Sections 3-2 and 3-3)

2 Electron notation (structure) of the elements (Section 4-6)

3 The relationship of electron structure to the periodic table (Section 4-6)

4 Trends in ionization energy and electron affinity (Sections 5-5 and 5-6)

6-1 Bond Formation and the Representative Elements

In a way, people are much like the atoms of which they are made. Both people and atoms are social creatures, forming bonds with others because of a mutual need for a more stable arrangement. There is an important difference in their behavior, however. The most content group of atoms are those that have little or no desire to join with others to form bonds. As mentioned previously, these are atoms of the noble gas elements. The noble gases all exist as individual gaseous atoms under normal conditions. Until recently these elements were also known as the "inert gases" because it was widely believed that they could never form bonds. Since 1962, however, several compounds of xenon, krypton, and radon have been prepared with the most reactive elements, fluorine and oxygen.

Since the formation of bonds necessitates an alteration of the element's electron structure, we can conclude that noble gases are stable with their electron arrangement as it is. The other representative elements (and some transition elements) prefer the electron structure of the noble gases and form bonds in such a manner as to achieve a similar arrangement. The electron structure of a noble gas is comprised of filled outer s and p subshells which hold a total of eight electrons (except He). *The* **octet rule** *states that atoms of the representative elements form bonds in such a manner as to have access to exactly eight outer s and p electrons (also known as the* **valence electrons***).* This goal can be achieved in three ways:

1 A *metal* may lose one to three electrons to form a cation with the structure of the previous noble gas.

2 A *nonmetal* may gain one to three electrons to form an anion with the structure of the next noble gas.

3 Atoms (usually two nonmetals) may share electrons with other atoms to attain the number of electrons in the next noble gas.

Cases 1 and 2 complement each other to form ionic compounds. Case 3 produces covalent compounds.

6-2 Lewis Dot Structures

The bonding of elements of the representative elements focuses mainly on the valence electrons, which are those electrons in the outer s and p sub-

Figure 6-1 Dot structures.

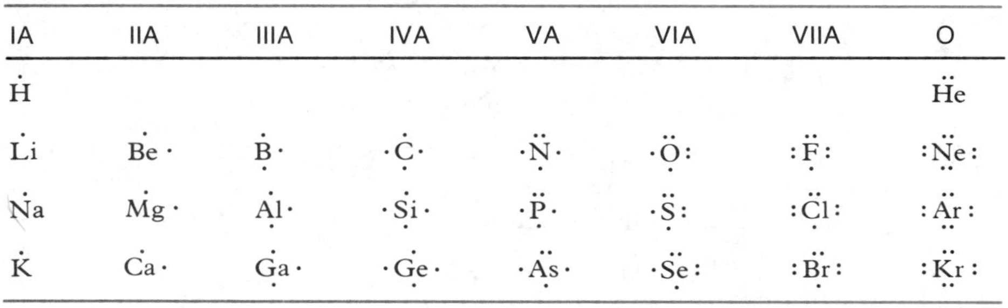

IA	IIA	IIIA	IVA	VA	VIA	VIIA	O
Ḣ							:He
Li	Be·	Ḃ·	·Ċ·	·N̈·	·Ö:	:F̈:	:N̈e:
Na	Mg·	Al·	·Si·	·P̈·	·S̈:	:C̈l:	:Ar̈:
K̇	Ca·	Ġa·	·Ge·	·As·	·S̈e:	:Br̈:	:K̈r:

shells. **Lewis dot structures*** *of these elements represent the valence electrons as dots around the symbol of the element.* Since bonding of these elements involves access to eight electrons (four pairs), the electrons are represented as one or two dots on the four sides of the element's symbol. Although the valence electrons come from two different subshells (s and p), only the total number of these electrons is important for discussions of bonding. Thus it is convenient to place one electron on each side of the symbol first (groups IA through IVA) and then represent pairs of electrons (group VA through 0).† In this manner, the Lewis dot structures of the first four periods are shown in Figure 6-1. Notice that the group number also represents the number of dots or outer electrons for a neutral atom of the representative elements.

See Problems 6-1, 6-2

6-3 The Formation of Ions

In Section 5-4 the formation of positive ions was discussed. When one electron is removed from sodium, it was found that the process does not require a large amount of energy compared to other ionizations such as the removal of an electron from a nonmetal or even a second electron from sodium. As shown below, using a dot structure for Na, the electron that is lost is the outer $3s^1$ electron, leaving a cation with the noble gas structure of Ne.

$$\dot{N}a \longrightarrow Na^+ + e^-$$
$$[Ne]3s^1 \qquad [Ne]$$

All other metals in group IA can lose one electron to form a $+1$ ion with the electron structure of the preceding noble gas.

---

* Named after the American chemist G. N. Lewis (1875–1946), who developed this theory of bonding.
† This procedure of representing the s electrons as unpaired for bonding purposes has a basis in fact. For example, although we learned in Chapter 4 that carbon has two paired electrons ($2s^2$) and two unpaired electrons (in $2p$ orbitals), it forms bonds as if all four valence electrons were in equivalent or what are referred to as "hybridized" orbitals.

The element magnesium can lose two electrons to leave an ion with the electron structure of Ne:

$$\overset{\cdot}{\text{Mg}}\cdot \longrightarrow \text{Mg}^{2+} + 2e^-$$
$$[\text{Ne}]3s^2 \qquad [\text{Ne}]$$

All other metals in group IIA can lose two electrons to form a $+2$ ion with the structure of the preceding noble gas.

Group IIIA *metals* can lose three electrons to form $+3$ ions:

$$\overset{\cdot}{\text{Al}}\cdot \longrightarrow \text{Al}^{3+} + 3e^-$$
$$[\text{Ne}]3s^23p^1 \qquad [\text{Ne}]$$

In this group, boron is *not* a metal and does not form a $+3$ ion. Its bonds are covalent in nature and will be discussed later. There is some deviation from the noble gas concept in this group as follows:

$$\overset{\cdot}{\text{Ga}}\cdot \longrightarrow \text{Ga}^{3+} + 3e^-$$
$$[\text{Ar}]4s^23d^{10}4p^1 \qquad [\text{Ar}]3d^{10}$$

Gallium loses three electrons (the outer $4s$ and $4p$ electrons) to form a $+3$ ion but maintains the $3d^{10}$ electrons. This is known as a pseudo-noble gas structure. *A* **pseudo-noble gas structure** *has a filled d subshell in addition to the electrons of the previous noble gas.* This electron arrangement is apparently a satisfactory substitute for a noble gas structure for Ga, In, and Tl.★

In group IVA, the metals Ge, Sn and Pb would have to lose four electrons to form a pseudo-noble gas electron structure. This doesn't happen, since it takes too much energy to remove four electrons from an atom. Like charging on a credit card, there is a limit to how far in the hole one can go and still be covered. Ge, Sn, and Pb bond either by electron sharing or by forming a $+2$ ion that does not have a pseudo-noble gas structure.

In summary, *metals of the representative element can attain a noble gas structure (or a pseudo-noble gas structure) by losing up to three electrons to form cations.*

Positive ions do not form independently, as the lost electrons must go somewhere. The electrons are added by atoms that form negative ions. In Chapter 5 we learned that this process is more favorable for nonmetals.

★ Group IB elements can lose one electron to form $+1$ ions, and group IIB elements can lose two electrons to form $+2$ ions. These ions also have pseudo-noble gas structures. The electrons lost to form these ions are the *ns* electrons and not the $(n-1)d^{10}$ electrons. Group IIIB elements lose three electrons $[ns^2(n-1)d^1]$ to form a $+3$ ion with a noble gas electron structure. Other transition metals form one or more cations (e.g., Fe^{2+}, Fe^{3+}) that have no relation to a noble gas structure. Some of these ions will be discussed in Chapter 7.

Group VIIA elements (including H) are all one electron short of a noble gas structure. If they add one electron they attain this noble gas structure and form a -1 ion:

$$e^- + \text{H} \cdot \longrightarrow \quad \text{H} \!:^-$$
$$1s^1 \qquad\qquad 1s^2 = [\text{He}]$$

$$e^- + \quad :\!\ddot{\text{Cl}}\!: \quad \longrightarrow \quad :\!\ddot{\text{Cl}}\!:^-$$
$$[\text{Ne}]3s^2 3p^5 \qquad [\text{Ne}]3s^2 3p^6 = [\text{Ar}]$$

Group VIA nonmetals can attain a noble gas structure by adding two electrons to form a -2 ion:

$$2e^- + \quad \cdot \ddot{\text{O}}\!: \quad \longrightarrow \quad :\!\ddot{\text{O}}\!:^{2-}$$
$$[\text{He}]2s^2 2p^4 \qquad [\text{He}]2s^2 2p^6 = [\text{Ne}]$$

In group VA, nitrogen and phosphorus can add three electrons to form -3 ions:

$$3e^- + \quad \cdot \ddot{\text{N}}\!\cdot \quad \longrightarrow \quad :\!\ddot{\text{N}}\!:^{3-}$$
$$[\text{He}]2s^2 2p^3 \qquad [\text{He}]2s^2 2p^6 = [\text{Ne}]$$

For the most part, group IVA nonmetals bond by electron sharing instead of ion formation. Although there is some evidence for the C^{4-} ion, formation of such a highly charged ion would be an energetically unfavorable process.

In summary, *nonmetals can attain a noble gas electron structure by adding up to three electrons to form anions.*

See Problems 6-3, 6-4

6-4 Formulas of Binary Ionic Compounds

Metals part with electrons with an input of energy (ionization energy) and nonmetals add an electron with the evolution of energy (electron affinity). Although the energy involved in the overall process involves more than just the ionization energy of the metal and the electron affinity of the nonmetal, you can appreciate the basis for a compatible marriage between metals and nonmetals. For example, the combination of sodium and chlorine produces a **binary** (*composed of two elements*) ionic compound:

$$\text{Na} \cdot \; + \; \cdot \ddot{\text{Cl}}\!: \; \longrightarrow \quad \text{Na}^+ :\!\ddot{\text{Cl}}\!:^-$$
$$\text{Formula} = \text{NaCl}$$

The transfer of the one electron results in two ions with opposite but equal charges. Both ions have noble gas structures. However, when lithium combines with oxygen, two lithium atoms are needed to satisfy the need for two electrons for one oxygen. Notice that the resulting charges cancel. The chemical formula for the compound formed is Li_2O.

$$\begin{array}{l} Li \cdot \\ \quad\quad + \cdot \ddot{O}: \longrightarrow (Li^+)_2 : \ddot{O} :^{2-} \\ Li \cdot \end{array}$$

Formula $= Li_2O$

When calcium combines with bromine the opposite situation exists. Two Br atoms are needed to take up the two electrons from one Ca.

$$Ca \cdot \longrightarrow \begin{array}{l} \cdot \ddot{Br}: \\ \\ \cdot \ddot{Br}: \end{array} \longrightarrow Ca^{2+} \left(: \ddot{Br}:^- \right)_2$$

Formula $= CaBr_2$

When aluminum combines with oxygen, it is somewhat more complex to follow the shuffle of the three electrons of aluminum onto the oxygens, which need two each. In this case two Al's and three O's are needed to achieve a charge balance of zero and to satisfy the octet rule.

$$\begin{array}{l} Al \quad\quad \ddot{O}: \\ \\ \quad\quad \ddot{O}: \longrightarrow (Al^{3+})_2 \left(: \ddot{O}:^{2-} \right)_3 \\ \\ Al \quad\quad \ddot{O}: \end{array}$$

Formula $= Al_2O_3$

To predict the formulas of binary ionic compounds, only the group number of each element is needed to give the charge on the respective ions. First, write the appropriate ions adjacent to each other (IA, +1; IIA, +2; IIIA, +3; VA, −3; VIA, −2; VIIA, −1). The number of the charge becomes the subscript of the other element.

$$Na^{1+} \quad N^{3-} = Na_3N$$
$$Ga^{3+} \quad S^{2-} = Ga_2S_3$$
$$Ca^{2+} \quad O^{2-} = Ca_2O_2 = CaO \quad \text{(write the simplest formula)}$$

Example 6-1
What is the formula of the ionic compound formed between

(a) aluminum and fluorine and (b) barium and sulfur?

Solution:

(a) Aluminum is in group IIIA and fluorine is in group VIIA, and they have the dot structures

$$\cdot \overset{\cdot}{Al} \cdot \qquad : \overset{\cdot}{\underset{\cdot\cdot}{F}} :$$

Three F atoms are needed to take up the three electrons from one Al, forming the compound

$$Al^{3+} \left(: \overset{\cdot\cdot}{\underset{\cdot\cdot}{F}} :^- \right)_3 = \underline{\underline{AlF_3}}$$

We could also determine the formula from the charges of the respective ions formed. Al becomes Al^{3+} (group IIIA) and F becomes F^{1-} (group VIIA).

$$Al \overset{+3}{\underset{\textstyle\frown}{}} F \overset{-1}{} = \underline{\underline{AlF_3}}$$

(b) Barium is in group IIA and sulfur is in group VIA, and they have the dot structures

$$\overset{\cdot}{Ba} \cdot \qquad \overset{\cdot\cdot}{\underset{\cdot\cdot}{S}} :$$

One Ba atom gives up two electrons and one S atom takes up two electrons, forming the compound

$$Ba^{2+} : \overset{\cdot\cdot}{\underset{\cdot\cdot}{S}} :^{2-} = \underline{\underline{BaS}}$$

We could also determine the formula from the charges of the respective ions formed. Ba becomes Ba^{2+} and S becomes S^{2-}.

$$Ba \overset{2+}{\underset{\textstyle\frown}{}} S \overset{2-}{} = Ba_2S_2 = \underline{\underline{BaS}}$$

Besides single atoms, two or more atoms can also bond together by electron sharing with the whole unit behaving as an ion (e.g., CO_3^{2-}). These are called **polyatomic ions** and are discussed later in this chapter.

See Problems 6-5, 6-6, 6-7

In Section 3-3 an example of an ionic crystal was pictured. It is important to recall that the formula for an ionic compound tells us the *ratio* of the ions present. In the crystal pictured in Figure 3-4, there are no discrete NaCl "molecules." *The Na^+ ions are surrounded by six Cl^- ions and the Cl^- ions by six Na^+ ions in a three-dimensional arrangement called a* **lattice.** The crystal is held together by the forces of attraction between the oppositely charged ions. *The energy holding the crystal together is determined mainly by Coulomb's law (Section 5-4) and is known as the* **lattice energy.** Different lattice arrangements occur for other ionic compounds with corresponding lattice energies.

See Problem 6-8

6-5 The Covalent Bond

As we mentioned, ionic bonds form between two elements that are complementary. One element has a strong attraction for electrons (high electron affinity) and the other has a weak grasp on its outer electrons (low ionization energy). The strong dominates the weak and the electrons are exchanged. But what if neither of the atoms gives up electrons easily? Bonds still form, but another suitable arrangement must be made to attain a noble gas structure for the elements involved. If one element is not strong enough to take the electrons away from the other, the alternative arrangement is to share electrons.

When fluorine combines with a metal, it can reach an octet by complete removal of an electron from the metal to form F^-. If it combines with another F atom with the same electron affinity and ionization energy, a sharing arrangement is necessary:

$$:\ddot{F}\!\overset{\frown}{\underset{\smile}{}}\!\ddot{F}: \longrightarrow \qquad :\ddot{F}\!\!\overset{..}{\odot}\!\!\ddot{F}:$$

Shared pair of electrons
(one from each F)

The two fluorine atoms have achieved noble gas electron structures by means of a **covalent bond,** *which is a shared pair of electrons (in this case, one electron from each F).*★

There is a simple analogy to the covalent bond. A husband and wife each have $7, but both feel the need to tell the world that they have access to $8, no more, no less. They will have to cooperate to accomplish the goal. Each must hold $6 in a private bank account and each must contribute $1 toward a $2 joint bank account which is shared. Both can then truthfully say that they *have attained access* to exactly $8 (see Figure 6-2).

Just as Bill and Sue each have access to $8, each fluorine atom has access to eight electrons—six of its own and two shared in the bond.

The concept of Lewis dot structures can now be extended to include molecules. There are several variations of how Lewis structures are repre-

---

★ A fluorine *atom* is *paramagnetic,* with one unpaired electron in a $2p$ orbital. Formation of a covalent bond between two fluorine atoms produces a *diamagnetic molecule,* since the two electrons that pair have opposite spins. Although most atoms of the representative elements are paramagnetic, most molecules of elements (e.g., F_2) or molecules of compounds (e.g., H_2O) are diamagnetic. Ionic compounds involving the representative elements are also diamagnetic, since unpaired electrons are the ones lost in forming cations or paired in forming anions.

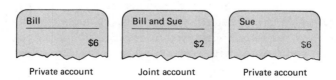

Bill		Bill and Sue		Sue	
	$6		$2		$6
Private account		Joint account		Private account	

Figure 6-2 INCREASE YOUR MONEY BY SHARING. The money in a joint account is analogous to electrons in a covalent bond.

sented for molecules. Pairs of electrons are sometimes represented as a pair of dots (:) or as a dash (—). In this text, we will use a pair of dots to represent unshared pairs (called lone pairs) of electrons on an atom and a dash to represent a pair of electrons shared between atoms. In this way the two different environments of electrons (shared and unshared) can be distinguished.

By writing Lewis structures in accordance with the octet rule, we will be able either to (1) justify the formula of a given compound or (2) predict the formula of a compound between two elements. (Although we will not pursue this topic to a large extent in this text, the shape or geometry of most simple molecules can also be predicted from their Lewis structures.)

The Lewis structure of F_2 is illustrated as follows.

Total of 14 outer electrons (7 from each F)

:F̈—F̈— Three lone pairs on each F

Two shared electrons in a covalent bond

All of the halogens exist as diatomic molecules for the same reason as does F_2. Hydrogen, which forms the simplest of all molecules, also exists as a diatomic gas with one covalent bond between atoms:

$$H—H$$

Notice that hydrogen does *not* follow the octet rule; its outer subshell holds only two electrons, which is the electronic structure of He.

How does the covalent bond hold the two hydrogen atoms together? Taken alone, the two nuclei (represented as H_2^{2+} in Figure 6-3) would repel each other because of their like charges. (→ represents a repulsive force). The presence of one electron (with a negative charge) between the two hydrogen nuclei (represented as H_2^+) changes the situation. The nuclei are now held together (at a certain distance) by the two nucleus–electron attractions. These two attractions are stronger than the nucleus–nucleus repulsion. (∿∿∿)

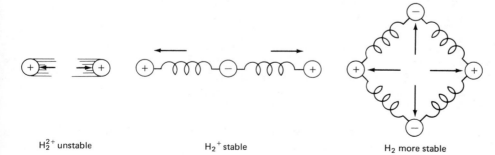

H_2^{2+} unstable H_2^+ stable H_2 more stable

Figure 6-3 THE COVALENT BOND. The presence of electrons between the two nuclei holds the atoms together.

represents an attractive force in Figure 6-3). If there are two electrons between the two hydrogen atoms as in the H_2 molecule, the nuclei are held together even tighter by the *four* nucleus–electron attractions even though the second electron adds an electron–electron repulsion.

6-6 Simple Binary Molecules

The formulas of some simple binary molecules can be predicted by writing out their Lewis structures and applying the octet rule. A compound between hydrogen and chlorine is correctly predicted to have the formula HCl, since each atom is only one electron short of a noble gas structure.

$$H \qquad :\ddot{C}l: \longrightarrow H-\ddot{C}l:$$

The simplest compound between hydrogen and oxygen has the formula H_2O:

$$\begin{array}{c} H \\ \\ H \end{array} \ddot{O}: \longrightarrow \begin{array}{c} H-\ddot{O}: \\ | \\ H \end{array}$$

In a similar manner, the Lewis structures of the simplest N, H and C, H compounds are shown below.

$$\begin{array}{ccc} H \cdot & & H \\ H \cdot \longrightarrow :N: \longrightarrow & H-N: \\ H \cdot & & H \end{array}$$

Ammonia (NH_3)

$$\begin{array}{ccc} & H & & H \\ & & & | \\ H \cdot :C: \cdot H \longrightarrow & H-C-H \\ & & & | \\ & H & & H \end{array}$$

Methane (CH_4)

Writing the Lewis structures of simple molecules is easy if you become familiar with the following facts. We will expand on these facts or "rules" later for more complex molecules.

1 Representative elements in the second period (Li to F) never attain more than eight outer electrons. There are some cases where they have fewer than eight. Nonmetals in the third, fourth, and fifth periods can have access to more than eight in certain compounds that will not be discussed here.

2 Hydrogen never has access to more than two electrons. Therefore, H bonds to only one atom at a time.

3 Generally, the atoms in a molecule have a very symmetrical and compact skeletal arrangement (i.e., SO_3 has a S surrounded by three O's,

$$O$$
$$S$$
$$O \quad O$$

rather than structures such as

$$S\ O\ O\ O \qquad O\ S\ O\ O \qquad \begin{matrix} S & O \\ O & O \end{matrix}$$

In most cases, the first atom in a formula is the central atom, with the other atoms bound to it.

Example 6-2
Write the Lewis structure for PF_3.

Procedure:
Arrange the atoms in a symmetrical manner. In this case the P will be surrounded by three F's:

$$F \quad P \quad F$$
$$F$$

Now add the outer electrons. The P has five (represented by X's) and each F has seven (represented by dots).

$$:\!\overset{\cdot\cdot}{F}\!\overset{\times\times}{\underset{\times}{\times}}\!\overset{\cdot\cdot}{P}\!\overset{\cdot\cdot}{\underset{\times}{\times}}\!\overset{\cdot\cdot}{F}\!:$$
$$:\!\overset{\cdot\cdot}{F}\!:$$

Finally, we can represent the bonded electrons as dashes.

$$:\!\overset{\cdot\cdot}{F}\!\!-\!\!\overset{\cdot\cdot}{P}\!\!-\!\!\overset{\cdot\cdot}{F}\!:$$
$$|$$
$$:\!\overset{\cdot\cdot}{F}\!:$$

Example 6-3
Write the Lewis structure for N_2H_4 (hydrazine).

Procedure:
Arrange the atoms in a symmetrical manner. Notice that all of the H's must be on the outside, since they bond to only one atom at a time.

$$H \quad N \quad N \quad H$$
$$H \quad H$$

Now add the electrons. Each H contributes one and each N contributes 5. The electrons from the H's are represented by X's and the electrons from N's are represented by dots for clarity. As far as the molecule is concerned, however, all electrons are identical.

$$H\!:\!\overset{\cdot\cdot}{\underset{\overset{\cdot\cdot}{\underset{H}{}}}{N}}\!:\!\overset{\cdot\cdot}{\underset{\overset{\cdot\cdot}{\underset{H}{}}}{N}}\!:\!H \longrightarrow H\!-\!\overset{\cdot\cdot}{\underset{\underset{H}{|}}{N}}\!-\!\overset{\cdot\cdot}{\underset{\underset{H}{|}}{N}}\!-\!H$$

See Problems 6-9, 6-10

More examples of Lewis structures will be worked in Section 6-9 after we have developed the concept of the multiple bond.

6-7 The Multiple Covalent Bond

In Figure 6-2 an analogy to the covalent bond illustrated how two people could each claim access to \$8 with only total of \$14 between them. What if they only had \$12 between them? In this case they could still claim \$8 each if they shared \$4 in the joint account and kept \$4 each in private accounts.

Two atoms may also share four electrons (two pairs) to achieve an octet of valence electrons. *The sharing of two pairs of electrons (between the same two atoms) is known as a* **double bond.** An example of a molecule containing a double bond is illustrated by the Lewis structure of SO_2:

$$\overset{\cdot\cdot}{\underset{:\overset{..}{O}}{S}}\diagdown\!\!\!\diagup_{\overset{..}{O}:}$$

Notice that all atoms in the molecule have access to eight electrons.*

The N_2 molecule has a total of 10 outer electrons available to achieve octets. In this case, each N can hold only one lone pair to itself and must share six electrons (three from each N). *When three pairs of electrons are shared between atoms it is known as a* **triple bond.** The Lewis structure for N_2 is

$$:N\!\equiv\!N:$$

* It would seem that O_2, with 12 electrons, would make an excellent example of the simplest molecule with a double bond as follows:

$$\overset{\cdot\cdot}{:}O\!=\!O\overset{\cdot\cdot}{:}$$

This Lewis structure implies that all of the electrons in O_2 are paired and that the molecule is diamagnetic. However, experiments show that O_2 is paramagnetic with two unpaired electrons. Although writing Lewis structures works very well in explaining the bonding in most simple molecules, it should be kept in mind that it is simply the representation of a theory. In this case, the theory just doesn't work. Thus O_2 is usually not represented by a Lewis structure. Other theories on bonding work well to explain the paramagnetism and bonding of O_2 but these will not be discussed here.

Only C, N, and O among the representative elements are ever involved in triple bonds. Triple and quadruple bonds (eight shared electrons) also exist in certain bonds between transition metals.

6-8 Polyatomic Ions

As we mentioned earlier, *two or more atoms can bond together by covalent bonds to form what is known as a* **polyatomic ion.** The charge on the ion is not located on any particular atom in the ion but belongs to the entire structure. A simple example of a polyatomic ion is the hypochlorite ion, ClO^-. The existence of the -1 charge tells us that the ion contains one electron *in addition* to those from neutral Cl and O atoms. The total number of outer electrons involved in the ClO^- ion is calculated as follows:

From a neutral Cl	7
From a neutral O	6
Additional electrons indicated by charge	1
Total number of electrons	14

Two atoms bonded together with 14 electrons has a Lewis structure like F_2 with a single covalent bond:

$$\left[:\ddot{C}l - \ddot{O}: \right]^-$$

Notice that the brackets indicate that the total ion has a -1 charge. The extra electron has not been specifically identified, since electrons are all identical and the charge belongs to the ion as a whole.

See Problem 6-11

6-9 Writing Lewis Structures

Writing correct Lewis structures is an important skill in basic chemistry, since it is necessary in later courses such as organic chemistry and biochemistry. Fortunately, the whole process becomes straightforward (with some practice, of course) if one follows a few basic rules. These rules that follow have been expanded from those mentioned in page 118.

1   Check to see whether any ions are involved in the compounds (see Section 6-4). Write out any ions present.

2   For a molecule, add up all of the outer (or valence) electrons of the neutral atoms. For an ion, add (if negative) or subtract (if positive) the number of electrons indicated by the charge.

3   Arrange the atoms of the molecule or ion in a skeletal arrangement. As mentioned, this is usually the most symmetrical and compact arrangement. Remember that H can bond to only one atom at a time.

4 Put a dash representing a shared pair of electrons between adjacent atoms that have covalent bonds (not between ions). Subtract the electrons used for this (two for each bond) from the total calculated in step 2.

5 Distribute the remaining electrons among the atoms so that no atom has more than eight electrons.

6 Check all atoms for an octet (except H). If an atom has access to fewer than eight electrons, put an electron pair from an adjacent atom into a double bond. Each double bond increases by two the number of electrons available to the atom needing electrons.

Example 6-4
Write the Lewis structure for NCl_3.

1 This is a binary compound between two nonmetals. Therefore, it is not ionic.

2 The total number of electrons available for bonding is

$$
\begin{array}{lll}
\text{N} & 1 \times 5 = & 5 \\
\text{Cl} & 3 \times 7 = & 21 \\
\end{array}
$$
$$\text{Total outer electrons} = \overline{26}$$

3 The skeletal arrangement is

$$\text{Cl} \quad \text{N} \quad \text{Cl}$$
$$\text{Cl}$$

4 Use six electrons to form bonds:

$$\text{Cl}-\text{N}-\text{Cl}$$
$$|$$
$$\text{Cl}$$

5 Distribute the remaining 20 electrons ($26 - 6 = 20$):

$$:\ddot{\text{C}}\text{l}-\overset{..}{\text{N}}-\ddot{\text{C}}\text{l}:$$
$$|$$
$$:\ddot{\text{C}}\text{l}:$$

6 Check to make sure that all atoms satisfy the octet rule.

$$:\ddot{\text{C}}\text{l}\!-\!\overset{..}{\text{N}}\!-\!\ddot{\text{C}}\text{l}:$$
$$:\ddot{\text{C}}\text{l}:$$

Example 6-5
Write the Lewis structure for K_2S.

1 This is a binary compound between a group IA metal and a group VIA nonmetal. It is therefore ionic. Since K is a $+1$ ion, the S is a -2 ion.

The Lewis structure is simply that of an ionic compound (no dashes between ions).

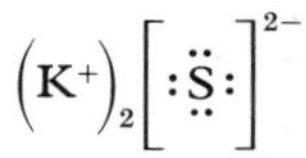

$$\left(K^+\right)_2 \left[:\ddot{S}: \right]^{2-}$$

Example 6-6

Write the Lewis structure for N_2O_4.

1 This is not ionic.

2 The total number of outer electrons available is

$$
\begin{array}{lll}
N & 2 \times 5 = & 10 \\
O & 4 \times 6 = & \underline{24} \\
\text{Total} & & 34
\end{array}
$$

3 The symmetrical skeletal structure is

4 Add 10 electrons for the five bonds:

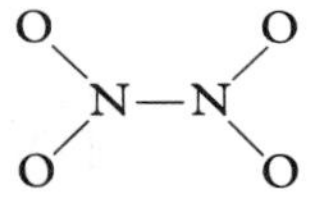

5 Add the remaining 24 electrons ($34 - 10 = 24$):

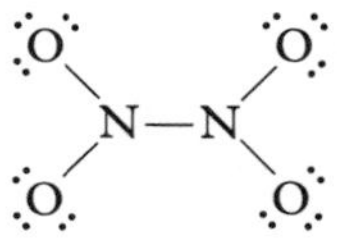

6 Check for octets. The oxygens are all right, but both N's need access to one more pair of electrons each. Therefore, make one double bond for each N using a lone pair of electrons from an adjacent O.

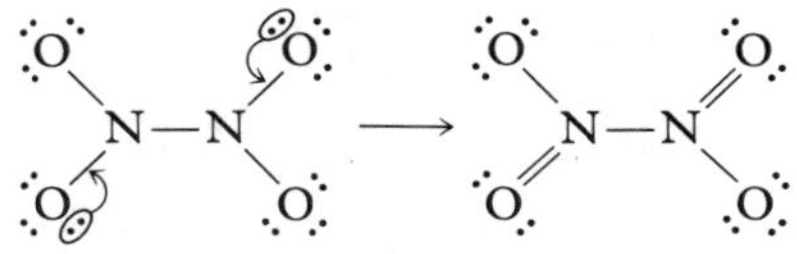

Example 6-7

Write the Lewis structure for $CaCO_3$.

1 This is an ionic compound composed of Ca^{2+} and CO_3^{2-} ions (since you know that Ca is in group IIA, it must have a $+2$ charge, therefore the polyatomic anion must be -2). A Lewis structure can be written for CO_3^{2-}.

2 For the CO_3^{2-} ion the total number of outer electrons available is:

$$
\begin{array}{ll}
C & 1 \times 4 = 4 \\
O & 3 \times 6 = 18 \\
\text{From charge} = 2 \\
\text{Total electrons} = 24
\end{array}
$$

3, 4 The skeletal structure with bonds is

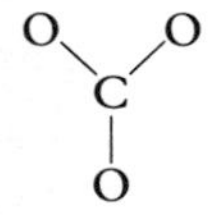

5 Add the remaining 18 electrons ($24 - 6 = 18$):

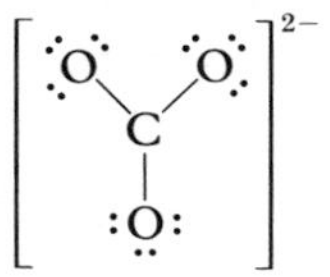

6 The C needs two more electrons, so one double bond is added using one lone pair from one oxygen:

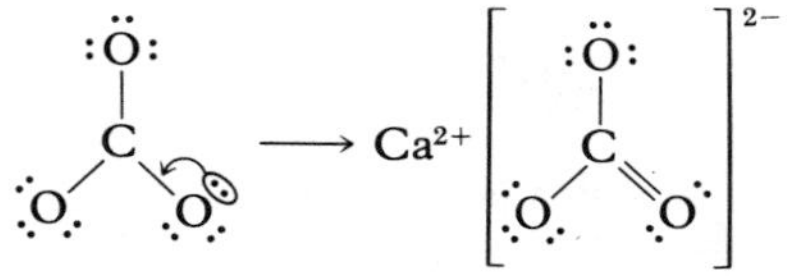

Example 6-8
Write the Lewis structure for H_2SO_4.

1 All three atoms are nonmetals, which means that all are covalent.

2 The total number of outer electrons available is

$$
\begin{array}{ll}
H & 2 \times 1 = 2 \\
S & 1 \times 6 = 6 \\
O & 4 \times 6 = \underline{24} \\
\text{Total} & 32
\end{array}
$$

3 In most molecules containing H and O the H is bound to an O and the O to some other atom, which in this case is S. The skeletal structure is

$$
\begin{array}{ccccc}
 & & O & & \\
H & O & S & O & H \\
 & & O & &
\end{array}
$$

4 Add 12 electrons for the six bonds:

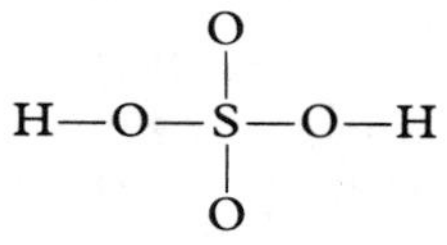

$$
\begin{array}{c}
O \\
| \\
H-O-S-O-H \\
| \\
O
\end{array}
$$

5 Add the remaining 20 electrons (32 − 12 = 20):

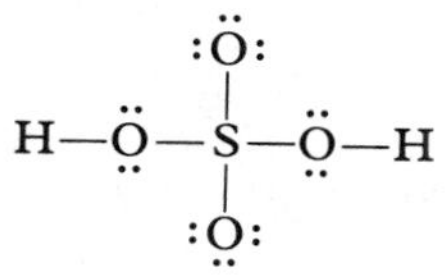

$$
\begin{array}{c}
:\ddot{O}: \\
| \\
H-\ddot{O}-S-\ddot{O}-H \\
| \\
:\ddot{O}:
\end{array}
$$

6 All octets are satisfied.

Example 6-9
Write the Lewis structure for NO.

    1 It is not ionic.

    2 There are 5 + 6 = 11 outer electrons.

3, 4, 5 With a bond and the other electrons the structure is

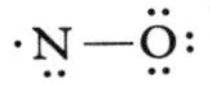

$$\cdot\ddot{N}-\ddot{O}:$$

    6 As written, the N has access to five electrons. By making one double bond, the nitrogen can be brought up to seven electrons. However, there is no way that an odd number of electrons can be arranged so that all atoms have access to the even number of eight electrons. In this case one of the atoms does not have an octet, as shown below:*

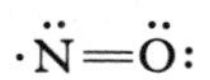

$$\cdot\ddot{N}=\ddot{O}:$$

See Problems 6-12 through 6-16

6-10 Resonance Hybrids

The Lewis structure for SO_2 is

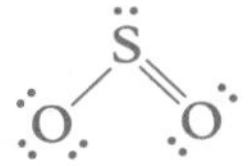

Experiments indicate that the length and strength of a double bond are different than those of a single bond between the same two elements. However, the above structure by itself implies something that isn't true according to available evidence. It implies that one S—O bond is different than the other.

* This would be a paramagnetic compound with one unpaired electron. Such compounds occur but are rare among the representative elements.

In fact, both S—O bonds are known to be identical and have length and strength halfway between a single and a double bond. Since there is no way to illustrate this fact with one Lewis structure, it must be done with two. The two structures below (connected by a double-headed arrow) are known as **resonance structures.** The actual structure is a **hybrid** of the resonance structures.

(1)   (2)   (1)   (2)

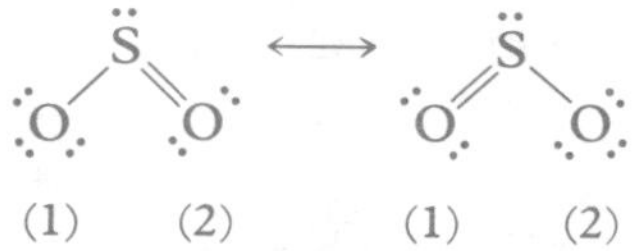

Resonance structures can be written for compounds when equally correct Lewis structures can be written without changing the basic skeletal geometry.

Example 6-10

Write resonance hybrid structures for the CO_3^{2-} ion.

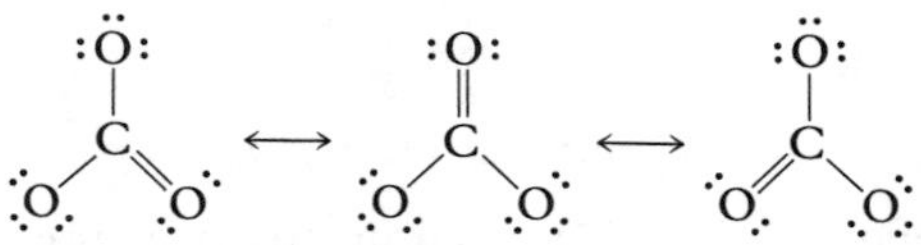

The three resonance structures indicate that the C—O bond has one-third double bond properties and two-thirds single bond properties.

The word "resonance" is an unfortunate choice. It implies that the S—O bond is changing (reasonating) between a double and a single bond and that what is seen is actually just an average of the two extremes. This is not the case; the S—O bond actually *exists* full-time as half double and half single. In this sense a resonance hybrid is like the hybrid tomato that you may grow in the garden or buy in the grocery store. The tomato has a sweet taste and large size but not because it is changing rapidly back and forth between a large tomato and a small, sweet tomato. It exists full-time with **See Problem 6-17**   properties of its two parents.

6-11 Polarity and Electronegativity

When two atoms compete for a pair of electrons in a bond, there are three things that can happen:

1   Both atoms share the electrons equally.

2   One atom takes the pair of electrons completely to itself and there is no sharing.

3   The two atoms share the electrons but not equally.

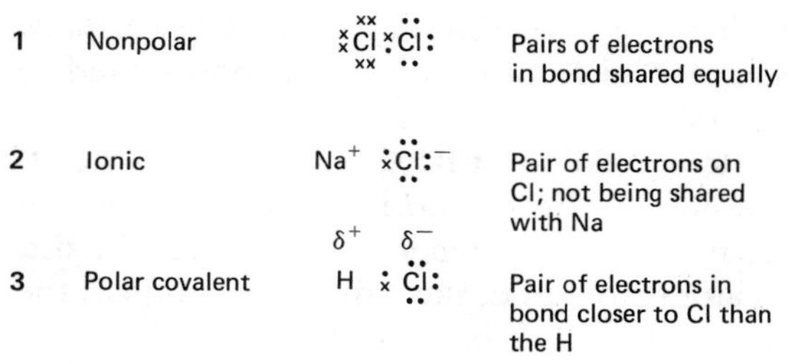

1	Nonpolar	$\overset{xx}{\underset{xx}{x}}Cl\overset{\cdot\cdot}{\underset{\cdot\cdot}{x}}Cl\overset{\cdot\cdot}{\underset{\cdot\cdot}{:}}$	Pairs of electrons in bond shared equally
2	Ionic	$Na^+ \ \overset{\cdot\cdot}{\underset{\cdot\cdot}{x}}Cl\overset{\cdot\cdot}{\underset{\cdot\cdot}{:}}^-$	Pair of electrons on Cl; not being shared with Na
3	Polar covalent	$\overset{\delta^+}{H} \ \overset{\delta^-}{\underset{\cdot\cdot}{x}Cl\overset{\cdot\cdot}{:}}$	Pair of electrons in bond closer to Cl than the H

Figure 6-4 NONPOLAR, IONIC, AND POLAR COVALENT BONDS.

The molecule Cl_2 illustrates case 1, as shown in Figure 6-4. Since both atoms are identical, the electron pair in the bond is shared equally between the two Cl atoms. Each Cl has access to its three lone pairs of electrons plus exactly one-half of the electron pair in the bond.

Case 1:   *Each Cl*   Unshared electrons = 6
                    Portion of shared electrons ($\frac{1}{2} \times 2$) = 1
                                        Total electrons = 7

The seven electrons leave each Cl (group VIIA) *exactly neutral. Such a bond is said to be* **nonpolar.** This is one extreme case.

The ionic compound NaCl illustrates the other extreme case as shown in Figure 6-4 (case 2). The pair of electrons that could have been shared is transferred completely to the Cl. The Cl has three lone pairs of electrons plus the two electrons in question.

Case 2:   *Each Cl*   Unshared electrons = 6
                    Portion of shared electrons (1×2) = 2
                                        Total electrons = 8

Since seven electrons leave a Cl atom neutral, eight gives the Cl a charge of -1. The Na has an equal but opposite charge of $+1$.

In between these extremes lies the vast number of bonds formed between two atoms that do not share electrons equally. In HCl (Case 3), the Cl has a greater attraction for the pair of electrons in the bond than H but not enough to take them completely to form an ion. The Cl has its three lone pairs of electrons plus *more than* ($>$) *one-half* of the electrons in the bond, which gives the Cl a *partial* negative charge (symbolized by δ^-, the Greek letter delta).

Case 3:   *Each Cl*   Unshared electrons =   6
                    Portion of shared electrons ($>\frac{1}{2} \times 2$) = >1
                                        Total electrons = >7

This leaves the H with an equal and opposite partial positive charge (symbolized by δ^+). *A chemical bond that has a partial separation of charge because of unequal sharing of electrons is known as a* **polar covalent bond.**

A molecule with a positive end and a negative end separated by a distance is said to contain a **dipole** (*two poles*).* The dipole of a bond is represented by an arrow going from positive to negative (↦).

Since two atoms of different elements do not have exactly the same attraction for electrons in a bond, all bonds between unlike atoms are polar to some extent. The amount of polarity of a bond depends mainly on the difference in the tendency of the two atoms to attract the pair of electrons in the bond. *The ability of the atoms of an element to attract the electrons in the bond can be expressed as a numerical quantity and is known as the element's* **electronegativity.** Electronegativity is a periodic property with much the same trend as nonmetal behavior. That is, electronegativity increases up and to the right in the periodic table. Some values for electronegativity are listed in Table 6-1. They were calculated by Linus Pauling (winner of two Nobel Prizes and of vitamin C fame) but have since been refined by others. In any case, the actual values are not as important as how the electronegativity of one element compares to another. Notice that the element fluorine has the highest electronegativity. This means that of all the elements, fluorine has the strongest attraction for electrons in a bond. As a result, fluorine is *always*

* The earth is polar (contains a dipole), with a north and a south magnetic pole.

Table 6-1   ELECTRONEGATIVITY

1 H 2.1																	
3 Li 1.0	4 Be 1.5											5 B 2.0	6 C 2.5	7 N 3.0	8 O 3.5	9 F 4.0
11 Na 0.9	12 Mg 1.2											13 Al 1.5	14 Si 1.8	15 P 2.1	16 S 2.5	17 Cl 3.0
19 K 0.8	20 Ca 1.0	21 Sc 1.3	22 Ti 1.5	23 V 1.6	24 Cr 1.6	25 Mn 1.5	26 Fe 1.8	27 Co 1.8	28 Ni 1.8	29 Cu 1.9	30 Zn 1.6	31 Ga 1.6	32 Ge 1.8	33 As 2.0	34 Se 2.4	35 Br 2.8
37 Rb 0.8	38 Sr 1.0	39 Y 1.2	40 Zr 1.4	41 Nb 1.6	42 Mo 1.8	43 Tc 1.5	44 Ru 2.2	45 Rh 2.2	46 Pd 2.2	47 Ag 2.4	48 Cd 1.7	49 In 1.7	50 Sn 1.8	51 Sb 1.9	52 Te 2.1	53 I 2.5
55 Cs 0.7	56 Ba 0.9	57 La 1.1	72 Hf 1.3	73 Ta 1.5	74 W 1.7	75 Re 1.9	76 Os 2.2	77 Ir 2.2	78 Pt 2.2	79 Au 2.4	80 Hg 1.9	81 Ti 1.8	82 Pb 1.8	83 Bi 1.9	84 Po 2.0	85 At 2.2
87 Fr 0.7	88 Ra 0.9	89 Ac 1.1														

the negative end of the dipole. Oxygen is the negative end of the dipole in all bonds except those with fluorine.

   The greater the difference of electronegativity between two atoms, the greater the polarity (the bond has a larger dipole). As a matter of fact, if the difference between two atoms is around 1.9 to 2.0, the bond is generally ionic, meaning that one atom has gained complete control of the electron pair in the bond.

See Problems 6-18, 6-19

6-12 Molecular Polarity

The polarity of molecules is determined by:

1   The polarity of the bonds in the molecule and

2   The geometry of the molecule.*

   In a molecule such as CO_2, both C—O bonds have a dipole as shown below. (Since O is more electronegative than C, the C is the positive end and the O negative.)

$$\ddot{O}\!=\!C\!=\!\ddot{O}$$

The size of each dipole is the same, but the direction† is exactly opposite since CO_2 is a linear molecule. *Thus the bond dipoles cancel, leaving the molecule nonpolar.* On the other hand, COS is also linear, but the dipoles in opposite directions are unequal (in fact, Table 6-1 predicts that the C—S bond is nonpolar since C and S have the same electronegativity). Since the bond dipoles do not cancel, there is a resultant molecular dipole. *The* **molecular dipole** *is the net or resultant effect of all of the individual bond dipoles.*

$$S\!=\!C\!=\!O$$

Bond dipoles

↙

↖ Resultant molecular dipole

   The relation of individual dipoles to the molecular dipole is analogous to the effect of two men pulling on a block of concrete as illustrated in Figure 6-5. The force exerted by each man is analogous to an individual bond dipole and the net effect (the resultant) of their efforts is analogous to the molecular dipole.

---

* The geometry of simple molecules can be predicted from either the Lewis structure or a knowledge of the shapes of what are called "hybridized" orbitals. This topic is not covered in this text, so the geometry of the sample molecules should be accepted as fact.
† The size and direction together make up what is known as the dipole moment of the bond.

No
movement

←— Direction of
movement

Figure 6-5 FORCES AT 180°. If forces are equal but opposite (180°) they cancel. If the forces are not equal there is a resultant force.

Other examples of molecules that are nonpolar although the bonds are polar include:

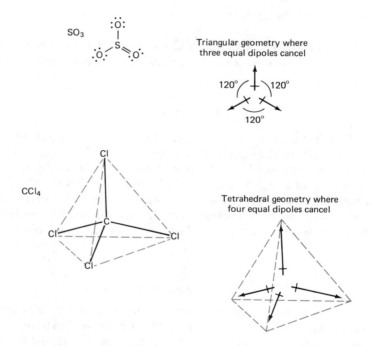

SO_3

Triangular geometry where three equal dipoles cancel

120°   120°

120°

CCl_4

Tetrahedral geometry where four equal dipoles cancel

Many molecules exist in which the equal bond dipoles do not cancel. Ordinary water is an important example of this. Although an H_2O molecule has two equal bond dipoles, they are not at an angle of 180° and thus do not cancel. If H_2O was a linear molecule, it would be nonpolar and have vastly different properties than those discussed in Chapter 11.

Bond dipoles at an angle less than 180° is a situation analogous to two men pulling on a block of concrete at an angle as illustrated in Figure 6-6. Both the direction and resultant force depend on the strength of the two men and the angle between them. (The resultant force and direction of movement also depend on their comparative strengths if different.) In a similar manner,

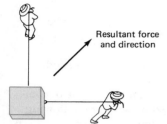

Resultant force and direction

Figure 6-6 FORCES AT LESS THAN 180°. Both the direction and resultant force depends on the angle between the two men.

the resultant dipole for the H_2O molecule is determined by the size of each H—O dipole and the angle between the two dipoles:

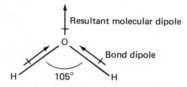

Resultant molecular dipole

Bond dipole

In later chapters the importance of the polarity of molecules will become evident. The physical state and other physical properties as well as chemical properties of compounds are determined to a large extent by the polarity of the molecules. We will refer back to this discussion at that time.   **See Problem 6-20**

Review Questions

1   What is meant by the octet rule?
2   Why don't noble gases form many compounds?
3   What are the dot structures of the representative elements?
4   Define pseudo-noble gas structure.
5   What general classes of compounds form binary ionic compounds?
6   What are the charges on the common ions of the representative elements?
7   Define a covalent bond. How does it differ from an ionic bond?
8   How is the Lewis structure of a compound determined?
9   Why doesn't hydrogen seek an "octet" of electrons?

10   Describe a double bond. Describe a triple bond.
11   What is a polyatomic ion?
12   What is meant by resonance hybrids?
13   When is a bond nonpolar?
14   What is meant by a polar covalent bond? How does it differ from an ionic bond?
15   What type of bonds have a dipole?
16   Define electronegativity. What are the periodic trends?
17   What determines molecular polarity? How can polar bonds form nonpolar molecules?
18   Why is water known to be a nonlinear molecule?

Problems

Dot Structures of Elements

6-1 Write Lewis dot structures for
(a) Ca (b) Sb (c) SN (d) I
■ (e) Ne ■ (f) Bi
■ (g) All group VIA elements

6-2 Identify the group from the given dot structure:

(a) $\cdot \overset{\cdot}{M} \cdot$ (b) $\cdot \overset{\cdot\cdot}{\underset{\cdot\cdot}{X}} \cdot$ (c) $\overset{\cdot}{A} \cdot$

Binary Ionic Compounds

6-3 Which of the following ions do *not* have a noble gas structure?
(a) Sr^{2+} (b) S^- (c) Cr^{2+} (d) Te^{2-} (e) In^+
■ (f) Pb^{2+} ■ (g) Ba^{2+} ■ (h) Tl^{3+}

6-4 Write all of the ions that would have the noble gas structure of Kr.

6-5 Complete the following table with formulas of the ionic compounds that form between the anion and cation shown.

Cation/ Anion	Br^-	S^{2-}	N^{3-}
Cs^+	CsBr		
Ba^{2+}			
In^{3+}			

6-6 Determine the charge on the cation from the charge on the anion for the following compounds. The cation is the metal.
(a) Cr_2O_3 (b) FeF_3 (c) MnS (d) CoO
■ (e) $NiBr_2$ ■ (f) VN

6-7 Which of the following two atoms could combine to form an ionic compound?
(a) Mg and S (b) As and O (c) Cr and Cl
(d) I and F
■ (e) Mg and Al ■ (f) B and C

6-8 Why isn't an ionic compound such as $BaCl_2$ referred to as a molecule?

Lewis Structures of Compounds

6-9 Write Lewis structures for the following:
(a) C_2H_6 (b) H_2O_2 (c) NF_3

■ (d) SCl_2 ■ (e) C_2H_6O (there are two correct answers; both have the lone pairs on the oxygen)

*6-10 Refer to Problem 5-14. Use the periodic table from the planet Zerk.
(1) What are the simplest formulas of compounds formed between the following elements? (Example: between 7 and 7 is 7_2.)
(a) 1 and 7 (b) 1 and 3 (c) 1 and 5
(d) 7 and 9 (e) 7 and 13 (f) 10 and 13
(g) 6 and 7 (h) 3 and 6
(2) Write Lewis structures for all of the above. Indicate which are ionic. (Remember that on Zerk there will be something different than an octet rule.)

6-11 Determine the charge on each of the following polyatomic anions from the charge on the cation.
(a) K_2SO_4 (b) $Ca(ClO_3)_2$ (c) $Al_2(SO_3)_3$
(d) $Ca_3(PO_4)_2$
■ (e) CaC_2 ■ (f) $NaBrO$ ■ (g) $SrSeO_3$

6-12 Write Lewis structures for:
(a) CO (b) SO_3 (c) KCN (d) H_2SO_3 (both H's on different O's)

6-13 Write Lewis structures for:
(a) N_2O (b) $Ca(NO_2)_2$ (c) $AsCl_3$ (d) H_2S
(e) CH_2Cl_2 (f) NH_4^+

6-14 Write Lewis structures for:
(a) Cl_2O (b) SO_3^{2-} (c) C_2H_4 (d) H_2CO
(e) BF_3 (f) NO^+

■ **6-15** Write Lewis structures for:
(a) CO_2 (b) $BaCl_2$ (c) NO_3^- (d) $SiCl_4$
(e) C_2H_2 (f) O_3

■ **6-16** Write Lewis structures for:
(a) Cs_2Se (b) $CH_3CO_2^-$ (all H's on one C, both O's on the other C) (c) $LiClO_3$
(d) N_2O_3 (e) PBr_3

Resonance

6-17 Write all resonance structures (if any) for each of the following:
(a) SO_3 ■ (b) NO_2^- (c) SO_3^{2-}

Electronegativity and Polarity

6-18 Which of the following bonds would be nonpolar?
(a) I—F (b) I—I (c) C—H (d) N—N

6-19 Rank the following bonds in order of increasing polarity using Table 6-1. Indicate with a dipole arrow (↦) the direction of the dipole. Does the difference in electronegativity in any of the following indicate strictly ionic character for the bond?
(a) Al—Cl (b) I—B (c) Ca—Cl (d) I—F
(e) O—N (f) Se—Br (g) K—O (h) Al—F
(i) B—H (j) N—H

6-20 From the geometry of the following, determine which molecules will be polar.
(a) CS₂   S—C—S   linear
(b) COS   S—C—O   linear

(c) SF₆   perfect
        octahedron

■ (d) SF₂   bent

■ (e) CCl₄   tetrahedron

(f) CHCl₃   tetrahedron

(g) BF₃   triangular

■ (h) BF₂Cl   triangular
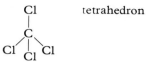

The Naming of Compounds

The listings of contents on labels indicate that the naming of compounds is a significant endeavor for the chemist.

It seems hardly a day goes by that we aren't asked to make a judgment about certain chemicals in our environment. Sometimes the initials of the chemicals read like government agencies such as 2,4,5-T, PCB, PVC, and DDT. Besides these there are sulfur dioxide and carbon monoxide gases in our atmosphere, nitrates and nitrites in our meats, as well as phosphates and acids in our lakes. To the average citizen the names of chemical compounds must seem overwhelming and complex, with very little order to the whole process.

A hundred years ago there were not all that many known compounds, so the citizen as well as the chemist did not need much order in nomenclature. Thus many compounds, such as lye, baking soda, ammonia, and water, have *common* names dating from long ago. These names tell little, if anything, about the elements of which they are composed. (The common names of some well-known compounds are given in Problem 7-7.) This, of course, would never do in the naming of the millions of compounds that have now been identified. Obviously, a more orderly approach was needed. Since 1921, a group of chemists who make up an organization called IUPAC (International Union of Pure and Applied Chemistry) has met regularly to discuss the naming of compounds. From this group comes *systematic* methods of chemical nomenclature.

This chapter involves the naming of inorganic compounds, which includes all those compounds except organic (carbon–hydrogen) compounds. This is essentially the language of chemistry. Like any language, a good deal of memorization is necessary. With some basic ground rules, however, what might otherwise be an impossible task becomes quite reasonable and orderly.

As background to this chapter you should be familiar with the differences between metals and nonmetals and between ionic and covalent compounds as discussed in the preceding chapter.

7-1 Oxidation States

Since two elements can form more than one compound (e.g., FeO and Fe_2O_3), a method of keeping track of the electrons used in bond formation is needed. This is done by reference to the **oxidation states** of the elements. *The oxidation state* of an element is the charge the atoms of that element would have if the electrons in each bond are assigned to the more electronegative element.* A familiarity with oxidation states will not only help in the naming of compounds in this chapter but will also provide a basis for our discussion of chemical reactions involving changes in oxidation states in Chapter 13.

The calculation of oxidation states of elements in compounds can be simplified by a knowledge of the following rules.

1 The oxidation state of an element in its natural state is zero (e.g., O_2, Br_2, P_4).

2 The oxidation state of a monatomic ion is the same as the charge (e.g., $Na^+ = +1$ oxidation state, $O^{2-} = -2$, $Al^{3+} = +3$).
   (a) Alkali metals are always $+1$.
   (b) Alkaline earth metals are always $+2$.

3 The halogens have a -1 oxidation state in binary (two-element) compounds whether ionic or covalent *when bound to a less electronegative element.*

4 Oxygen is usually -2. Certain compounds (which are rare), called peroxides or superoxides, contain O in a lower negative oxidation state. Oxygen is positive when bound to F.

---

* Also referred to as the **oxidation number.**

5 Hydrogen is usually $+1$. When combined with a less electronegative element, usually a metal, H has a -1 oxidation state (e.g., LiH).

6 The oxidation states of all of the elements in a neutral compound add to zero. In the case of polyatomic ions, the sum of the oxidation states equals the charge on the ion.

The following examples illustrate the assignment of oxidation states.

Example 7-1
What is the oxidation states of the elements in FeO?
Rule 6: The oxidation states of the two elements add to zero.

$$(\text{ox. state Fe}) + (\text{ox. state O}) = 0$$

Rule 4: Since O is -2,

$$\text{Fe} + (-2) = 0$$

$$\text{Fe} = \underline{\underline{+2}} \qquad (\text{This is an ionic compound.})$$

Example 7-2
What is the oxidation state of the N in N_2O_5?
Rule 6: The oxidation states add to zero, as shown by the equation

$$2(\text{ox. state N}) + 5(\text{ox. state O}) = 0$$

Rule 4: Since O is -2,

$$2N + 5(-2) = 0$$

$$2N = +10$$

$$N = \underline{\underline{+5}} \qquad (\text{This is a covalent compound.})$$

Example 7-3
What is the oxidation state of the S in H_2SO_3?
Rule 6: The oxidation states add to zero.

$$2(\text{ox. state H}) + (\text{ox. state S}) + 3(\text{ox. state O}) = 0$$

Rules 4 and 5: H is usually $+1$ and O is usually -2.

$$2(+1) + S + 3(-2) = 0$$

$$S = \underline{\underline{+4}}$$

Example 7-4
What is the oxidation state of the As in AsO_4^{3-}?
Rule 6: The oxidation states add to the charge on the ion.

$$(\text{ox. state As}) + 4(\text{ox. state O}) = -3$$

$$As + 4(-2) = -3$$

$$As = \underline{\underline{+5}}$$

See Problems 7-1, 7-2

7-2 Naming Metal–Nonmetal Binary Compounds

Most metal–nonmetal binary compounds are classified as ionic compounds. There are two kinds of metals, however. In the first kind, such as the metals in group IA, IIA, and Al in IIIA, the metal forms only one oxidation state. The second kind of metals form more than one oxidation state and will be discussed shortly.

In both writing and naming a compound, the metal comes first and the nonmetal second. *Name the metal using its full English name and the anion with its English root plus* ide. Table 7-1 lists all of the monatomic anions and their names.

Example 7-5

Name the following binary ionic compounds: KCl, Li_2S, and Mg_3N_2.

Answer:

KCl	Potassium chloride
Li_2S	Lithium sulfide
Mg_3N_2	Magnesium nitride

Iron (in group VIII) forms $FeCl_2$ and $FeCl_3$. As a result, the name iron chloride does not distinguish between the two compounds. Currently, the most common method used to differentiate between the two compounds is called the **Stock method.** *In this method the oxidation state of the metal is listed with Roman numerals in parentheses after the name of the element. The* **older method** *of naming these ions uses the root of the English name of the metal plus* ous *for the lower oxidation state and* ic *for the higher oxidation state. If its symbol is derived from a Latin name, the Latin root is used.* Several metals with variable oxidation states are shown using both methods of nomenclature in Table 7-2.

The Stock method is the more convenient of the two, since it is not necessary to memorize the oxidation state corresponding to each name.

Table 7-1   Monatomic Anions

Element	Ion	Name of Anion	Element	Ion	Name of Anion
Hydrogen	H^-	Hydride	Sulfur	S^{2-}	Sulfide
Fluorine	F^-	Fluoride	Tellurium	Te^{2-}	Telluride
Chlorine	Cl^-	Chloride	Selenium	Se^{2-}	Selenide
Bromine	Br^-	Bromide	Nitrogen	N^{3-}	Nitride
Iodine	I^-	Iodide	Phosphorus	P^{3-}	Phosphide
Oxygen	O^{2-}	Oxide	Carbon	C^{4-}	Carbide

Table 7-2   Metals with Variable Oxidation States

Metal	Stock Method	Old Method	Metal	Stock Method	Old Method
Thallium	Thallium(I)	Thallous	Lead	Lead(II)	Plumbous
	Thallium(III)	Thallic		Lead (IV)	Plumbic
Iron	Iron(II)	Ferrous	Tin	Tin(II)	Stannous
	Iron(III)	Ferric		Tin(IV)	Stannic
Chromium	Chromium(II)	Chromous	Copper	Copper(I)	Cuprous
	Chromium(III)	Chromic		Copper(II)	Cupric
Cobalt	Cobalt(II)	Cobaltous	Gold	Gold(I)	Aurous
	Cobalt(III)	Cobaltic		Gold(III)	Auric

Metals whose oxidation state is $+4$ or greater are not generally considered to be ionic compounds. For example, $SnBr_4$ has properties more typical of a covalent compound than of an ionic compound. Formation of a $+4$ ion would require a large amount of energy (see Section 5-4). It should be noted then that an oxidation state for a metal does not necessarily indicate the presence of an ion.

Example 7-6
Name the following compounds: $SnBr_4$, CoN, and Au_2O_3.

Answer:

$SnBr_4$	Tin(IV) bromide* or stannic bromide
CoN	Cobalt(III) nitride or cobaltic nitride
Au_2O_3	Gold(III) oxide or auric oxide

(*Pronounced tin-four bromide)

Example 7-7
Write formulas for lead(IV) oxide and iron(II) chloride.

Answer:

Lead(IV) oxide	PbO_2
Iron(II) chloride	$FeCl_2$

See Problems 7-3, 7-4, 7-5

7-3 Naming Compounds with Polyatomic Ions

In naming metal–polyatomic anion compounds, we follow essentially the same rules as before. The metal is named first (followed by the oxidation state if necessary) and then the name of the anion is given.

The anions (and one cation) listed in Table 7-3 are some of the common polyatomic ions. The name and the corresponding formula (including charge) should be memorized for each.

Table 7-3   Polyatomic Ions

Ion	Name	Ion	Name
$C_2H_3O_2^-$	Acetate	HSO_3^-	Hydrogen sulfite or bisulfite
NH_4^+	Ammonium	OH^-	Hydroxide
CO_3^{2-}	Carbonate	ClO^-	Hypochlorite
ClO_3^-	Chlorate	NO_3^-	Nitrate
ClO_2^-	Chlorite	NO_2^-	Nitrite
CrO_4^{2-}	Chromate	$C_2O_4^{2-}$	Oxalate
CN^-	Cyanide	ClO_4^-	Perchlorate
$Cr_2O_7^{2-}$	Dichromate	MnO_4^-	Permanganate
HCO_3^-	Hydrogen carbonate or bicarbonate	PO_4^{3-}	Phosphate
HSO_4^-	Hydrogen sulfate or bisulfate	SO_4^{2-} SO_3^{2-}	Sulfate Sulfite

There is some systemization possible that will help in learning Table 7-3. In many cases *the anions are composed of oxygen and one other element. These anions are thus called* **oxyanions.** When there are two oxyanions of the same element (e.g., SO_3^{2-} and SO_4^{2-}) they, of course, have different names. The anion with the element other than oxygen in the *higher* oxidation state uses the root of the element plus *ate*. The anion with the element in the *lower* oxidation state uses the root of the element plus *ite*. For example:

$$SO_4^{2-} \quad \text{Sulf}\underline{ate} \quad (\text{S is in} +6 \text{ ox. state})$$

$$SO_3^{2-} \quad \text{Sulf}\underline{ite} \quad (\text{S is in} +4 \text{ ox. state})$$

The four oxyanions for Cl are

$$ClO_4^- \quad \text{Perchlor}\underline{ate} \quad (\text{Cl is in} +7 \text{ ox. state})$$

$$ClO_3^- \quad \text{Chlor}\underline{ate} \quad (\text{Cl is in} +5 \text{ ox. state})$$

$$ClO_2^- \quad \text{Chlor}\underline{ite} \quad (\text{Cl is in} +3 \text{ ox. state})$$

$$ClO^- \quad \text{Hypochlor}\underline{ite} \quad (\text{Cl is in} +1 \text{ ox. state})$$

Notice that the two oxyanions of intermediate oxidation states are assigned the *ate* (+5) and the *ite* (+3) endings. The highest oxidation state (+7) also received a prefix *per* in addition to the *ate* ending. The lowest oxidation state (+1) receives the prefix *hypo* in addition to the *ite* ending. Oxyanions of Br and I follow the same pattern.

Example 7-8
Name the following compounds: K_2CO_3 and $Fe_2(SO_4)_3$.

Answer:

$$K_2CO_3 \qquad \underline{\text{Potassium carbonate}}$$
$$Fe_2(SO_4)_3 \qquad \underline{\text{Iron(III) sulfate}}$$

Example 7-9
Give formulas for barium acetate and ammonium sulfate.

Answer:

$$\text{Barium acetate} \qquad \underline{Ba(C_2H_3O_2)_2}$$
$$\text{Ammonium sulfate} \qquad \underline{(NH_4)_2SO_4}$$

Example 7-10
Arsenic forms two oxyanions with the formulas $AsO_3{}^{3-}$ and $AsO_4{}^{3-}$. Name the two anions using the root *arsen*.

Answer:

$AsO_3{}^{3-}$ is arsen<u>ite</u> (As is in the $+3$ oxidation state).
$AsO_4{}^{3-}$ is arsen<u>ate</u> (As is in the $+5$ oxidation state).

Example 7-11
Referring to the nomenclature for Cl, name the following: $NaBrO_4$ and $Ca(IO)_2$.

Answer:
$NaBrO_4$
The Br is in the $+7$ oxidation state. Thus it is named in an analogous manner to Cl in the $+7$ oxidation state. Since $ClO_4{}^-$ is the *perchlorate* ion, the $BrO_4{}^-$ ion is the *perbromate* ion.

$$\underline{\text{Sodium perbromate}}$$

$Ca(IO)_2$
The I is in the $+1$ oxidation state. Thus the IO^- ion is analogous to the ClO^- ion.

$$\underline{\text{Calcium hypoiodite}}$$

**See Problems 7-6,
7-7, 7-8**

7-4 Naming Nonmetal–Nonmetal Binary Compounds

In the case of all binary compounds, the procedure is to write the element that is the metal or is the closest to being a metal (the least electronegative) first and the other element second. For example, we write CO rather than OC and H_2O rather than OH_2. Exceptions to this rule are ammonia, which is written NH_3, and methane (and other hydrogen–carbon compounds), which is written CH_4. This is simply because it's always been done that way for these well-known compounds.

The compounds are also named with the least electronegative element (most metallic) first when its English name is used. The more nonmetallic element is second, with its root plus *ide* as discussed before. Since nonmetals

Table 7-4   Greek Prefixes

Number	Prefix	Number	Prefix	Number	Prefix
1	Mono	5	Penta	8	Octa
2	Di	6	Hexa	9	Nona
3	Tri	7	Hepta	10	Deca
4	Tetra				

(except F) have variable oxidation states, some means must be provided to specify particular compounds. The Stock system is *not* preferred in this case, because it can still be ambiguous. For example, both NO_2 and N_2O_4 would have the Stock names of nitrogen(IV) oxide. In these compounds use is made of the Greek prefixes shown in Table 7-4 to indicate numbers of atoms in a compound.

The use of prefixes is illustrated by the names of the oxides of nitrogen listed in Table 7-5.

7-5 Naming Acids

Acids are so named because of the particular type of chemical reaction they undergo particularly in water solution (Chapter 12). For our present purposes, acids are *covalent compounds* formed from the anions combined with H^+.

Binary acids *consist of H^+ plus a monatomic anion.* They are privilaged to have two names, with the first as discussed for other binary compounds. The second is the acid name and is obtained by adding *hydro* to the anion root and changing the *ide* ending to *ic* followed by the word *acid*. This is illustrated in Table 7-6.

The following are *not* generally considered binary acids: H_2O, NH_3, CH_4, and PH_3.

In addition to binary acids, acids can form from *a combination of hy-*

Table 7-5   The Oxides of Nitrogen

Formula	Name	Oxidation State of N
N_2O	Dinitrogen monoxide (sometimes referred to as nitrous oxide)	+1
NO	Nitrogen monoxide (sometimes referred to as nitric oxide)	+2
N_2O_3	Dinitrogen trioxide	+3
NO_2	Nitrogen dioxide	+4
N_2O_4	Dinitrogen tetroxide[a]	+4
N_2O_5	Dinitrogen pentoxide[a]	+5

[a] The "a" is left off of *tetra* and *penta* for ease in pronunciation.

See Problems
7-9, 7-10, 7-11

Table 7-6 Binary Acids

Anion	Formula of Acid	Compound Name	Acid Name
F^-	HF	Hydrogen fluoride	Hydrofluoric acid
I^-	HI	Hydrogen iodide	Hydroiodic acid
S^{-2}	H_2S	Hydrogen sulfide	Hydrosulfuric acid

drogen with many of the oxanions discussed previously. These are known as **oxyacids.** To name an oxyacid, simply use the root of the anion to form the name of the acid. If the name of the oxyanion ends in *ate*, it is changed to *ic* followed by the word *acid*. If the name of the anion ends in *ite*, it is changed to *ous* plus the word *acid*. Hydrogen compounds of oxyanions are usually named only as an acid (e.g., HNO_3 is referred to only as *nitric acid*, not as hydrogen nitrate). Naming of oxyacids is illustrated in Table 7-7.

Table 7-7 Oxyacids

Anion	Formula of Acid	Name of Acid
$C_2H_3O_2^-$	$HC_2H_3O_2$	Acetic acid
CO_3^{2-}	H_2CO_3	Carbonic acid
ClO_2^-	$HClO_2$	Chlorous acid
ClO_4^-	$HClO_4$	Perchloric acid
SO_4^{2-}	H_2SO_4	Sulfuric acid
SO_3^{2-}	H_2SO_3	Sulfurous acid

Example 7-12
Name the following acids: H_2Se, $H_2C_2O_4$, and HClO.

Answer:

H_2Se	Hydroselenic acid
$H_2C_2O_4$	Oxalic acid
HClO	Hypochlorous acid

Example 7-13
Give formulas for the following: permanganic acid, chromic acid, and acetic acid.

Answer:

Permanganic acid	$HMnO_4$
Chromic acid	H_2CrO_4
Acetic acid	$HC_2H_3O_2$

See Problems
7-12, 7-13, 7-14

Review Questions

1   Recite the rules used to compute oxidation states.
2   Which metals have only one oxidation state?
3   Discuss the Stock method of naming metal ions with variable oxidation states.
4   How are metal ions named by the older method?
5   From the names of the polyatomic anions in Table 7-3, give the formula and the charge of the ion.

6   If both elements in a binary compound are nonmetals, how does one determine which is written and named first?
7   When are Greek prefixes used in the naming of binary compounds?
8   What is a binary acid, and how is it named?
9   What is an oxyacid, and how is it named?

Problems

Oxidation States

7-1   What are the oxidation states of the elements in each of the following compounds?
(a) PbO_2 (b) P_4O_{10} (c) C_2H_2 (d) N_2H_4
(e) LiH (f) BCl_3 (g) Rb_2Se (h) Bi_2S_3 (i) ClO_2
■ (j) XeF_2 ■ (k) CO ■ (l) O_2F_2
■ (m) Mn_2O_7

7-2   What is the oxidation state of each of the following?
(a) The P in H_3PO_4
(b) The C in $H_2C_2O_4$
(c) The Cl in ClO_4^-
(d) The Cr in $CaCr_2O_7$
(e) The S in SF_6
(f) The N in $CsNO_3$
(g) The Mn in $KMnO_4$
■ (h) The Se in SeO_3^{2-}
■ (i) The I in H_5IO_6
■ (j) The S in $Al_2(SO_3)_3$

Metal-Nonmetal Binary Compounds

7-3   Name the following. Use the Stock method where appropriate.
(a) LiF (b) SnO_2 (c) $CaSe$ (d) Mn_2O_7
■ (e) Sr_3N_2 ■ (f) Bi_2O_5

7-4   Which of the compounds in Problem 7-3 are primarily covalent?

7-5   Write formulas for each of the following:
(a) Copper(I) sulfide

(b) Vanadium(III) oxide
(c) Potassium phosphide
■ (d) Aluminum carbide
■ (e) Silver(I) bromide

Polyatomic Ions

7-6   Name the following. Use the Stock method where appropriate.
(a) $CrSO_4$ (b) $Al_2(SO_3)_3$ (c) $Fe(CN)_2$
(d) $RbHCO_3$
■ (e) $(NH_4)_2CO_3$ ■ (f) $KBrO$ ■ (g) NH_4NO_3

Compounds with Polyatomic Ions

7-7   Give the systematic name for each of the following:

Common Name	Systematic Name
(a) Table salt	$NaCl$
(b) Baking soda	$NaHCO_3$
(c) Marble or limestone	$CaCO_3$
(d) Lye	$NaOH$
(e) Chile saltpeter	$NaNO_3$
(f) Sal ammoniac	NH_4Cl
(g) Alumina	Al_2O_3
(h) Slaked lime	$Ca(OH)_2$
(i) Caustic potash	KOH

7-8 Give formulas for the following:
    (a) Magnesium permanganate
    (b) Cobalt(II) cyanide
    (c) Strontium hydroxide
    (d) Thallium(I) sulfite
    (e) Indium(III) bisulfate
    (f) Barium periodate
    ■ (g) Iron(III) oxalate
    ■ (h) Ammonium dichromate
    ■ (i) Mercury(I) acetate [The mercury(I) ion exists as Hg_2^{2+}.]

Binary Compounds from Two Nonmetals

7-9 The following two elements combine to make binary compounds. Which element should be written and named first?
    (a) Si and S   (e) H and F
    (b) F and I   ■ (f) S and P
    (c) H and B   ■ (g) O and Cl
    (d) Kr and F   ■ (h) O and F

7-10 Name the following:
    (a) CS_2 (b) BF_3 (c) P_4O_{10} (d) Br_2O_3 (e) SO_3
    ■ (f) Cl_2O ■ (g) PCl_5 ■ (h) SF_6

7-11 Write formulas for each of the following:
    (a) Tetraphosphorus hexoxide
    (b) Carbon tetrabromide
    (c) Iodine trifluoride
    (d) Dichlorine heptoxide
    ■ (e) Sulfur hexafluoride
    ■ (f) Xenon dioxide

Acids

7-12 Name the following acids:
    (a) HCl (b) HNO_3 (c) HClO (d) $HMnO_4$
    ■ (e) HIO_4 ■ (f) HBr

7-13 Write formulas for each of the following acids:
    (a) Oxalic acid
    (b) Nitrous acid
    (c) Dichromic acid
    (d) Phosphoric acid
    ■ (e) Hypobromous acid
    ■ (f) Hydrotelluric acid

7-14 Write the formulas and name the acids formed from the arsenite (AsO_3^{3-}) ion and the arsenate ion (AsO_4^{3-}).

Quantitative Relationships: The Mole

There are more atoms in a single drop of water than grains of sand in the picture.

Thus far, our efforts have been directed toward a description of the most basic forms of matter. Now we turn our attention to the quantitative relations of chemistry. For this effort, the material in Chapter 2 and the mathematical review available in the Appendix will help to make this work surprisingly straightforward. In this chapter, you will find that it is especially helpful to work problems assigned in the margin as you proceed. Throughout the chapter, one concept builds directly on another.

In Chapter 3, the small size of the atom was discussed. Even with the most powerful microscope, atoms cannot be seen except for fuzzy images of the larger atoms. Indeed, in order to have enough carbon atoms to be able to see a small, black smudge the size of the period at end of this sentence, it would take about 10^{15} atoms (100 trillion atoms). You would not be able to count that far in a lifetime.

The atomic weights of the elements were also discussed in Chapter 3. The mass or weight of *one* atom of $^{12}_{6}C$ was defined as exactly 12 amu (atomic mass units). From this the mass of other isotopes could be determined by comparison to $^{12}_{6}C$ and from an element's natural abundance of isotopes, the average atomic weight of the element was obtained. The weight unit *amu*, which was introduced at that time, is of no use in laboratory situations, however, because it represents such a tiny weight. For example, the weight of *one* hydrogen atom, which is about 1 amu, in grams is

$$1.00 \text{ amu} = 1.66 \times 10^{-24} \text{ g}$$

Obviously, in order to deal with enough atoms to see and weigh, a convenient unit, such as a dozen or gross but one representing a very large number, is needed.

8-1 The Mole

It would be convenient if one of the familiar units such as dozen or gross could be used with atoms. Unfortunately, they are of little help. Instead of about 10^{14} atoms, there would be about 10^{13} dozen atoms, or about 10^{12} gross of atoms. Obviously, this hardly simplifies the numbers involved. Thus it is necessary to introduce a new unit to deal with very large numbers. This unit is called the **mole** (the SI symbol is "mol"). The value of 1 mol is

$$1.00 \text{ mol} = 6.02 \times 10^{23} \text{ objects or particles}$$

One mole of iron atoms in the form of nails weighs about 2 oz, which is plenty to deal with in a laboratory situation.

The **molar number,** *6.02×10^{23}, is also known as* **Avogadro's number.** It is not an exact and defined number such as 12 in one dozen or 144 in one gross. It is actually known to more significant figures than the three that are shown and used here.

8-2 The Molar Mass of the Elements

The mole is defined by a mass as well as a number. *The mass of 1 mol of an element is the atomic weight expressed in grams. This is known as the* **molar mass** *of an element.* From the periodic table you can see that the mass of 1 mol of oxygen atoms is 16.0 g (to three significant figures), the mass of 1 mol of helium atoms is 4.00 g, and the mass of 1 mol of uranium atoms is 238 g.

One mole of atoms implies two things: the molar *mass*, which is different for each element, and the molar *number*, which is the same for each element.

$$24.3 \text{ g} \qquad\qquad 19.0 \text{ g}$$

1.00 mol of Mg $\longleftrightarrow$ 6.02×10^{23} atoms $\longleftrightarrow$ 1.00 mol of F

Table 8-1 Weight Relation of C and He

C	He	Number of Atoms of Each Element Present
12.0 amu	4.00 amu	1
24.0 amu	8.00 amu	2
360 amu	120 amu	30
12.0 g	4.00 g	6.02×10^{23}
12.0 lb	4.00 lb	2.73×10^{26}
24.0 ton	8.00 ton	1.09×10^{30}

At this point it is important to be convinced that the molar mass of each element actually represents the *same number of atoms*. When you are convinced of this, the actual number of atoms in 1 mol becomes less important. As an example from the periodic table, notice that one helium atom (4.00 amu) weighs one-third as much as one carbon atom (12.0 amu). As shown in Table 8-1, whenever helium and carbon are present in a 4:12 (1:3) weight ratio, regardless of the units, the same number of atoms are present in each case.

Example 8-1

If we have 46.0 lb of magnesium, how much does the same number of sulfur atoms weigh?

Procedure:

From the atomic weights in the periodic table, notice that there are the same number of magnesium atoms in 24.3 g of magnesium as there are sulfur atoms in 32.1 g of sulfur. There is also the same number of atoms in 24.3 lb of magnesium as there are in 32.1 lb of sulfur. This statement can be converted into two conversion factors, which we can use to convert a weight of one element to an equivalent weight of the other.

$$(1) \; \frac{24.3 \text{ lb of Mg}}{32.1 \text{ lb of S}} \qquad (2) \; \frac{32.1 \text{ lb of S}}{24.3 \text{ lb of Mg}}$$

Use factor 2 to convert weight of Mg to an equivalent weight of S:

$$46.0 \; \cancel{\text{lb of Mg}} \times \frac{32.1 \text{ lb of S}}{24.3 \; \cancel{\text{lb of Mg}}} = \underline{\underline{60.8 \text{ lb of S}}}$$

See Problems 8-1, 8-2

As you can see, there are a number of relationships that are implied with the mole. Before illustrating the large number of problems possible from these relationships, let's preview just how familiar the problems will be with a simple, everyday, dozen oranges. We will then work problems involving the less familiar mole of atoms. In these problems the value of the unit-factor method of working problems will become apparent.

Review the unit-factor method in Appendix D and Chapter 2.

In these problems, assume that there is such a thing as a standard orange (all oranges are exactly the same). The "dozen mass" (the weight of 12) of those particular oranges is 3.60 lb. The relationships given or implied by this statement are as follows:

$$\text{1 doz oranges} \approx \text{12 oranges} \qquad \text{1 doz oranges} \approx \text{3.60 lb}$$

From these relationships, four conversion factors can be written:

$$(1) \ \frac{\text{1 doz oranges}}{\text{12 oranges}} \qquad (2) \ \frac{\text{12 oranges}}{\text{doz oranges}}$$

$$(3) \ \frac{\text{3.60 lb}}{\text{doz oranges}} \qquad (4) \ \frac{\text{1 doz oranges}}{\text{3.60 lb}}$$

Example 8-2a

What is the weight of exactly 3 dozen oranges?

Procedure:

Convert dozen to weight. Use a conversion factor that has what you want (lb) in the numerator and what you want to cancel (doz) in the denominator. This is factor 3. In shorthand this can be written:

$$\text{doz} \xrightarrow{\text{factor 3}} \text{wt}$$

Solution:

$$3.00 \ \cancel{\text{doz oranges}} \times \frac{\text{3.60 lb}}{\cancel{\text{doz oranges}}} = \underline{\underline{\text{10.8 lb}}}$$

Example 8-3a

How many dozens of oranges are there in 18.0 lb?

Procedure:

This is the reverse of the previous problem, so use the inverse of the previous factor which is factor 4.

$$\text{wt} \xrightarrow{\text{factor 4}} \text{doz}$$

Solution:

$$18.0 \ \cancel{\text{lb}} \times \frac{\text{1 doz}}{3.60 \ \cancel{\text{lb}}} = \underline{\underline{\text{5.00 doz oranges}}}$$

Example 8-4a

What is the weight of 142 oranges?

Procedure:

Since a factor that relates number to weight is not available, use two factors in a two-step process. Note that you can go from number to dozens (factor 1) and from dozens to weight (factor 3).

$$\text{number} \xrightarrow{\text{factor 1}} \text{doz} \xrightarrow{\text{factor 3}} \text{wt}$$

Solution:

$$142 \text{ oranges} \times \frac{1 \text{ doz oranges}}{12 \text{ oranges}} \times \frac{3.60 \text{ lb}}{\text{doz oranges}} = \underline{\underline{42.6 \text{ lb}}}$$

Example 8-5a

How many individual oranges are there in 73.5 lb?

Procedure:

This is the reverse of the previous problem, so convert what's given to dozens (factor 4) and then dozens to individual oranges (factor2).

$$\text{wt} \xrightarrow{\text{factor 4}} \text{doz} \xrightarrow{\text{factor 2}} \text{number}$$

Solution:

$$73.5 \text{ lb} \times \frac{1 \text{ doz oranges}}{3.60 \text{ lb}} \times \frac{12 \text{ oranges}}{\text{doz oranges}} = \underline{\underline{245 \text{ oranges}}}$$

Now, instead of working with dozens and "dozen mass," let's work some very similar problems with moles and molar mass. Review scientific notation in Appendix C if necessary. Notice that Example 8-2a is the analog of Example 8-2b, and so forth.

Review scientific notation in Appendix C

In these problems you will work with sodium atoms instead of oranges. The molar mass (the weight of 6.02×10^{23} atoms) of Na is 23.0 g. The relationships given or implied by this statement are as follows:

$$1 \text{ mol of Na} \approx 6.02 \times 10^{23} \text{ atoms} \qquad 1 \text{ mol of Na} \approx 23.0 \text{ g}$$

From these relationships, four conversion factors can be written:

(1) $\dfrac{1 \text{ mol of Na}}{6.02 \times 10^{23} \text{ atoms of Na}}$

(2) $\dfrac{6.02 \times 10^{23} \text{ atoms of Na}}{\text{mol of Na}}$

(3) $\dfrac{23.0 \text{ g}}{\text{mol of Na}}$

(4) $\dfrac{1 \text{ mol of Na}}{23.0 \text{ g}}$

Example 8-2b

What is the weight of exactly 3 mol of Na?

Procedure:

Convert moles to weight. Use a conversion factor that has what you want (g) in the numerator and what you want to cancel (mol) in the denominator. This is factor 3. In shorthand, this can be written:

$$\text{mol} \xrightarrow{\text{factor 3}} \text{wt}$$

Solution:

$$3.00 \text{ mol of Na} \times \frac{23.0 \text{ g}}{\text{mol of Na}} = \underline{\underline{69.0 \text{ g}}}$$

Example 8-3b
How many moles are there in 34.5 g of Na?

Procedure:
This is the reverse of the previous problem, so use the inverse of the previous factor, which is factor 4.

$$\text{wt} \xrightarrow{\text{factor 4}} \text{mol}$$

Solution:

$$34.5 \; \cancel{g} \times \frac{1 \text{ mol of Na}}{23.0 \; \cancel{g}} = \underline{\underline{1.50 \text{ mol of Na}}}$$

Example 8-4b
What is the weight of 1.20×10^{24} atoms of Na?

Procedure:
Since a factor that relates number to weight is not available, use two factors in a two-step process. Note that you can go from number to moles (factor 1) and from moles to weight (factor 3).

$$\text{number} \xrightarrow{\text{factor 1}} \text{mol} \xrightarrow{\text{factor 3}} \text{wt}$$

Solution:

$$1.20 \times 10^{24} \; \cancel{\text{atoms of Na}} \times \frac{1 \; \cancel{\text{mol of Na}}}{6.02 \times 10^{23} \; \cancel{\text{atoms of Na}}} \times \frac{23.0 \text{ g}}{\cancel{\text{mol of Na}}} = \underline{\underline{46.0 \text{ g}}}$$

Example 8-5b
How many individual atoms are there in 11.5 g of Na?

Procedure:
This is the reverse of the previous problem, so convert what's given to moles (factor 4) and then moles to individual atoms (factor 2).

$$\text{wt} \xrightarrow{\text{factor 4}} \text{mol} \xrightarrow{\text{factor 2}} \text{number}$$

Solution:

$$11.5 \; \cancel{g} \times \frac{1 \; \cancel{\text{mol of Na}}}{23.0 \; \cancel{g}} \times \frac{6.02 \times 10^{23} \text{ atoms of Na}}{\cancel{\text{mol of Na}}}$$

$$= \underline{\underline{3.01 \times 10^{23} \text{ atoms of Na}}}$$

At this point you should pay special attention to the problems at the end of this chapter indicated in the margin. It is essential that you be able to work these types of problems backwards and forwards. If you master these problems, the rest of this chapter and the next should flow smoothly. For additional practice on problems of this type, refer to Appendix D, especially Examples D-12, D-13, D-18, D-19, D-21, and D-22.

**See Problems 8-3
through 8-6**

8-3 The Molar Mass of Compounds

*The **formula weight** of a compound is determined from the number of atoms and the atomic weight of each element indicated by the formula.* As mentioned in previous chapters, the formula of a compound may represent discrete molecular units* or ratios of ions in ionic compounds (referred to as a formula unit). In this text we will use formula weight to cover either possibility. The following examples illustrate the computation of formula weights.

Example 8-6

What is the formula weight of CO_2?

Atom	Number of Atoms in Molecule	Atomic Weight	Total Weight of Atom in Molecule
C	1	× 12.0 amu	= 12.0 amu
O	2	× 16.0 amu	= 32.0 amu
			44.0 amu

The formula weight of CO_2 is

$$44.0 \text{ amu}$$

Example 8-7

What is the formula weight of $Fe_2(SO_4)_3$?

Atom	Number of Atoms in Formula Unit	Atomic Weight	Total Weight of Atom in Formula Unit
Fe	2	× 55.8 amu	= 112 amu
S	3	× 32.1 amu	= 96.3 amu
O	12	× 16.0 amu	= 192 amu
			400 amu

The formula weight of $Fe_2(SO_4)_3$ is

$$400 \text{ amu}$$

See Problem 8-7

*The weight of 1 mol of molecules or ionic formula units is referred to as the **molar mass** of a compound. The molar mass is the formula weight expressed in grams and is the weight of 6.02×10^{23} molecules or ionic formula units.*

To illustrate the rather significant amount of information that has been covered so far, let's look at a sample compound. A molecule of sulfuric acid

---

* Where molecules are present, the formula weight is often referred to as the **molecular weight**.

Total		Components	
1.00 mol of molecules 6.02 × 10²³ molecules 98.1 g	H_2SO_4	2 mol of H atoms 1 mol of S atoms 4 mol of O atoms	$\begin{cases} 2.02 \text{ g} \\ 1.20 \times 10^{24} \text{ H atoms} \end{cases}$ $\begin{cases} 32.1 \text{ g} \\ 6.02 \times 10^{23} \text{ S atoms} \end{cases}$ $\begin{cases} 64.0 \text{ g} \\ 2.41 \times 10^{24} \text{ O atoms} \end{cases}$

Figure 8-1 ONE MOLE OF H_2SO_4.

(H_2SO_4) is composed of two atoms of hydrogen, one atom of sulfur, and four atoms of oxygen. One mole of H_2SO_4 has the composition shown in Figure 8-1.

Example 8-8
How many moles of each atom are present in 0.345 mol of $Al_2(CO_3)_3$? What is the total number of moles of atoms present?

Procedure:
In this problem, notice that there are 2 Al atoms, 3 C atoms, and 9 O atoms in each formula unit of compound. Therefore, in 1 mol of compound, there are 2 mol of Al, 3 mol of C, and 9 mol of O atoms. This can be expressed with conversion factors as

$$\frac{2 \text{ mol of Al}}{\text{mol of Al}_2(CO_3)_3} \qquad \frac{3 \text{ mol of C}}{\text{mol of Al}_2(CO_3)_3} \qquad \frac{9 \text{ mol of O}}{\text{mol of Al}_2(CO_3)_3}$$

Solution:

Al $0.345 \text{ mol of Al}_2(CO_3)_3 \times \dfrac{2 \text{ mol of Al}}{\text{mol of Al}_2(CO_3)_3} = \underline{\underline{0.690 \text{ mol of Al}}}$

C $0.345 \text{ mol of Al}_2(CO_3)_3 \times \dfrac{3 \text{ mol of C}}{\text{mol of Al}_2(CO_3)_3} = \underline{\underline{1.04 \text{ mol of C}}}$

O $0.345 \text{ mol of Al}_2(CO_3)_3 \times \dfrac{9 \text{ mol of O}}{\text{mol of Al}_2(CO_3)_3} = \underline{\underline{3.10 \text{ mol of O}}}$

Total moles of atoms present:

   0.690 mol of Al + 1.04 mol of C + 3.10 mol of O = 4.83 mol of atoms

Example 8-9
What is the weight of 0.345 mol of $Al_2(CO_3)_3$, and how many individual ionic formula units does this amount represent?

Procedure:
This problem is worked much like the examples in which we were dealing with moles of atoms rather than compounds. First, find the molar mass of the compound as illustrated in Examples 8-6 and 8-7. The formula weight of the compound is 234 amu, so the molar mass is 234 g/mol.

Solution:

$$0.345 \text{ mol of } Al_2(CO_3)_3 \times \frac{234 \text{ g}}{\text{mol of } Al_2(CO_3)_3} = \underline{\underline{80.7 \text{ g}}}$$

$$0.345 \text{ mol of } Al_2(CO_3)_3 \times \frac{6.02 \times 10^{23} \text{ formula units}}{\text{mol of } Al_2(CO_3)_3}$$

$$= \underline{\underline{2.08 \times 10^{23} \text{ formula units}}}$$

It is important to be specific when discussing moles. For example, with the element chlorine be careful to distinguish whether you are discussing a mole of atoms (Cl) or a mole of molecules (Cl_2). Notice that there are two Cl atoms per Cl_2 molecule and 2 mol of Cl atoms per mole of Cl_2 molecules.

$$1 \text{ mol of Cl atoms} \begin{cases} 35.5 \text{ g} \\ 6.02 \times 10^{23} \text{ atoms} \end{cases}$$

$$1 \text{ mol of } Cl_2 \text{ molecules} \begin{cases} 71.0 \text{ g } (35.5 \text{ g} + 35.5 \text{ g}) \\ 6.02 \times 10^{23} \text{ molecules } (1.20 \times 10^{24} \text{ atoms}) \end{cases}$$

This situation is analogous to the difference between a dozen sneakers and a dozen *pairs* of sneakers. Both the weight and the actual number of the dozen pair of sneakers is double the dozen sneakers.

See Problems 8-8 through 8-11

8-4 Percent Composition, Empirical and Molecular Formulas

It is rather easy to express the *percent by weight* (which is known as **percent composition**) *of the elements in a compound* if the molecular formula is given. For example, to obtain the molar mass of CO_2 (44.0 g/mol), 12.0 g of C were added to 32.0 g of O (from 1 mol of carbon atoms and 2 mol of oxygen atoms). The percent composition is calculated like any other percent, which is the component part over the total times 100.

$$\frac{\text{Total wt of component atom}}{\text{Total wt (molar mass)}} \times 100 = \text{percent composition}$$

In CO_2, the percent composition of C is

$$\frac{12.0 \text{ g of C}}{44.0 \text{ g of } CO_2} \times 100 = \underline{\underline{27.3\% \text{ C}}}$$

and the percent composition of O is

$$\frac{32.0 \text{ g of O}}{44.0 \text{ g of } CO_2} \times 100 = \underline{\underline{72.7\% \text{ O}}}$$

(or, since there are only two components, if one is 27.3% the other must be $100 - 27.3 = 72.7\%$. Right?)

Example 8-10

What is the percent composition of all of the elements in limestone, $CaCO_3$?

Procedure:

Find the molar mass and convert the *total* weight of each element to percent of the molar mass.

Solution:

For $CaCO_3$,

$$Ca \quad 1 \text{ mol of Ca} \times \frac{40.1 \text{ g}}{\text{mol of Ca}} = 40.1 \text{ g}$$

$$C \quad 1 \text{ mol of C} \times \frac{12.0 \text{ g}}{\text{mol of C}} = 12.0 \text{ g}$$

$$O \quad 3 \text{ mol of O} \times \frac{16.0 \text{ g}}{\text{mol of O}} = 48.0 \text{ g}$$

$$\text{Molar mass} = 100.1 \text{ g/mol of } CaCO_3$$

$$\%Ca \quad \frac{40.1 \text{ g of Ca}}{100.1 \text{ g of } CaCO_3} \times 100 = \underline{\underline{40.0\%}}$$

$$\%C \quad \frac{12.0 \text{ g of C}}{100.1 \text{ g of } CaCO_3} \times 100 = \underline{\underline{12.0\%}}$$

$$\%O \quad \frac{48.0 \text{ g of O}}{100.1 \text{ g of } CaCO_3} \times 100 = \underline{\underline{48.0\%}}$$

Example 8-11

What is the percent composition of all the elements in borazine, $B_3N_3H_6$?

Procedure:

Find the molar mass and convert the *total* weight of each element to percent of the molar mass.

Solution:

For $B_3N_3H_6$,

$$B \quad 3 \text{ mol of B} \times \frac{10.8 \text{ g}}{\text{mol of B}} = 32.4 \text{ g}$$

$$N \quad 3 \text{ mol of N} \times \frac{14.0 \text{ g}}{\text{mol of N}} = 42.0 \text{ g}$$

$$H \quad 6 \text{ mol of H} \times \frac{1.01 \text{ g}}{\text{mol of H}} = 6.1 \text{ g}$$

$$\text{Molar mass} = 80.5 \text{ g/mol of } B_3N_3H_6$$

$$\%B \quad \frac{32.4 \text{ g of B}}{80.5 \text{ g of } B_3N_3H_6} \times 100 = \underline{\underline{40.2\%}}$$

$$\%N \quad \frac{42.0 \text{ g of N}}{80.5 \text{ g of } B_3N_3H_6} \times 100 = \underline{\underline{52.2\%}}$$

$$\%H \qquad \frac{6.1 \text{ g of H}}{80.5 \text{ g of B}_3\text{N}_3\text{H}_6} \times 100 = \underline{\underline{7.6\%}}$$

See Problems 8-12 through 8-15

In the preceding problem, notice that the molecular formula of borazine, $B_3N_3H_6$, is not the simplest whole-number ratio of atoms. *The simplest whole-number ratio of atoms in a compound is called the* **empirical formula.** In the case of borazine, the numerical subscripts can be divided through by 3 to leave the empirical formula, BNH_2. Of course, the percent composition of the elements is the *same* for an empirical formula as for a molecular formula.

If the percent composition is given for a compound, the empirical formula can be calculated. To calculate the empirical formula, three steps are necessary:

1 Convert percent composition to an actual weight.

2 Convert weight to moles.

3 Find the whole-number ratio of the moles of different atoms. This is best illustrated by the following examples.

Example 8-12

What is the empirical formula of laughing gas, which is 63.6% nitrogen and 36.4% oxygen?

Procedure:

The easiest way to convert percent to a weight is to assume a 100-g quantity of the original compound. (Actually, you can assume any quantity; you will get the same answer.) Find the weight of nitrogen and oxygen in the 100 g by multiplying 100 g times the percent expressed in decimal form (i.e., $63.6\% = 63.6/100 = 0.636$).

$$100 \text{ g} \times 0.636 = 63.6 \text{ g of N in 100 g of compound}$$

$$100 \text{ g} \times 0.364 = 36.4 \text{ g of O in 100 g of compound}$$

Convert the weight to moles:

$$63.6 \text{ g of N} \times \frac{1 \text{ mol of N}}{14.0 \text{ g of N}} = 4.54 \text{ mol of N in 100 g}$$

$$36.4 \text{ g of O} \times \frac{1 \text{ mol of O}}{16.0 \text{ g of O}} = 2.28 \text{ mol of O in 100 g}$$

The ratio of N to O atoms will be the same as the ratio of the moles of N to O (*or the ratio of the number of dozens or any other unit*). The formula cannot remain fractional, since only whole numbers of atoms are present in a compound. To find the whole-number ratio of moles, divide through by the smallest number of moles, which in this case is 2.28 mol of O.

$$N \qquad \frac{4.54}{2.28} = 2.0 \qquad\qquad O \qquad \frac{2.28}{2.28} = 1.0$$

The empirical formula of the compound is

$$N_2O$$

Example 8-13

Ordinary rust is composed of 30.1% oxygen and 69.9% iron. What is the empirical formula of rust?

Procedure:

Convert the percent to a weight by assuming 100 g of compound.

Fe   100 g × 0.699 = 69.9 g of Fe in 100 g of compound

O   100 g × 0.301 = 30.1 g of O in 100 g of compound

Convert weight to moles:

$$69.9 \; \cancel{\text{g of Fe}} \times \frac{1 \; mol}{55.8 \; \cancel{\text{g of Fe}}} = 1.25 \; mol \; of \; Fe$$

$$30.1 \; \cancel{\text{g of O}} \times \frac{1 \; mol}{16.0 \; \cancel{\text{g of O}}} = 1.88 \; mol \; of \; O$$

Divide through by the smallest number of moles:

$$Fe \quad \frac{1.25}{1.25} = 1.0 \qquad O \quad \frac{1.88}{1.25} = 1.5$$

This time we're not quite finished, since $FeO_{1.5}$ still has a fractional number that must be cleared. (You should keep at least two significant figures in these numbers, so as not to round off a number like 1.5 to 2.) This fractional number can be cleared by multiplying both subsubscripts by a number that produces whole numbers only. In this case, multiply both subscripts by "2" and you have the empirical formula:

$$Fe_{(1\times2)}O_{(1.5\times2)} = \underline{\underline{Fe_2O_3}}$$

See Problems 8-16 through 8-20

In order to determine the molecular formula of a compound, the molar mass (g/mol) and the mass of one empirical unit (g/emp. unit) are needed. The ratio of these two quantities must be a whole number. This whole number ("a") is the number of empirical units in 1 mol.

$$\text{"a"} = \frac{Molar \; mass}{Emp. \; mass} = \frac{X \; g/mol}{Y \; g/emp. \; unit} = 1, 2, 3, \text{etc., emp. unit/mol}$$

Multiply the subscripts of the atoms in the empirical formula by "a" and you have the molecular formula. For example, the empirical formula of benzene is CH and its molar mass is 78 g/mol. The empirical mass (formula mass of CH) is 13.0 g/emp. unit.

$$\text{"a"} = \frac{78 \; g/mol}{13 \; g/emp. \; unit} = 6 \; emp. \; unit/mol$$

The molecular formula is

$$C_{(1\times6)}H_{(1\times6)} = C_6H_6$$

Example 8-14

A pure phosphorous–oxygen compound is 43.7% phosphorus and the remainder oxygen. The molar mass is 284 g/mol. What are the empirical and molecular formulas of this compound?

Procedure:

First find the empirical formula. In 100 g of this compound there are

$$100 \text{ g} \times 0.437 = 43.7 \text{ g of P} \qquad (\%0 \text{ is } 100 - 43.7 = 56.3\%)$$

$$100 \text{ g} \times 0.563 = 56.3 \text{ g of O}$$

or, expressed in moles,

$$43.7 \text{ g of P} \times \frac{1 \text{ mol of P}}{31.0 \text{ g of P}} = 1.41 \text{ mol of P}$$

$$56.3 \text{ g of P} \times \frac{1 \text{ mol of O}}{16.0 \text{ g of O}} = 3.52 \text{ mol of O}$$

Divide through by the smallest number of moles:

$$\text{P} \quad \frac{1.41}{1.41} = 1.0 \qquad \text{O} \quad \frac{3.52}{1.41} = 2.5$$

Multiply both numbers by "2" to remove the fraction:

$$\text{PO}_{2.5} = \text{P}_{(1\times2)}\text{O}_{(2.5\times2)} = \underline{\text{P}_2\text{O}_5}$$

To find the molecular formula, first compute the empirical mass, which is

$$
\begin{array}{lll}
\text{P} & 2 \times 31.0 \text{ g} = & 62.0 \text{ g} \\
\text{O} & 5 \times 16.0 \text{ g} = & \underline{80.0 \text{ g}} \\
& & 142 \text{ g/emp. unit}
\end{array}
$$

$$\text{"a"} = \frac{284 \text{ g/mol}}{142 \text{ g/emp. unit}} \qquad \text{"a"} = 2 \text{ emp. unit/mol}$$

The molecular formula is

$$\text{P}_{(2\times2)}\text{O}_{(2\times5)} = \underline{\text{P}_4\text{O}_{10}}$$

See Problems 8-21 through 8-24

8-5 The Use of Empirical and Molecular Formulas

In 1978 about 350,000 new and unique chemical compounds were prepared, identified, and reported in the chemical literature (see Figure 8-2). Obviously, since the molecular formula is of primary importance in the identification of a new compound, the types of calculations performed in this chapter are frequently carried out. Before a research chemist reports the discovery of a new compound, several items must be firmly established: (1) purity, (2) empirical formula, (3) molecular formula, and (4) molecular structure (order and geometry of the atoms in the molecule). A listing of some physical and chemical properties of the new substance may also be reported.

Figure 8-2 A RESEARCH LABORATORY. The preparation and identification of new compounds occurs frequently in chemical laboratories.

Before we move on, let's discuss a little more about how (1) through (4) are established.

1 *Is it a pure substance?* This can be determined by examination of physical properties. A mixture of salt and sugar looks pure (as any joker who has put salt in the sugar jar knows), but heating the mixture soon reveals the truth. The sugar crystals soon melt and decompose, but the salt crystals do not. Pure substances have definite and sharp melting and boiling points, mixtures do not (see Chapter 1).

If the substance is pure and the chemists have established the melting and/or boiling point, they should then take a look in a handbook listing physical properties of known compounds to make sure they haven't just rediscovered something old. If, as often happens, they have, then back to the lab. If the compound still appears to be original, then on to the next step.

2 *What is the empirical formula of the new compound?* Most chemists take the easy but smart way out and send a sample of their new pure substance to a commerical analytical laboratory, which reports back on percent composition of the elements requested. To actually obtain percent composition, many methods are used depending on the particular element; but the most commonly requested elements, carbon and hydrogen, are determined by analysis of combustion products. Such a determination is illustrated in Problems 8-20 and 8-24.

3 *What is the molecular formula of the new compound?* With the empirical formula now available, one next needs the molar mass to obtain the molecular formula. The molar mass can be obtained commercially like percent composition, or it may be obtained by several straightforward laboratory experiments. The determination of the molar mass of an unknown substance is usually performed in the general chemistry laboratory.

4 *What is the structure of the molecule?* Molecules can have the same molecular formula but different structures. For example, C_2H_6O is the formula for both dimethyl ether and ethyl alcohol (see Problem 6-9). Determination of structure can take a few minutes or a few years depending on the nature of the molecule, its complexity, and the instruments necessary. Nobel Prizes have been awarded for determination of structure of molecules (e.g., Watson and Crick for the structure of DNA).

With all of the above information in hand, chemists should feel confident enough to face their peers with their discovery.

Review Questions

1 Define the mole. Why is such a unit needed in Chemistry?
2 Give the value of Avogadro's number.
3 What is the molar mass of an element?
4 Discuss what is meant by the formula weight of a compound.
5 What is the molar mass of a compound?

6 Is there a difference between a mole of oxygen atoms and a mole of oxygen molecules?
7 Define percent composition.
8 What is an empirical formula of a compound?
9 Discuss the relationship between an empirical formula and a molecular formula. Can they be the same for a compound?

Problems

Relative Numbers of Atoms
8-1 A piece of pure gold weighs 145 g. What would the same number of silver atoms weigh?

■ **8-2** Some copper pennies weigh 16.0 g, the same number of atoms of a precious metal weigh 49.1 g. What is the metal?

Moles of Atoms
8-3 Which has more atoms, 50.0 g of Al or 50.0 g of Fe?
8-4 Which contains more Ni, 20.0 g, 2.85 × 10^{23} atoms, or 0.450 mol?
8-5 Fill in the blanks in the table at the top of the next page; use the unit-factor method to determine the answers.

8-6 What is the weight in grams of each of the following?
(a) 1.00 mol of Cu
(b) 0.50 mol of S
(c) 6.02 × 10^{23} atoms of Ca
■ (d) 6.02 × 10^{22} atoms of Ca
■ (e) 15.0 mol of Li

Formula Weight
8-7 What is the formula weight of each of the following (to three significant figures)?
(a) $KClO_2$ (e) Na_2CO_3
(b) SO_3 ■ (f) CH_3COOH
(c) N_2O_5 ■ (g) $CuSO_4 \cdot 6H_2O$
(d) H_2SO_4 ■ (h) B_4H_{10}

Element	Weight in Grams	Number of Moles	Number of Atoms
S	8.00	0.250	1.50×10^{23}
(a) P	14.5		
(b) Rb		1.75	
(c) Al			6.02×10^{23}
(d)	363		3.01×10^{24}
(e) Ti			1
■ (f) Na		3.01×10^{22}	
■ (g) K	5.75×10^{-12}		

Moles of Molecules

8-8 How many moles of each type of atom are present in 2.55 mol of grain alcohol, C_2H_6O? What is the total number of moles of atoms present? What is the weight of each element present? The total weight?

■ **8-9** What weight of each element is present in 8.95 mol of boric acid (H_3BO_3)?

8-10 How many moles of molecules are there in 1.20×10^{22} O_2 molecules? How many moles of oxygen atoms? What is the weight of oxygen molecules? What is the weight of oxygen atoms?

8-11 Fill in the blanks in the table at the bottom of this page; use the unit-factor method to determine the answers.

Percent Composition

8-12 What is the percent composition of all of the elements in

Molecules	Weight in Grams	Number of Moles	Number of Molecules or Formula Units
(a) N_2O	23.8	0.542	3.26×10^{23}
(b) H_2O		10.5	
(c) BF_3			3.01×10^{21}
(d) SO_2	14.0		
(e) K_2SO_4		1.20×10^{-4}	
(f) SO_3			4.50×10^{24}
(g) $N(CH_3)_3$	0.450		
■ (h) UF_6			12
■ (i) O_3		0.0651	
■ (j) NO_2	23.8		

(a) C_2H_6O (b) C_3H_6 (c) C_9H_{18}
■ (d) Na_2SO_4 ■ (e) $(NH_4)_2CO_3$

8-13 What is the percent composition of all of the elements in $Na_2B_4O_7 \cdot 10H_2O$ (borax)?

■ **8-14** What is the percent composition of all of the elements in saccharin ($C_7H_5SNO_3$)?

8-15 What is the percent composition of a compound composed of 1.375 g of N and 3.935 g of O?

Empirical Formulas

8-16 What is the empirical formula of a compound that has the percent composition 63.1% oxygen and 36.8% nitrogen?

■ **8-17** What is the empirical formula of a compound that has the following percent composition: 41.0% K, 33.7% S, and 25.3% O?

8-18 In an experiment it was found that 8.25 g of potassium combines with 6.75 g of O_2. What is the empirical formula of the compound?

■ **8-19** Orlon is composed of very long molecules with a percent composition of 26.4% N, 5.66% H, and 67.9% C. What is the empirical formula for orlon?

***8-20** A hydrocarbon (a compound that contains only carbon and hydrogen) was burned and the products of the combustion were collected and weighed. All of the carbon present in the original compound is now present in 1.20 g of CO_2. All of the hydrogen in 0.489 g of H_2O. What is the empirical formula of the compound? (Hint: Remember that all of the moles of C atoms in CO_2 and H atoms in H_2O came from the original compound.)

Molecular Formulas

8-21 A compound has the following percent composition: 20.0% C, 2.2% H, and 77.8% Cl. The molar mass of the compound is 545 g/mol. What is the molecular formula of the compound?

■ **8-22** A compound is composed of 1.65 g of nitrogen and 3.78 g of sulfur. If its molar mass is 184 g, what is its molecular formula?

8-23 A compound has a percent composition of 14.4% B, 16.0% C, 64.2% O, and 5.35% H. Its molar mass is about 150 g/mol. What is its molecular formula?

★■ **8-24** A 0.500-g sample of a compound containing C, H, and O was burned and the products collected. The combustion produced 0.733 g of CO_2 and 0.302 g of H_2O. The molar mass of the compound is 60.0 g/mol. What is the molecular formula of this compound? (Hint: Find the weight of C and H in the original compound; the remainder of the 0.500 g will be O.)

9 Quantitative Relationships: The Chemical Equation

The amount of gas produced in this coal gasification plant is determined by the amount of coal that is processed.

How many eggs does it take to make one medium-size cake? How many gallons of gasoline can be produced from a barrel of crude oil? How much natural gas (CH_4) can be formed from a ton of coal? How much time will I have to put in to get an "A" in this course? All of these questions indicate that the end product of an endeavor is always limited by the input of ingredients. In other words, "you can't get something for nothing."

In a chemical reaction, the amount of products relates to the ingredients (reactants) by what is called "the chemical equation." The main purpose of this chapter is to explore and develop the extensive quantitative relationships implied by these equations. Before we begin the study of the quantitative aspects, however, we will carefully examine the construction and the balancing of an equation and then briefly discuss some of the types of chemical reactions that can be illustrated by an equation.

The background for this chapter mainly involves a thorough knowledge of Chapter 8 concerning mole relationships.

9-1 Chemical Equations

Gaseous hydrogen molecules combine with gaseous oxygen molecules in a 2:1 ratio to produce liquid water. A much simpler way to represent all of this information is with a chemical equation. *A* **chemical equation** *is the symbolic representation of a chemical reaction.* Our example appears as follows:

$$2H_2(g) + O_2(g) \longrightarrow 2H_2O(l)$$

Let's go back a few steps and evolve this chemical equation from the start in order to appreciate fully all that is shown. At first, to represent the word equation, hydrogen plus oxygen yields water, we write:

$$H + O \longrightarrow H_2O$$

(Yield is represented as → or =.) It is necessary, however, to represent the hydrogen and oxygen in the chemical state as they are found in nature at normal temperature and pressure or as found at the actual reaction conditions. As discussed previously, both of these elements exist as diatomic molecules. Including this information, we have

$$H_2 + O_2 \longrightarrow H_2O$$

At this point, recall that matter cannot be created or destroyed in a reaction. As you'll notice above, we apparently did just that, because one oxygen atom has been lost by the reaction. *Equations thus must be* **balanced,** *meaning that all atoms on the left of the yield or equal sign* (**the reactants**) *must be found on the right* (**the products**). The equation cannot be balanced by changing the subscripts, because that would be changing the identity of the compound. For example, the equation could be balanced by simply changing the H_2O to H_2O_2. However, H_2O_2 is hydrogen peroxide, which is not the same as water (even though some people wash their hair in it). *The equation can be balanced by introducing* **coefficients.** In this case, a 2 in front of the H_2O solves the original oxygen problem but unbalances the hydrogens:

$$H_2 + O_2 \longrightarrow 2H_2O$$

This problem can be solved easily. Simply return to the left and put a coefficient of 2 in front of the H_2 and the equation is balanced:

$$2H_2 + O_2 \longrightarrow 2H_2O$$

Finally, the physical states of the reactants and products under the reaction conditions are sometimes added in parentheses after the molecule. To indicate the states or reaction conditions, use the following:

(g) = gas $\qquad$ (l) = liquid

(s) = solid $\qquad$ (aq) = aqueous or water solution

So now we have

$$2H_2(g) + O_2(g) \longrightarrow 2H_2O(l)$$

Other symbols sometimes used in chemical equations are ↑ (an arrow up) for a gas that is evolved in the products and ↓ (an arrow down) for a solid formed in the products. The symbol Δ (delta) over the arrow indicates that the reactants must be heated before a reaction occurs. For example,

$$Ba(s) + H_2SO_4(aq) \longrightarrow BaSO_4 \downarrow + H_2 \uparrow$$

As you can see, the simple chemical equation contains a wealth of information in shorthand. When you practice balancing equations at the end of the chapter, remember the following:

1 The subscripts of a compound are fixed; they cannot be changed to balance an equation.

2 The coefficients used should be the smallest whole numbers possible.

3 The coefficient multiplies every number in the formula. For example, $2KClO_2$ contains two atoms of K, two atoms of Cl, and four atoms of O.

Example 9-1
Write a balanced chemical equation from the following word equation: Nitrogen gas reacts with hydrogen gas to produce ammonia gas.

Procedure:
The unbalanced chemical equation is

$$N_2(g) + H_2(g) \longrightarrow NH_3(g)$$

There are two N's on the left and only one on the right. Put a 2 before the NH_3 on the right to balance the nitrogens:

$$N_2(g) + H_2(g) \longrightarrow 2NH_3(g)$$

There are now six hydrogens on the right but only two on the left. Put a 3 before the H_2 on the left and the equation is balanced (see Figure 9-1).

Solution:

$$\underline{N_2(g) + 3H_2(g) \longrightarrow 2NH_3(g)}$$

Example 9-2
Balance the following equation:

$$B_2H_6(g) + H_2O(l) \longrightarrow H_3BO_3(aq) + H_2(g)$$

Procedure:
The best place to start is with an element that occurs only once on each side of the equation (e.g., B as opposed to H). Once that's done, return to the other side to balance an element that is now "locked in" (e.g., O). In this equation

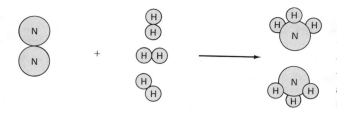

Figure 9-1 NITROGEN PLUS HYDROGEN YIELDS AMMONIA. In the chemical reaction, the atoms are simply re-arranged into different molecules.

then, start with B. Since there are two on the left, two are needed on the right:

$$B_2H_6(g) + H_2O(l) \longrightarrow 2H_3BO_3(aq) + H_2(g)$$

Now, notice that six oxygens on the right have been "locked in." Return to the left, where six H_2O's are needed:

$$B_2H_6(g) + 6H_2O(l) \longrightarrow 2H_3BO_3(aq) + H_2(g)$$

Only hydrogen remains. On the left, 18 hydrogens have been locked in. On the right, 6 hydrogens have been locked in with the $2H_3BO_3$. We will need 12 additional hydrogens on the right, which can be provided by placing a 6 before the H_2.

Solution:

$$\underline{B_2H_6(g) + 6H_2O(l) \longrightarrow 2H_3BO_3(aq) + 6H_2(g)}$$

Example 9-3

Balance the following equation:

$$H_3PO_4(aq) + KOH(aq) \longrightarrow K_2HPO_4(aq) + H_2O(l)$$

Procedure:

Notice that K and P appear only once on each side of the equation. The P is balanced but the K is not, so balance the K's on the left:

$$H_3PO_4(aq) + 2KOH(aq) \longrightarrow K_2HPO_4(aq) + H_2O(l)$$

Now balance the H's, saving the O's as a final check. There are five H's on the left, so a coefficient of 2 in front of the H_2O will provide five H's on the right:

$$\underline{H_3PO_4(aq) + 2KOH(aq) \longrightarrow K_2HPO_4(aq) + 2H_2O(l)}$$

Notice that there are six O's on both sides of the equation, indicating a correctly balanced equation.

See Problems 9-1, 9-2

Equations can be somewhat more complex than those illustrated by these examples and in the problems. In fact, balancing such equations by inspection is tedious if not impossible. However, in Chapter 13 we will find that there are systematic methods for balancing the more complicated equations.

9-2 Types of Chemical Reactions

Most of the chemical reactions that will be studied in this text can be classified into five basic types: (1) combination reactions, (2) decomposition reactions, (3) combustion reactions, (4) single replacement or substitution reactions, and (5) double displacement reactions (see Figure 9-2).

Combination Reactions Combination reactions concern the synthesis of
one compound from the elements or from a union of other compounds. In
both cases, however, only one product is formed.

$$2Na(s) + Cl_2(g) \longrightarrow 2NaCl(s)$$

$$C(s) + O_2(g) \longrightarrow CO_2(g)$$

$$CaO(s) + CO_2(g) \longrightarrow CaCO_3(s)$$

$$SO_2(g) + H_2O(l) \longrightarrow H_2SO_3(aq)$$

Decomposition Reactions Decomposition is simply the opposite of combination. In decomposition reactions one compound decomposes or breaks
down into two or more elements or compounds. In most cases decomposition reactions take place only at high temperature (indicated by Δ).

$$2HgO(s) \xrightarrow{\Delta} 2Hg(l) + O_2(g)$$

$$CaCO_3(s) \xrightarrow{\Delta} CaO(s) + CO_2(g)$$

$$2KClO_3(s) \xrightarrow{\Delta} 2KCl(s) + 3O_2(g)$$

$$H_2CO_3(aq) \longrightarrow CO_2(g) + H_2O(l)$$

Combustion Reactions Combustion, better known as burning, is the reaction of a compound or element with oxygen. Notice, however, that the combustion of elements can also be classified as combination reactions.

$$C(s) + O_2(g) \longrightarrow CO_2(g)$$

$$2Mg(s) + O_2(g) \longrightarrow 2MgO(s)$$

When compounds containing carbon and hydrogen burn in sufficient
oxygen, carbon dioxide and water are formed.

$$CH_4(g) + 2O_2(g) \longrightarrow CO_2(g) + 2H_2O(l)$$

$$2C_3H_8O(l) + 9O_2(g) \longrightarrow 6CO_2(g) + 8H_2O(l)$$

When insufficient oxygen is present (as in the combustion of gasoline in an
automobile engine), some carbon monoxide (CO) also forms.

Single Replacement or Substitution Reactions Single replacement or substitution reactions involve the substitution of one element in a compound for another. These reactions usually take place in aqueous solution. They will be discussed in more detail in Chapter 13.

$$Zn(s) + CuCl_2(aq) \longrightarrow ZnCl_2(aq) + Cu(s)$$
$$\text{(Zn in, Cu out)}$$

$$Mg(s) + 2HCl(aq) \longrightarrow MgCl_2(aq) + H_2(g)$$
$$\text{(Mg in, H}_2\text{ out)}$$

$$2Na(s) + 2H_2O(l) \longrightarrow 2NaOH(aq) + H_2(g)$$
$$\text{(Na in, H}_2\text{ out)}$$

Double Displacement Reactions Double displacement reactions also occur mostly in aqueous solution. In this case two compounds react to form two

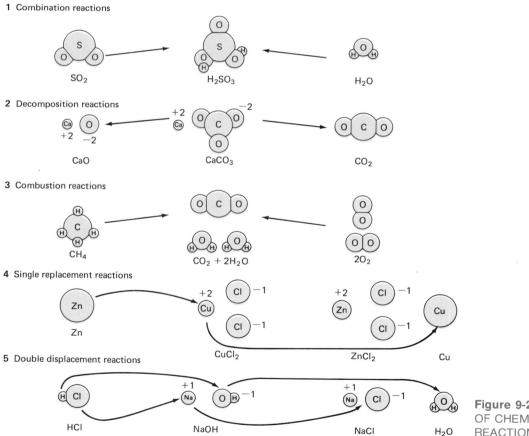

1 Combination reactions

SO$_2$ H$_2$SO$_3$ H$_2$O

2 Decomposition reactions

CaO   CaCO$_3$ CO$_2$

3 Combustion reactions

CH$_4$ CO$_2$ + 2H$_2$O   2O$_2$

4 Single replacement reactions

Zn   CuCl$_2$ ZnCl$_2$ Cu

5 Double displacement reactions

HCl   NaOH   NaCl   H$_2$O

Figure 9-2 TYPES OF CHEMICAL REACTIONS.

others. In a single replacement reaction we saw a situation analogous to a football player on the bench replacing a player on the field. A double replacement reaction is analogous to two players on the field changing positions. This type of reaction occurs because of the formation of a solid or gaseous product, as illustrated by the following two examples:

$$(NH_4)_2S(aq) + Pb(NO_3)_2(aq) \longrightarrow PbS(s) + 2NH_4NO_3(aq)$$

$$2HCl(aq) + K_2S(aq) \longrightarrow H_2S(g) + 2KCl(aq)$$

The first type of reaction is discussed in Chapter 11, the second in Chapter 12. Another example of a double displacement reaction involves the formation of a covalent compound from ions in solution. The following reaction is also classified as an acid–base reaction and is discussed in more detail in Chapter 12.

See Problems 9-3, 9-4

$$2HNO_3(aq) + Ca(OH)_2(aq) \longrightarrow Ca(NO_3)_2(aq) + 2H_2O(l)$$

9-3 Stoichiometry

Let us now look at the following balanced equation and list some of the quantitative relationships that are directly implied.

$N_2(g)$	$+ 3H_2(g)$	$\longrightarrow 2NH_3(g)$
(a) 1 molecule	+ 3 molecules	$\longrightarrow$ 2 molecules
(b) 12 molecules	+ 36 molecules	$\longrightarrow$ 24 molecules
(c) 1 dozen molecules	+ 3 dozen molecules	$\longrightarrow$ 2 dozen molecules
(d) 6.02×10^{23} molecules	+ 18.1×10^{23} molecules	$\longrightarrow 12.0 \times 10^{23}$ molecules
(e) 1 mol	+ 3 mol	$\longrightarrow$ 2 mol
(f) 28 g	+ 6 g	$\longrightarrow$ 34 g

We will be concerned primarily with the last three relationships, as they concern laboratory situations.

Balanced equations give the necessary relationships to convert moles, grams, or number of molecules of one reactant or product into the equivalent number of moles, grams, or number of molecules of another reactant or product. *The quantitative relationship among reactants and products is known as* **stoichiometry.**

First we will examine the relationship of moles in the above equation. The equation tells us:

1 mol of N_2 produces 2 mol of NH_3

3 mol of H_2 produces 2 mol of NH_3

1 mol of N_2 reacts with 3 mol of H_2

From these three relationships six conversion factors can be written. *These conversion factors that originate from a balanced chemical equation will be called* **reaction factors.** They are exact numbers.

(1) $\dfrac{1 \text{ mol of } N_2}{2 \text{ mol of } NH_3}$ (2) $\dfrac{2 \text{ mol of } NH_3}{1 \text{ mol of } N_2}$ (3) $\dfrac{1 \text{ mol of } N_2}{3 \text{ mol of } H_2}$

(4) $\dfrac{3 \text{ mol of } H_2}{1 \text{ mol of } N_2}$ (5) $\dfrac{3 \text{ mol of } H_2}{2 \text{ mol of } NH_3}$ (6) $\dfrac{2 \text{ mol of } NH_3}{3 \text{ mol of } H_2}$

The following examples illustrate the types of stoichiometry problems.   **See Problem 9-5**

Mole–Mole

Example 9-4
How many moles of NH_3 can be produced from 5.00 mol of H_2?

Procedure:
Convert moles of what's given (H_2) to moles of what's requested (NH_3). Use reaction factor 6, which has what's requested in the numerator and what's given in the denominator.

Scheme:

	Given			*Requested*	
	moles (H_2)		$\xrightarrow[\text{factor 6}]{\text{reaction}}$	moles (NH_3)	

Solution:

$$5.00 \; \cancel{\text{mol of } H_2} \times \frac{2 \text{ mol of } NH_3}{3 \; \cancel{\text{mol of } H_2}} = \underline{\underline{3.33 \text{ mol of } NH_3}}$$

Mole–Weight

Example 9-5
How many moles of NH_3 can be produced from 33.6 g of N_2?

Procedure:
First convert weight of what's given to moles (N_2) and then as in Example 9-4 convert to moles of what's requested (NH_3) using reaction factor 2.

Scheme:

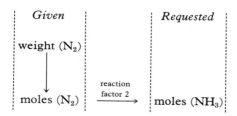

Solution:

$$33.6 \; \cancel{\text{g of N}_2} \times \frac{1 \; \cancel{\text{mol of N}_2}}{28.0 \; \cancel{\text{g of N}_2}} \times \frac{2 \; \text{mol of NH}_3}{1 \; \cancel{\text{mol of N}_2}} = \underline{\underline{2.40 \; \text{mol of NH}_3}}$$

Weight–Weight

Example 9-6

What weight of H_2 is needed to produce 119 g of NH_3?

Procedure:

As in Example 9-5, (1) convert weight to moles of what's given, (2) convert moles of what's given to moles of what's requested using reaction factor 5, and (3) convert moles of what's requested to weight.

Scheme:

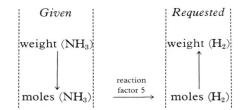

Solution:

$$119 \; \cancel{\text{g of NH}_3} \times \frac{1 \; \cancel{\text{mol of NH}_3}}{17.0 \; \cancel{\text{g of NH}_3}} \times \frac{3 \; \cancel{\text{mol of H}_2}}{2 \; \cancel{\text{mol of NH}_3}}$$

$$\times \frac{2.02 \; \text{g of H}_2}{\cancel{\text{mol of H}_2}} = \underline{\underline{21.2 \; \text{g of H}_2}}$$

Weight–Number

Example 9-7

How many individual molecules of N_2 are needed to react with 17.0 g of H_2?

Procedure:

This problem is similar to Example 9-6 except that in the final step we convert moles to number of molecules instead of a weight.

Scheme:

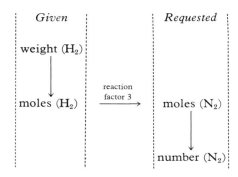

Solution:

$$17.0 \; \cancel{\text{g of H}_2} \times \frac{1 \; \cancel{\text{mol of H}_2}}{2.02 \; \cancel{\text{g of H}_2}} \times \frac{1 \; \cancel{\text{mol of N}_2}}{3 \; \cancel{\text{mol of H}_2}}$$

$$\times \frac{6.02 \times 10^{23} \; \text{molecules of N}_2}{\cancel{\text{mol of N}_2}} = \underline{\underline{1.69 \times 10^{24} \; \text{molecules of N}_2}}$$

Example 9-8

What weight of NH_3 is produced by 4.65×10^{22} molecules of H_2?

Procedure:

This example is similar to Example 9-7 except that in the first step we convert number of molecules to moles of what's given.

Scheme:

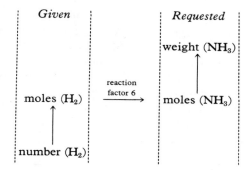

Solution:

$$4.65 \times 10^{22} \; \cancel{\text{molecules of H}_2} \times \frac{1 \; \cancel{\text{mol of H}_2}}{6.02 \times 10^{23} \; \cancel{\text{molecules of H}_2}}$$

$$\times \frac{2 \; \cancel{\text{mol of NH}_3}}{3 \; \cancel{\text{mol of H}_2}} \times \frac{17.0 \; \text{g of NH}_3}{\cancel{\text{mol of NH}_3}} = \underline{\underline{0.875 \; \text{g of NH}_3}}$$

The general procedure for working stoichiometry problems covering all of the possibilities discussed so far is shown in Figure 9-3. In words, the general procedure is as follows:

1 Write down (a) what's given and (b) what's requested.

2 If weight or a number of molecules is given, convert to moles in the first step.

3 Using the correct reaction factor from the *balanced equation*, convert moles of what's given to moles of what's requested in the second step.

4 If a weight or a number of molecules is requested, convert from moles of what's requested in the third step.

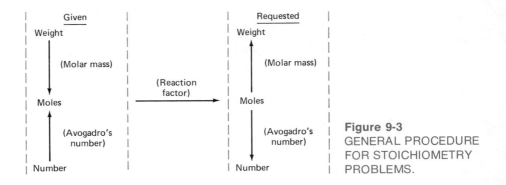

Figure 9-3
GENERAL PROCEDURE
FOR STOICHIOMETRY
PROBLEMS.

As you work the problems, first map out a procedure as shown in the examples. Set up the problems, making sure that all of the factors are correct and the appropriate units cancel, and finally *carefully* do the math. If you have worked stoichiometry problems previously using proportions, you may find the procedures outlined here slightly more time consuming; but you will also find these procedures more dependable and organized in obtaining the correct answer. In chemistry that's important.

Before proceeding, we will follow the general procedure with two more examples of stoichiometry problems.

Example 9-9

Some sulfur is present in coal in the form of pyrite, FeS_2 (also known as "fool's gold"). When it burns it pollutes the air with the combustion product SO_2, as shown by the following chemical equation:

$$4FeS_2(s) + 11O_2(g) \longrightarrow 2Fe_2O_3(s) + 8SO_2(g)$$

How many moles of Fe_2O_3 are produced from 145 g of O_2?

Procedure:
Given: 145 g of O_2;   Requested: ____?____ mol of Fe_2O_3.
Convert grams of O_2 to moles of Fe_2O_3.

Scheme:

$$\boxed{\text{wt } O_2}$$
$$\downarrow \text{32.0 g/mol}$$
$$\text{mol of } O_2 \longrightarrow \boxed{\text{mol of } Fe_2O_3}$$

From the equation, a reaction factor relating O_2 to Fe_2O_3 is needed. The factor will have what's requested in the numerator.

$$\frac{2 \text{ mol of } Fe_2O_3}{11 \text{ mol of } O_2}$$

Solution:

$$145 \text{ g of O}_2 \times \frac{1 \text{ mol of O}_2}{32.0 \text{ g of O}_2} \times \frac{2 \text{ mol of Fe}_2\text{O}_3}{11 \text{ mol of O}_2} = 0.824 \text{ mol of Fe}_2\text{O}_3$$

Example 9-10

From the equation used in the preceding example, determine the weight of SO_2 that is produced from the combustion of 38.8 g of FeS_2.

Procedure:

Given: 38.8 g of FeS_2; Requested: _____?_____ g of SO_2.
Convert grams of FeS_2 to grams of SO_2.

Scheme:

$$\boxed{\text{wt FeS}_2} \qquad\qquad \boxed{\text{wt SO}_2}$$

$$\downarrow 120 \text{ g/mol} \qquad\qquad \uparrow 64.1 \text{ g/mol}$$

$$\text{mol of FeS}_2 \longrightarrow \text{mol of SO}_2$$

The reaction factor from the equation that relates moles of what's given to moles of what's requested (in the numerator) is

$$\frac{8 \text{ mol of SO}_2}{4 \text{ mol of FeS}_2}$$

Solution:

$$38.8 \text{ g of FeS}_2 \times \frac{1 \text{ mol of FeS}_2}{120 \text{ g of FeS}_2} \times \frac{8 \text{ mol of SO}_2}{4 \text{ mol of FeS}_2}$$

$$\times \frac{64.1 \text{ g of SO}_2}{\text{mol of SO}_2} = 41.5 \text{ g of SO}_2$$

See Problems 9-6 through 9-11

9-4 Percent Yield

In the reaction between N_2 and H_2 to produce NH_3 we have treated the reaction as if all the N_2 and H_2 present reacts to form NH_3. In fact, this isn't the case. If a mixture of N_2 and H_2 is allowed to react until no additional NH_3 is formed, there will still be some unreacted H_2 and N_2 present. This phenomenon, called equilibrium, is discussed in more detail in Chapter 14. Thus, in many reactions only a certain portion of the reactants convert to products. *The amount of product that forms in an incomplete reaction is known as the* **actual yield.** *The amount of product that would form if at least one of the reactants is completely consumed is known as the* **theoretical yield.** *The* **percent yield** *is the actual yield in grams divided by the theoretical yield in grams times 100.*

$$\frac{\text{Actual yield}}{\text{Theoretical yield}} \times 100 = \text{percent yield}$$

Example 9-11

In a certain experiment a 4.70-g quantity of H_2 is allowed to react with N_2; a 12.5-g quantity of NH_3 is formed. What is the percent yield based on the H_2?

Procedure:

Find the weight of NH_3 that would form if all the 4.70 g of H_2 is converted to NH_3 (the theoretical yield). Using the actual yield (12.5 g), find the percent yield.

Scheme:

$$\boxed{\text{wt } H_2} \qquad\qquad \boxed{\text{wt } NH_3}$$

$$\downarrow \text{2.02 g/mol} \qquad\qquad \uparrow \text{17.0 g/mol}$$

$$\text{mol of } H_2 \longrightarrow \text{mol of } NH_3$$

The reaction factor needed is:

$$\frac{2 \text{ mol of } NH_3}{3 \text{ mol of } H_2}$$

Solution:

$$4.70 \text{ g of } H_2 \times \frac{1 \text{ mol of } H_2}{2.02 \text{ g of } H_2} \times \frac{2 \text{ mol of } NH_3}{3 \text{ mol of } H_2} \times \frac{17.0 \text{ g of } NH_3}{\text{mol of } NH_3}$$

$$= \underline{26.4 \text{ g of } NH_3} \quad \text{(theoretical yield)}$$

$$\frac{12.5 \text{ g}}{26.4 \text{ g}} \times 100 = \underline{47.3\% \text{ yield}}$$

In a similar problem, a reaction is complete but the compound in question is only part of the original sample. By analyzing the products, the percent purity of the original sample can be determined.

Example 9-12

A 125-g sample of impure limestone is heated to drive off all of the CO_2 according to the following equation:

$$CaCO_3(s) \longrightarrow CaO(s) + CO_2(g)$$

If 50.6 g of CO_2 is collected, what is the purity of the original sample?

Procedure:

Assume that all $CaCO_3$ in the impure sample decomposes to form CaO and CO_2.

1 Find the weight of $CaCO_3$ needed to produce 50.6 g of CO_2 using the following scheme:

$$\boxed{\text{wt } CO_2} \qquad\qquad \boxed{\text{wt } CaCO_3}$$

$$\downarrow \text{44.0 g/mol} \qquad\qquad \uparrow \text{100 g/mol}$$

$$\text{mol of } CO_2 \longrightarrow \text{mol of } CaCO_3$$

The reaction factor needed is

$$\frac{1 \text{ mol of CaCO}_3}{1 \text{ mol of CO}_2}$$

2 Find the percent of the original sample that the weight of $CaCO_3$ found in step 1 represents.

1 $50.6 \text{ g of CO}_2 \times \dfrac{1 \text{ mol of CO}_2}{44.0 \text{ g of CO}_2} \times \dfrac{1 \text{ mol of CaCO}_3}{1 \text{ mol of CO}_2}$

$$\times \frac{100 \text{ g of CaCO}_3}{\text{mol of CaCO}_3} = \underline{115 \text{ g of CaCO}_3}$$

2 $\dfrac{115 \text{ g of CaCO}_3}{125 \text{ g of sample}} \times 100 = \underline{92.0\% \text{ pure}}$

See Problems 9-12 through 9-16

9-5 Limiting Reactant

In the simple reaction

$$2H_2 + O_2 \longrightarrow 2H_2O$$

a 4.0-g quantity of H_2 reacts completely with 32.0 g of O_2. In fact, whenever H_2 and O_2 is mixed in a $4.0:32$ $(1:8)$ weight ratio and a reaction is initiated, all reactants are consumed and only products appear. (This reaction goes to completion, which is a 100% yield.) *When reactants are mixed in exactly the weight ratio determined from the balanced equation, the mixture is said to be* **stoichiometric.**

$$4.0 \text{ g of } H_2 + 32.0 \text{ g of } O_2 \longrightarrow 36.0 \text{ g of } H_2O \quad \text{(stoichiometric)}$$

If a 6.0-g quantity of H_2 and a 32.0-g quantity of O_2 are mixed, there is still only 4.0 g of H_2 used so that 2.0 g of H_2 remains after the reaction is complete. Thus H_2 is in excess and the amount of H_2O formed is limited by the amount of O_2 originally present. The O_2 is the limiting reactant. *The* **limiting reactant** *is completely converted to products by the reaction and therefore determines the amount of products that are formed* (if the reaction goes 100% to the right).

$$6.0 \text{ g of } H_2 + 32.0 \text{ g of } O_2 \longrightarrow 36.0 \text{ g of } H_2O + \boxed{2.0 \text{ g of } H_2 \text{ unreacted}}$$
$$(H_2 \text{ in excess, } O_2 \text{ limiting reactant})$$

If a 4.0-g quantity of H_2 and a 38.0-g quantity of O_2 is mixed, 36.0 g of H_2O will be produced. Now O_2 is in excess and H_2 is the limiting reactant:

$$4.0 \text{ g of } H_2 + 38.0 \text{ g of } O_2 \longrightarrow 36.0 \text{ g of } H_2O + \boxed{6.0 \text{ g of } O_2 \text{ unreacted}}$$
$$(O_2 \text{ in excess, } H_2 \text{ limiting reactant})$$

When quantities of two or more reactants are given, it is necessary to determine which is the limiting reactant (unless they are mixed in exactly stoichiometric amounts). This can be simplified by the following procedures.

1   Determine the number of moles of a product produced by each reactant using the general procedure discussed earlier.

2   The reactant producing the *smallest* yield is the limiting reactant.

Example 9-13

Silver tarnishes (turns black) in homes because of the presence of small amounts of H_2S (a rotten-smelling gas which originates from the decay of food). The reaction is as follows:

$$4Ag(s) + 2H_2S(g) + O_2(g) \longrightarrow 2Ag_2S(s) + 2H_2O(l)$$
$$\text{(black)}$$

If 18.5 g of Ag is present with 4.55 g of H_2S, which is the limiting reactant?

Procedure:

Convert the weight of Ag and H_2S to moles of Ag_2S produced by each. The smallest yield is the limiting reactant.

$$\text{wt of reactant}$$
$$\Big| \text{ Molar mass}$$
$$\downarrow$$
$$\text{mol of reactant} \longrightarrow \text{mol of } Ag_2S$$

Solution:

Ag $18.5 \text{ g of Ag} \times \dfrac{1 \text{ mol of Ag}}{108 \text{ g of Ag}} \times \dfrac{2 \text{ mol of } Ag_2S}{4 \text{ mol of Ag}}$

$$= 0.0856 \text{ mol of } Ag_2S$$

H_2S $4.55 \text{ g of } H_2S \times \dfrac{1 \text{ mol of } H_2S}{34.1 \text{ g of } H_2S} \times \dfrac{2 \text{ mol of } Ag_2S}{2 \text{ mol of } H_2S}$

$$= 0.133 \text{ mol of } Ag_2S$$

Therefore, <u>Ag is the limiting reactant</u> and H_2S is in excess.

Example 9-14

Methanol (CH_3OH) is used as a fuel for racing cars. It burns in the engine according to the following equation:

$$2CH_3OH(l) + 3O_2(g) \longrightarrow 2CO_2(g) + 4H_2O(g)$$

If 40.0 g of methanol is mixed with 46.0 g of O_2:

(a) What is the limiting reactant?

(b) What is the theoretical yield of CO_2?

(c) If 38.0 g of CO_2 is actually produced, what is the percent yield?

Procedure:

(a) Determine the moles of CO_2 formed from each of the reactants.

$$\text{wt of reactant}$$
$$\Big\downarrow \text{Molar mass}$$
$$\text{mol of reactant} \longrightarrow \text{mol of } CO_2$$

$CH_3OH \qquad 40.0 \ \text{g of } CH_3OH \times \dfrac{1 \ \text{mol of } CH_3OH}{32.0 \ \text{g of } CH_3OH} \times \dfrac{2 \ \text{mol of } CO_2}{2 \ \text{mol of } CH_3OH}$

$$= 1.25 \ \text{mol of } CO_2$$

$O_2 \qquad 46.0 \ \text{g of } O_2 \times \dfrac{1 \ \text{mol of } O_2}{32.0 \ \text{g of } O_2} \times \dfrac{2 \ \text{mol of } CO_2}{3 \ \text{mol of } O_2}$

$$= 0.958 \ \text{mol of } CO_2$$

Therefore, O_2 is the limiting reactant

(b) The theoretical yield is determined from the amount of product formed *from the limiting reactant.* Thus we simply convert the 0.958 mol of CO_2 produced by the O_2 to grams.

$$0.958 \ \text{mol of } CO_2 \times \frac{44.0 \ \text{g of } CO_2}{\text{mol of } CO_2} = 42.2 \ \text{g of } CO_2$$

(c) From the theoretical yield and the actual yield determine percent yield.

$$\frac{38.0 \ \text{g}}{42.2 \ \text{g}} \times 100 = 90.0\% \ \text{yield of } CO_2$$

See Problems 9-17 through 9-22

Review Questions

1 Discuss the significance of a chemical equation. What is represented by a chemical equation?

2 What is meant by a *balanced* chemical equation?

3 To what do the following refer in a chemical equation: (s), (l), (g), (aq), ↑, ↓, and Δ?

4 Discuss the five types of chemical reactions mentioned in this chapter.

5 Define what is meant by stoichiometry.

6 What is a reaction factor? What is its purpose?

7 Give the general procedure for solving stoichiometry problems.

8 What is meant by the theoretical yield of a reaction? What is the percent yield?

9 What is meant by a stoichiometric mixture?

10 When is a compound the limiting reactant?

11 When is a reactant present in excess?

Problems

Chemical Equations

9-1 Write balanced chemical equations from the following word equations.

(a) Sodium metal plus water yields hydrogen gas and an aqueous sodium hydroxide solution.

(b) Potassium chlorate when heated yields potassium chloride plus oxygen gas. (Ionic compounds are solids.)

(c) An aqueous sodium chloride solution plus an aqueous silver nitrate solution yields a silver chloride precipitate (solid) and a sodium nitrate solution.

(d) An aqueous phosphoric acid solution plus an aqueous calcium hydroxide solution yields water and a calcium phosphate solution.

■ (e) Liquid propyl alcohol (C_3H_7OH) burns (combines with oxygen) to produce liquid water and gaseous carbon dioxide.

■ (f) An aqueous calcium hydroxide solution reacts with gaseous sulfur dioxide to produce a calcium sulfite precipitate and water.

9-2 Balance the following chemical equations.

(a) $CaCO_3 \xrightarrow{\Delta} CaO + CO_2$

(b) $Na + O_2 \rightarrow Na_2O$

(c) $H_2SO_4 + NaOH \rightarrow Na_2SO_4 + H_2O$

(d) $H_2O_2 \rightarrow H_2O + O_2$

(e) $Si_2H_6 + H_2O \rightarrow Si(OH)_4 + H_2$

(f) $Al + H_3PO_4 \rightarrow AlPO_4 + H_2$

(g) $Ca(OH)_2 + HCl \rightarrow CaCl_2 + H_2O$

(h) $Na_2NH + H_2O \rightarrow NH_3 + NaOH$

■ (i) $Mg + N_2 \rightarrow Mg_3N_2$

■ (j) $CaC_2 + H_2O \rightarrow C_2H_2 + Ca(OH)_2$

■ (k) $C_2H_6 + O_2 \rightarrow CO_2 + H_2O$

Types of Chemical Reactions

9-3 Which of the five basic types of reactions is each reaction in Problem 9-1?

■ **9-4** Which of the five basic types of reactions is each reaction in Problem 9-2?

Reaction Factors

9-5 Write one reaction factor illustrating the relationship between each reactant and all other reactants and products.

(a) $Mg + 2HCl \rightarrow MgCl_2 + H_2$

■ (b) $Cu + 4HNO_3 \rightarrow$
$$Cu(NO_3)_2 + 2NO_2 + 2H_2O$$

(c) $2C_4H_{10} + 13O_2 \rightarrow 8CO_2 + 10H_2O$

Stoichiometry

9-6 Given the reaction

$$2H_2 + O_2 \longrightarrow 2H_2O$$

answer the following.

(a) How many moles of H_2 are needed to produce 0.400 mol of H_2O?

(b) How many moles of H_2O will be produced from 0.640 g of O_2?

■ (c) How many moles of H_2 are needed to react with 0.032 g of O_2?

■ (d) What weight of H_2O would be produced from 0.400 g of H_2?

9-7 Propane burns according to the following equation:

$$C_3H_8 + 5O_2 \longrightarrow 3CO_2 + 4H_2O$$

(a) How many moles of CO_2 are produced from the combustion of 0.450 mol of C_3H_8? How many moles of H_2O? How many moles of O_2 are needed?

(b) What weight of H_2O is produced if 0.200 mol of CO_2 is also produced?

(c) What weight of C_3H_8 is required to produce 1.80 g of H_2O?

■ (d) What weight of C_3H_8 is required to react with 160 g of O_2?

■ (e) What weight of CO_2 is produced by the reaction of 1.20×10^{23} molecules of O_2?

■ (f) How many moles of H_2O are produced if 4.50×10^{22} molecules of CO_2 are produced?

9-8 The alcohol component of "gasohol" burns according to the following equation:

$$C_2H_6O(l) + 3O_2(g) \longrightarrow 2CO_2(g) + 3H_2O(g)$$

(a) What weight of O_2 is needed to produce 5.45 mol of CO_2?

(b) What weight of H_2O is produced from 4.58×10^{-4} mol of C_2H_6O?

(c) What weight of H_2O is produced from 125 g of C_2H_6O?

(d) What weight of CO_2 is produced if 50.0 g of H_2O is produced?

(e) What weight of C_2H_6O reacts with 8.54×10^{25} molecules of O_2?

9-9 In the atmosphere N_2 and O_2 do not react. In the high temperatures of an automobile engine, however, the following reaction does occur:

$$N_2(g) + O_2(g) \longrightarrow 2NO(g)$$

When the NO reaches the atmosphere through the engine exhaust, a second reaction takes place:

$$2NO(g) + O_2(g) \longrightarrow 2NO_2(g)$$

The NO_2 is a somewhat brownish gas that contributes to the haze of smog and is irritating to the nasal passages and lungs. What weight of N_2 is required to produce 155 g of NO_2?

■ 9-10 Manganese metal can be prepared by a reaction called a thermite reaction because it gives off a great amount of heat. In fact, the manganese metal is formed in the molten or liquid state:

$$4Al(s) + 3MnO_2(s) \longrightarrow 3Mn(l) + 2Al_2O_3(s)$$

From the reaction, determine what weight of Al is needed to produce 750 g of Mn. How many molecules of MnO_2 are used in the process?

*9-11 Liquid iron is made from iron ore (Fe_2O_3) in a three-step process in a blast furnace:

1 $3Fe_2O_3(s) + CO(g) \rightarrow$
$$2Fe_3O_4(s) + CO_2(g)$$

2 $Fe_3O_4(s) + CO(g) \rightarrow 3FeO(s) + CO_2(g)$

3 $FeO(s) + CO(g) \rightarrow Fe(l) + CO_2(g)$

What weight of iron would eventually be produced from 125 g of Fe_2O_3?

Theoretical and Percent Yield

9-12 Sulfur trioxide (SO_3) is prepared from SO_2 according to the following equation:

$$2SO_2(g) + O_2(g) \longrightarrow 2SO_3(g)$$

In this reaction not all SO_2 is converted to SO_3 even with excess O_2 present. In a certain experiment, 21.2 g of SO_3 was produced from 24.0 g of SO_2. What is the theoretical yield of SO_3 and the percent yield?

■ 9-13 A 50.0-g sample of impure $KClO_3$ is decomposed to KCl and O_2. If a 12.0 g quantity of O_2 is produced, what percent of the sample is $KClO_3$? (Assume that all of the $KClO_3$ present decomposes.)

9-14 Octane in gasoline burns in the automobile engine according to the following equation:

$$2C_8H_{18}(l) + 25O_2(g) \longrightarrow$$
$$16CO_2(g) + 18H_2O(g)$$

If a 57.0 g sample of octane is burned, 152 g of CO_2 is formed. What is the percent yield of CO_2?

*9-15 In Problem 9-14, the C_8H_{18} that is *not* converted to CO_2 forms CO. What is the weight of CO formed? (CO is a poisonous pollutant which is converted to CO_2 in the car's catalytic converter.)

■ 9-16 In Example 9-9, SO_2 was formed from the burning of pyrite (FeS_2) in coal. If a 312 g quantity of SO_2 was collected from the burning of 6.50 kg of coal, what percent of the original sample was pyrite?

Limiting Reactant

9-17 Given the following equation:

$$4NH_3(g) + 3O_2(g) \longrightarrow 2N_2(g) + 6H_2O(l)$$

If a 40.0 g sample of O_2 is mixed with 1.50 mol of NH_3, which is the limiting reactant? What is the theoretical yield (in moles) of N_2?

9-18 Given the following equation:

$$2AgNO_3(aq) + CaCl_2(aq) \longrightarrow$$
$$2AgCl(s) + Ca(NO_3)_2(aq)$$

If a solution containing 20.0 g of $AgNO_3$ is mixed with a solution containing 10.0 g of $CaCl_2$, which compound is the limiting reactant? What is the theoretical yield (in grams) of AgCl?

■ **9-19** Limestone ($CaCO_3$) dissolves in hydrochloric acid as shown by the equation:

$$CaCO_3(s) + 2HCl(aq) \longrightarrow \\ CaCl_2(aq) + CO_2(g) + H_2O(l)$$

If a 20.0-g sample of $CaCO_3$ and 10.0 g of HCl are mixed, what weight of CO_2 is produced?

9-20 Given the balanced equation:

$$2HNO_3(aq) + 3H_2S(aq) \longrightarrow \\ 2NO(g) + 4H_2O(l) + 3S(s)$$

If a 10.0 g quantity of HNO_3 is mixed with 5.00 g of H_2S, what weight of H_2O is produced?

9-21 Given the following equation:

$$4NH_3(g) + 5O_2(g) \longrightarrow 4NO(g) + 6H_2O(l)$$

When a 80.0 g sample of NH_3 is mixed with 200 g of O_2, a 40.0-g quantity of NO is formed. What is the percent yield?

*■ **9-22** Given the following equation:

$$3K_2MnO_4 + 4CO_2 + 2H_2O \longrightarrow \\ 2KMnO_4 + 4KHCO_3 + MnO_2$$

How many moles of MnO_2 will be produced if 9.50 mol of K_2MnO_4, 6.02×10^{24} molecules of CO_2, and 90.0 g of H_2O are mixed?

The Nature of Gases

Our atmosphere is a gigantic ocean of gases. We live at the bottom of this ocean.

If there is one thing that held up the development of modern chemistry, it was the ignorance that scientists had of the nature of gases. Even air, the sea of gases in which we live, was only vaguely understood until the experiments of Antoine Lavoisier in France and Joseph Priestly in England in the late 1700s solved much of the mystery. These scientists proved that air is actually a mixture of elements and that only one of these elements, oxygen, plays a role in combustion. Their research not only enlightened science on the nature of air but also laid the foundation for modern quantitative chemistry.

There is a disadvantage and an advantage in the study of gases. The disadvantage is that most gases are invisible and thus we must infer their presence indirectly from measurements of their properties. At least you can see what you are working with when working with solids and liquids. The advantage is that many of the properties that we will study are the same for all gases. This permits many convenient generalizations or "laws" concerning their behavior. On the other hand, few generalizations apply to all solids or liquids. As a result, they must be studied individually.

In this chapter, we will first examine how the volume of a quantity of gas is affected by external conditions such as pressure and temperature. As it turns out, the observed effect of changes in these conditions (now known as "laws") have been known for centuries. However, if we jump to the present and list what is known about the nature of gases (called the kinetic theory of gases), we find these and other "gas laws" to be natural consequences of this nature. Finally, since gases are involved in chemical reactions, we will expand on the scheme for working stoichiometry problems (introduced in Chapter 9) to include the laws governing this state of matter.

As background for this chapter you should be familiar with the concept of the mole as discussed in Chapter 8 and the procedure for solving stoichiometry problems discussed in Chapter 9.

10-1 The Pressure of a Gas

In 1643, an Italian scientist named Evangelista Torricelli experimented with an apparatus that is now known as a **barometer.** Torricelli filled a long glass tube with the dense, liquid metal, mercury, and inverted the tube into a bowl of mercury so that no air could enter the tube. Torricelli found that the mercury in the tube would stay suspended to a height of about 76 cm no matter how long or wide the tube (see Figure 10-1).

At the time scientists thought that since "nature abhors a vacuum," it was a vacuum in the top of the tube that held up the column of mercury. Torricelli had a different idea. He suggested that it was actually the weight of the air on the outside that pushed up the level of mercury. Otherwise, since tubes 2 and 3 in Figure 10-1 would have "more vacuum" than tube 1, the level would rise higher in those tubes. To prove his point, Torricelli took his

Figure 10-1 THE BAROMETER. When a long tube is filled with mercury and inverted in a bowl of mercury, the atmosphere supports the column to a height of 76.0 cm.

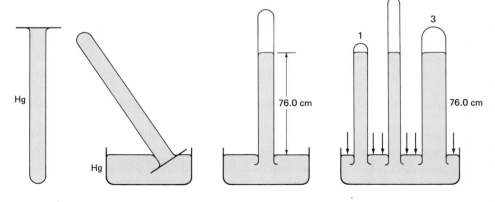

barometer up a mountain, where he reasoned that the thinner air would support less mercury in the tube. He was correct; the level of the barometer fell as he ascended the mountain.

In this discussion, we will define the total weight of a quantity of matter pressing on a surface as its **force** (force has a more formal definition in physics). Obviously, in a barometer the force of the atmosphere (the weight of the whole world's air) is not the same as the force of one small column of mercury. What is the same inside and outside of the tube is the pressure. **Pressure is defined as the force (in this case the weight) applied per unit area.** Mathematically, this definition is expressed as

$$P \text{ (pressure)} = \frac{F \text{ (force)}}{A \text{ (area)}}$$

In the barometer,

$$P_{air \, (outside)} = P_{Hg \, (inside)}$$

$$P_{Hg} = P_{air} = \frac{\text{weight of Hg}}{\text{area}}$$

$$\text{Weight of Hg} = \text{volume} \times \text{density}$$

$$\text{Volume} = \text{height} \times \text{area of opening}$$

In tube 1 the area of the opening is 1.00 cm².

$$\text{Volume} = 76.0 \text{ cm} \times 1.00 \text{ cm}^2 = 76.0 \text{ cm}^3$$

$$\text{Weight} = 76.0 \text{ cm}^3 \times 13.6 \text{ g/cm}^3 = 1030 \text{ g}$$

$$\text{Pressure} = 1030 \text{ g}/1.00 \text{ cm}^2 = 1030 \text{ g/cm}^2$$

Notice that the width of the tube in Torricelli's experiments does not affect the pressure. In tube 3, if the total area is 10.0 cm², the total force is

$$(76.0 \text{ cm} \times 10.0 \text{ cm}^2) \times 13.6 \text{ g/cm}^3 = 10,300 \text{ g}$$

The pressure, however, is the same as in tube 1.

$$P = \frac{F}{A} = \frac{10,300 \text{ g}}{10.0 \text{ cm}^2} = 1030 \text{ g/cm}^2$$

The human body has roughly 10,000 cm² of surface area (with, obviously, much individual variation). This means that we feel about 10^7 g (about 20,000 lb) of force from the air under normal atmospheric conditions. On days when it feels like that whole force is only on our heads it seems like we are "under a lot of pressure." However, despite how it may feel at times, the force is spread out over the total body area, so we should feel quite comfortable at the bottom of the ocean of gas where we live.

The normal pressure of the atmosphere at sea level is defined as exactly one atmosphere (1 atm) and is used as the standard. As we have seen, this is equivalent to the pressure exerted by a column of mercury 76.0 cm (760 mm) high.

Table 10-1   One Atmosphere

Unit	Special Use
760 mm Hg or 760 torr	Most chemistry laboratory measurements for pressures in the neighborhood of one atmosphere.
14.7 lb/in.²	U.S. pressure gauges
29.9 in. of Hg	U.S. weather reports
101.375 kPa (kilopascals)	The SI unit of pressure (1 newton/m²)
1.013 bars	Used in physics and astronomy mainly for very low pressures (millibars) or very high pressures (kilobars)

$$1.00 \text{ atm} = 76.0 \text{ cm Hg} = 760 \text{ mm Hg} = 760 \text{ torr}$$

(The unit of mm Hg is defined as torr in honor of Evangelista Torricelli.)

In addition to torr, there are several other units of pressure that have special uses. The special use of the unit and its relation to 1 atm is listed in Table 10-1.

See Problems 10-1, 10-2, 10-3

Example 10-1
What is 485 torr expressed in atmospheres?

Procedure:
The conversion factors are

$$\frac{760 \text{ torr}}{\text{atm}} \quad \text{and} \quad \frac{1 \text{ atm}}{760 \text{ torr}}$$

Solution:

$$485 \text{ torr} \times \frac{1 \text{ atm}}{760 \text{ torr}} = \underline{\underline{0.638 \text{ atm}}}$$

10-2 Volume and Pressure

The first quantitative relationship concerning gases was suggested in 1660 by the British chemist, Sir Robert Boyle. **Boyle's law** *states that there is an inverse relationship between the pressure exerted on a quantity of gas and its volume if the temperature is held constant.* (See Appendix B for a discussion of direct and inverse relationships and Appendix F for graphs of these relationships.)

Review Appendix B on proportionalities and Appendix F on graphs.

Boyle's law can be illustrated quite simply with an apparatus as shown in Figure 10-2. In experiment 1, a certain quantity of gas ($V_1 = 10.0$ ml) is trapped in a U-shaped tube by some mercury. Since the level of mercury is the same in both sides of the tube and the right side is open to the atmos-

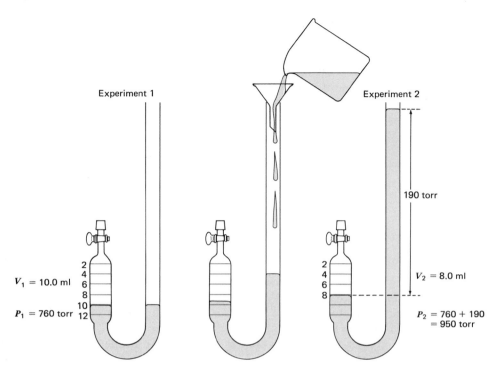

Experiment 1

Experiment 2

$V_1 = 10.0$ ml

$P_1 = 760$ torr

190 torr

$V_2 = 8.0$ ml

$P_2 = 760 + 190$
$= 950$ torr

Figure 10-2
BOYLE'S LAW
APPARATUS. Ad-
dition of mercury
in the apparatus
causes an increase
in pressure on the
trapped gas. This
leads to a reduction
in the volume.

phere, the pressure on the trapped gas is the same as the atmospheric pressure ($P_1 = 760$ torr).

When mercury is added to the tube, the pressure on the trapped gas is increased to 950 torr (760 torr originally plus 190 torr from the added mercury). Notice in experiment 2 that the *increase in pressure* has caused a *decrease in volume* to 8.0 ml.

The inverse relation of Boyle is represented as

$$V \propto \frac{1}{P}$$

or as an equality with k the constant of proportionality:

$$V = \frac{k}{P} \quad \text{or} \quad PV = k$$

Notice in experiment 1 in Figure 10-2 that

$$P_1V_1 = 760 \text{ torr} \times 10.0 \text{ ml} = 7600 \text{ torr} \cdot \text{ml} = k$$

indicates one set of conditions

In experiment 2,

$$P_2V_2 = 950 \text{ torr} \times 8.0 \text{ ml} = 7600 \text{ torr} \cdot \text{ml} = k$$

indicates a second set of conditions

As predicted by Boyle's law, notice that in both experiments PV equals the same value. Therefore, for a quantity of gas under two sets of conditions at the same temperature,

$$P_1V_1 = P_2V_2 = k$$

We can use this equation to calculate how a volume of a gas changes when the pressure changes. For example, if V_2 is a new volume that is to be found at a certain new pressure, P_2, the equation becomes

$$V_2 = V_1 \times \frac{P_1}{P_2}$$

(Final volume) = (initial volume) × (pressure correction factor)

In working problems of this nature, first determine whether V_2 (the final volume) is smaller or larger than V_1 (the initial volume). If V_2 is smaller, the pressure correction factor must be less than one, which means that the numerator is smaller than the denominator. If V_2 is larger than V_1, the pressure factor must be greater than one, which means that the numerator is larger than the denominator. The following two examples illustrate the use of Boyle's law.

Example 10-2
Inside a certain automobile engine, the volume of a cylinder is 475 ml when the pressure is 1.05 atm. When the gas is compressed, the pressure increases to 5.65 atm at the same temperature. What is the volume of the compressed gas?

Procedure:
The final volume equals the initial volume times the pressure factor. Since the pressure increases, the volume decreases and the pressure factor must be less than one.

Solution:

$$V_2 = V_1 \times \frac{P_1}{P_2}$$

$$V_2 = 475 \text{ ml} \times \frac{1.05 \text{ atm}}{5.65 \text{ atm}} = \underline{\underline{88.3 \text{ ml}}}$$

(Notice that the units of pressure are the same. If the initial and final pressures are given in different units, one must be converted to the other.)

Example 10-3

If the volume of a gas is 3420 ml at a pressure of 2.17 atm, what is the pressure if the gas expands to 8.75 liters?

Procedure:

The Boyle's law relationship can be solved for P_2:

$$P_2 = P_1 \times \frac{V_1}{V_2}$$

(Final pressure) = (initial pressure) × (volume correction factor)

In this case, the volume increases from 3420 ml (3.42 liters) to 8.75 liters. An increase in volume means a decrease in pressure, so the volume factor is less than one.

Solution:

$$P_2 = P_1 \times \frac{V_1}{V_2}$$

$$P_2 = 2.17 \text{ atm} \times \frac{3.42 \text{ liters}}{8.75 \text{ liters}} = \underline{\underline{0.848 \text{ atm}}}$$

See Problems 10-4 through 10-8

10-3 Volume and Temperature

It was more than a century before any further development in the quantitative understanding of gases occurred. However, in 1787 a French scientist, Jacques Charles, showed that there was a mathematical relationship between the volume of a gas and the temperature if the pressure is held constant. Charles showed that any gas expands by a definite fraction as the temperature rises. He found that the volume increases by a fraction of $\frac{1}{273}$ for each 1°C rise in the temperature. This observation is illustrated by the following example.

Example 10-4

If a gas has a volume of 10.0 liters at 0.0°C, what is its volume at 50.0°C if the pressure is held constant?

Solution:

The gas expands $\frac{1}{273}$ for each 1°C rise in temperature. For a 50.0°C rise the expansion is 50.0/273 times original volume. Therefore, the amount of expansion is

$$10.0 \text{ liters} \times \frac{50.0}{273} = 1.83 \text{ liters}$$

The new volume is the original volume plus the amount of expansion.

$$10.0 \text{ liters} + 1.83 \text{ liters} = \underline{\underline{11.8 \text{ liters}}}$$

See Figure 10-3.

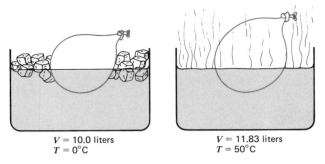

Figure 10-3 THE EFFECT OF TEMPERATURE ON VOLUME. When the temperature increases, the volume of the balloon increases.

V = 10.0 liters
T = 0°C

V = 11.83 liters
T = 50°C

In Figure 10-4, the data from the four experiments on the left is plotted in a graph on the right. Notice that this is a linear relationship. (Review Appendix F for a discussion of linear relationships.)

Notice from the graph in Figure 10-4 that the straight line is extrapolated (extended beyond the experimentally measured points) to where the gas would *theoretically* have zero volume. Obviously, it is impossible for matter to have zero volume, but since all gases condense to form solids or liquids at low temperatures anyway, there is no such thing as a "gas law" at very low temperatures. The temperature at zero volume (−273°C) is significant, however, as this turns out to be the *lowest possible temperature* (−273.16°C, to be more precise). The behavior of matter at this temperature is discussed shortly. *A temperature scale that assigns zero as the lowest possible temperature is known as the* **absolute** *or* **Kelvin** *scale.* Since the magnitude of the Celsius and Kelvin degrees is the same, we have the following simple relationship between scales:

$$K = °C + 273$$

We can now restate more simply the observations of Charles, which is known as **Charles' law:** *The volume of a gas is directly proportional to the Kelvin temperature at constant pressure.* Mathematically, this is

$$V \propto T$$

Exp.	T(°C)	V (liters)
1	100	136.7
2	50	118.3
3	10	103.7
4	0	100.0

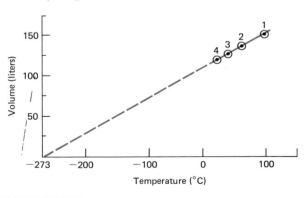

Figure 10-4 VOLUME AND TEMPERATURE.

As an equality with k as the constant of proportionality,

$$V = kT \quad \text{or} \quad \frac{V}{T} = k$$

For a quantity of gas under two sets of conditions at the same pressure,

$$\frac{V_1}{T_1} = \frac{V_2}{T_2}$$

This equation can be used to calculate how a volume of gas changes when the temperature changes. For example, if V_2 is a new volume that we are to find at a certain temperature, T_2, the equation becomes

$$V_2 = V_1 \times \frac{T_2}{T_1}$$

(Final volume) = (initial volume) × (temperature correction factor)

In this case, remember that if the temperature decreases the volume must also decrease, which means that the temperature factor is less than one. On the other hand, if the temperature increases the volume must increase, which means that the temperature factor is more than one.

Example 10-5
A certain quantity of gas in a balloon has a volume of 185 ml at a temperature of 50°C. What is the volume of the balloon if the temperature is lowered to −17°C? Assume that the pressure remains constant.

Procedure:
Since the temperature decreases the volume decreases. The temperature factor must therefore be less than one.

Solution:

$$V_2 = V_1 \times \frac{T_2}{T_1} \qquad \begin{array}{l} T_2 = -17°C + 273 = 256 \text{ K}\star \\ T_1 = 50°C + 273 = 323 \text{ K} \end{array}$$

$$V_2 = 185 \text{ ml} \times \frac{256 \text{ K}}{323 \text{ K}} = \underline{\underline{147 \text{ ml}}}$$

(Notice that temperature *must* be expressed in Kelvin degrees.)

See Problems 10-9 through 10-13

10-4 The Combined Gas Law

Boyle's law and Charles' law can be combined into one relationship, which is appropriately called the **Combined gas law:**

$$\frac{PV}{T} = k$$

⋆ Notice that a Celsius reading with two significant figures (e.g., −17°C) becomes a Kelvin reading with three significant figures (e.g., 256 K).

For two sets of conditions, the combined gas law is

$$\frac{P_1V_1}{T_1} = \frac{P_2V_2}{T_2}$$

Notice that if T is constant ($T_1 = T_2$) while P and V are changing, T cancels and the combined gas law becomes Boyle's law as shown below.

Since $T_1 = T_2$, substitute T_1 for T_2 and then cancel:

$$\frac{P_1V_1}{\cancel{T_1}} = \frac{P_2V_2}{\cancel{T_1}} \quad \text{or} \quad \underline{P_1V_1 = P_2V_2}$$

If P is constant ($P_1 = P_2$), P_1 can be substituted for P_2, P cancels, and Charles' law remains.

$$\frac{\cancel{P_1}V_1}{T_1} = \frac{\cancel{P_1}V_2}{T_2} \quad \text{or} \quad \frac{V_1}{T_1} = \frac{V_2}{T_2}$$

If the volume remains constant ($V_1 = V_2$) while P and T are changing, V cancels and the following relation is left:

$$\frac{P_1\cancel{V_1}}{T_1} = \frac{P_2\cancel{V_1}}{T_2} \quad \text{or} \quad \frac{P_1}{T_1} = \frac{P_2}{T_2}$$

See Problems 10-14 through 10-17

This is commonly referred to as **Gay-Lussac's law.** *It states that at constant volume, the pressure of a quantity of gas is directly proportional to the Kelvin temperature.* In fact, it is simply a direct result of Boyle's and Charles' laws.

It is apparent that the volume of a gas is very much dependent on conditions. This is unlike the volumes of liquids and solids, which are almost incompressible by comparison. When discussing gas volumes, then, it is useful to define certain conditions that are universally accepted as standard. These conditions are called **standard temperature and pressure (STP).** *These conditions are*

Standard temperature:	*0°C or 273 K*
Standard pressure:	*760 torr or 1 atm*

The following examples illustrate the use of the combined gas law and STP.

Example 10-6
A 25.8 liter quantity of gas has a pressure of 690 torr and a temperature of 17°C. What is the volume if the pressure is changed to 1.85 atm and the temperature to 345 K?

Solution:

Initial Conditions	Final Conditions

$V_1 = 25.8$ liters $V_2 = ?$

$P_1 = 690$ ~~torr~~ $\times \dfrac{1 \text{ atm}}{760 \text{ torr}}$ $P_2 = 1.85$ atm

$\quad = 0.908$ atm

$T_1 = 17°C + 273 = 290$ K $T_2 = 345$ K

$$\frac{P_1 V_1}{T_1} = \frac{P_2 V_2}{T_2} \qquad V_2 = V_1 \times \frac{P_1}{P_2} \times \frac{T_2}{T_1}$$

$$\text{(Final volume)} = \text{(initial volume)} \times \left(\begin{array}{c} \text{pressure} \\ \text{correction} \\ \text{factor} \end{array} \right) \times \left(\begin{array}{c} \text{temperature} \\ \text{correction} \\ \text{factor} \end{array} \right)$$

In this problem notice that the pressure increases, which decreases the volume. The pressure factor should therefore be less than one. On the other hand, the temperature increases, which increases the volume. The temperature factor is greater than one.

$$V_2 = 25.8 \text{ liters} \times \frac{0.908 \text{ atm}}{1.85 \text{ atm}} \times \frac{345 \text{ K}}{290 \text{ K}} = \underline{\underline{15.1 \text{ liters}}}$$

Example 10-7

A 5850-ft³ quantity of natural gas measured at STP was purchased from the Gas Company. Only 5625 ft³ was received at the house. Assuming that all of the gas was delivered, what was the temperature at the house if the delivery pressure was 1.10 atm?

Solution:

Initial Conditions	Final Conditions

$V_1 = 5850 \text{ ft}^3$ $V_2 = 5625 \text{ ft}^3$

$P_1 = 1.00 \text{ atm}$ $P_2 = 1.10 \text{ atm}$

$T_1 = 273 \text{ K}$ $T_2 = ?$

$$\frac{P_1 V_1}{T_1} = \frac{P_2 V_2}{T_2} \qquad T_2 = T_1 \times \frac{P_2}{P_1} \times \frac{V_2}{V_1}$$

In this case, the final temperature is corrected by a pressure and a volume correction factor. Since the final pressure is higher, the pressure factor must be greater than one to increase the temperature. The final volume is lower, so the volume correction factor must be less than one to decrease the temperature.

$$T_2 = 273 \text{ K} \times \frac{1.10 \text{ atm}}{1.00 \text{ atm}} \times \frac{5625 \text{ ft}^3}{5850 \text{ ft}^3} = 289 \text{ K}$$

$$289 \text{ K} - 273 = \underline{\underline{16°C}}$$

See Problems 10-18 through 10-22

10-5 The Nature of Gases

The gas laws discussed so far were accepted well before the nature of gases themselves were understood. As we will see, Boyle's, Charles', and other gas laws are obvious results of the **kinetic theory of gases,** *which summarizes our current understanding of the basic nature of gases.* The important assumptions of this theory are as follows:

1 A gas is composed of very small particles called molecules (atoms in the case of noble gases), which are widely spaced and occupy negligible volume. A gas is thus mostly empty space.

2 The molecules of the gas are in rapid, random motion, colliding with each other and the sides of the container. Pressure is a result of these collisions with the sides of the container.

3 All collisions involving gas molecules are elastic. (The total energy of two colliding molecules is conserved. A ball bouncing off of the pavement undergoes inelastic collisions, since it does not bounce as high each time.)

4 The gas molecules have negligible attractive (or repulsive) forces between them.

5 The temperature of a gas is related to average kinetic energy of the gas molecules. Also, at the same temperature different gases have the same average kinetic energy. Kinetic energy (K.E.) is given by

$$\text{K.E.} = \tfrac{1}{2}mv^2 \qquad (m = \text{mass}, \ v = \text{velocity})$$

The higher the temperature, the higher the average kinetic energy of the molecules of the gas. *Theoretically, all molecular motion stops at absolute zero.*

Let's examine briefly how the gas laws are predicted by kinetic theory. In Figure 10-5 we have, for simplicity, just four molecules of a gas in a large container on the left. The molecules travel a certain distance in a unit of time and eventually collide with the sides of the container, thus exerting pressure. Now let's confine the four molecules in a smaller container on the right at the same temperature. Since the temperature is the same, the molecules travel the same distance (on the average) per unit time. This is illustrated in the figure by the length of the lines of the paths of the molecules. In both containers the length of the lines are the same. In the smaller container, notice that collisions with the walls are more frequent. More frequent collisions in a certain area means a higher pressure. This, of course, is Boyle's law.

In Figure 10-6 we again have a container in the middle with four molecules. Now we will raise the temperature, which means that the average velocity of the molecules increases. A higher velocity means that the molecules travel farther in a unit of time as illustrated by the longer lines of travel. To the right of the middle container, the volume has been held constant. In

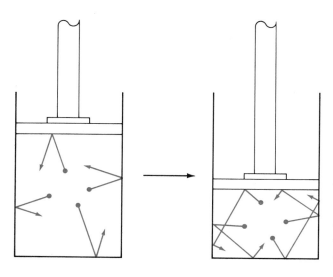

Figure 10-5 BOYLE'S LAW AND KINETIC THEORY. When the volume decreases, the pressure increases because of the more frequent collisions with the walls of the container.

this case, notice that the molecules collide more frequently and thus exert a greater pressure.* This relation of temperature and pressure at constant volume is Gay-Lussac's law. On the left of the middle container, the volume can change but the pressure remains constant. In order for the frequency of collisions (pressure) to remain the same, notice that the volume (the area of the container) must increase. Now we have Charles' law.

Other characteristics of gases can be easily expected by the kinetic theory.

* Since the molecules travel faster at higher temperatures, collisions with the walls are also more powerful. This also contributes to the greater pressure.

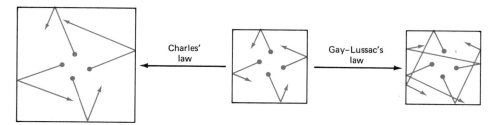

Figure 10-6 GAY-LUSSAC'S AND CHARLES' LAWS AND KINETIC THEORY. When the temperature increases, the molecules travel faster. If the volume is constant, the pressure increases as a result of the more frequent collisions. If the pressure is constant, the volume increases in order that the frequency of collisions remains the same.

1 Since gases are mostly empty space, they are highly *compressible*.

2 The small density of a gas can also be explained by the small amount of matter in the gas. For example, 1.0 g of *liquid* water has a volume of 1.0 ml. As *steam* at 100°C and 1 atm pressure, the 1.0 g of water occupies 1700 ml.

3 Because of the rapid motion of the molecules, gases mix thoroughly. In a chemistry building it doesn't take long for the whole campus to know that hydrogen sulfide experiments are on the schedule. The noxious smell of rotten eggs soon permeates the area.

4 The rapid motion of gas molecules also explains why gases expand to fill the space of the entire container.

The kinetic theory explains the gas laws that we have discussed so far. In addition, it opens the door to an understanding of the logic of several other laws that we will now discuss.

10-6 Temperature and Velocity

According to the kinetic theory, the molecules of two different gases at the same temperature have the same average kinetic energy.

Gas 1   K.E. $= \frac{1}{2}m_1v_1^2$ (m_1 and v_1 = mass and velocity of gas 1)

Gas 2   K.E. $= \frac{1}{2}m_2v_2^2$ (m_2 and v_2 = mass and velocity of gas 2)

Since $(K.E.)_1 = (K.E.)_2$, we have the following relationship:

$$\frac{1}{2}m_1v_1^2 = \frac{1}{2}m_2v_2^2$$

$$\frac{v_1}{v_2} = \sqrt{\frac{m_2}{m_1}}$$

This is known as **Graham's law of diffusion.** *It means that at the same temperature, the heavier the gas molecules the slower is its speed or velocity.* Thus lighter gases diffuse or mix faster than heavier gases (see Figure 10-7).

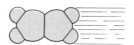

Figure 10-7 KINETIC ENERGY. To have the same kinetic energy, a light molecule travels at a higher velocity than a heavy molecule.

Example 10-8

How much faster does a He atom travel relative to an O_2 molecule at the same temperature?

Solution:

At the same temperature, $(K.E.)_{He} = (K.E.)_{O_2}$. Therefore,

$$m_{He} = 4.00 \text{ amu} \qquad m_{O_2} = 32.0 \text{ amu}$$

$$\frac{v_{He}}{v_{O_2}} = \sqrt{\frac{m_{O_2}}{m_{He}}} = \sqrt{\frac{32.0 \text{ amu}}{4.00 \text{ amu}}} = \sqrt{8.00} = 2.83$$

$$v_{He} = 2.83 v_{O_2}$$

On the average, He atoms travel 2.83 times faster than O_2 molecules.

Example 10-9

Nitrogen dioxide (NO_2) diffuses at a rate 1.73 times as fast as an unknown gas. What is the molar mass of the unknown gas?

Procedure:

The rate of diffusion is a function of the average velocity of the molecules. Also, molar mass (in g/mol) can be used in place of formula weight (in amu). Thus Graham's law is often expressed as

$$\frac{r_1}{r_2} = \sqrt{\frac{MM_2}{MM_1}} \qquad (r = \text{rate of diffusion}, MM = \text{molar mass})$$

Solution:

Let r_1 = rate of diffusion of NO_2 $(r_1 = r_{NO_2})$
  r_2 = rate of diffusion of the unknown gas
From the problem:

$$r_{NO_2} = 1.73 r_2 \qquad MM_{NO_2} = 46.0 \text{ g/mol}$$

$$\frac{r_{NO_2}}{r_2} = \frac{1.73 \, r_2}{r_2} = \sqrt{\frac{MM_2}{MM_{NO_2}}}$$

$$1.73 = \sqrt{\frac{MM_2}{46.0 \text{ g/mol}}}$$

Square both sides of the equation to remove the square root sign on the right.

$$(1.73)^2 = \left(\sqrt{\frac{MM_2}{46.0 \text{ g/mol}}} \right)^2 \qquad 2.99 = \frac{MM_2}{46.0 \text{ g/mol}}$$

$$MM_2 = 138 \text{ g/mol}$$

See Problems 10-23 through 10-26

10-7 The Nature of Mixtures of Gases

According to the kinetic theory, gas molecules have negligible interaction with each other. As a result, the pressure within a container is simply a function of the *number of gas molecules* present. Their individual nature is not relevant. In Figure 10-8, container 1 holds 1 mol of pure N_2 at a certain temper-

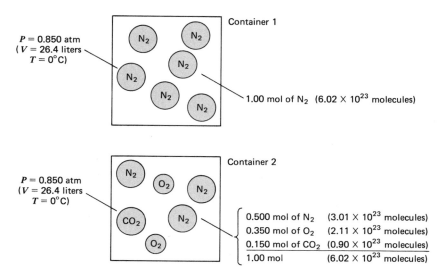

Figure 10-8 PRESSURES OF A PURE GAS AND A MIXTURE OF GASES.

ature. The pressure in the container is 0.850 atm. Container 2 holds a total of 1 mol of a mixture of gases. However, since the number of molecules, the volumes, and the temperatures are the same in both containers, *the pressures are also the same.*

In container 2 the total pressure (P_T = 0.850 atm) is the sum of the pressures of the individual component gases. This principle is given by **Dalton's law,** *which states: The total pressure in a system is the sum of the partial pressures of each component gas.*

$$P_T = P_1 + P_2 + P_3, \text{ etc.}$$

See Problems 10-27 through 10-29

Review Appendix A on expressing fractions.

In Figure 10-8, 50.0% of the molecules of gas present in the mixture are N_2 molecules. According to Dalton's law, 50.0% of the pressure must be due to N_2. Therefore the partial pressure of N_2 is 0.425 atm. A more direct way of approaching this problem is by use of the mole fraction of N_2 (same as molecule fraction). **Mole fraction** (X) *is the mole percent expressed in decimal form.* Like any other fraction, it is calculated by dividing the component part by the total amount. (For a review, refer to Appendix A, Section 6. Notice Examples A-9, A-10 and B-15, B-16.)

$$X_1 = \frac{n_1}{n_1 + n_2 + n_3 + \text{ etc.}} = \frac{\text{moles of gas 1}}{\text{total moles of gas}}$$

$$X_{N_2} = \frac{0.500}{0.500 + 0.350 + 0.150} = 0.500$$

$$X_{O_2} = \frac{0.350}{0.500 + 0.350 + 0.150} = 0.350$$

$$X_{CO_2} = \frac{0.150}{0.500 + 0.350 + 0.150} = 0.150$$

To find the partial pressure of a component, multiply the mole fraction of the component times the total pressure.

$$P_{N_2} = P_T X_{N_2} = 0.850 \text{ atm} \times 0.500 = \underline{0.425 \text{ atm}}$$

The pressure due to the other two components is calculated in the same way.

$$P_{O_2} = P_T X_{O_2} = 0.850 \text{ atm} \times 0.350 = \underline{0.298 \text{ atm}}$$

$$P_{CO_2} = P_T X_{CO_2} = 0.850 \text{ atm} \times 0.150 = \underline{0.128 \text{ atm}}$$

$$P_T = P_{N_2} + P_{O_2} + P_{CO_2} = 0.425 + 0.298 + 0.128 = 0.851 \text{ atm}*$$

Example 10-10

Three gases are mixed in a 5.00-liter container. Gas A has a pressure of 250 torr, gas B has a pressure of 0.300 atm, and gas C has a pressure of 750 torr. What is the total pressure?

Solution:

$$P_T = P_A + P_B + P_C$$

$$P_A = 250 \text{ torr}$$

$$P_B = 0.300 \text{ atm} \times \frac{760 \text{ torr}}{\text{atm}} = 228 \text{ torr}$$

$$P_C = 750 \text{ torr}$$

$$P_T = 250 + 228 + 750 = \underline{1228 \text{ torr}}$$

Example 10-11

A mixture of 52.0 g of N_2 and 48.5 g of O_2 is present in a container with a total pressure of 580 torr. What is the partial pressure of each gas?

Procedure:

1 Find the number of moles of each gas.

2 Find the mole fraction of one of the gases (X_1)

3 Find the partial pressure of that gas from $P_1 = X_1 P_T$.

4 Subtract P_1 from the total pressure to find the partial pressure of the other gas.

Solution:

1

$$52.0 \text{ g of } N_2 \times \frac{1 \text{ mol of } N_2}{28.0 \text{ g of } N_2} = 1.86 \text{ mol of } N_2$$

$$48.5 \text{ g of } O_2 \times \frac{1 \text{ mol of } O_2}{32.0 \text{ g of } O_2} = 1.52 \text{ mol of } O_2$$

---

* The pressures do not add to 0.850 atm because of rounding off.

2
$$X_{N_2} = \frac{n_{N_2}}{n_{N_2} + n_{O_2}} = \frac{1.86}{1.86 + 1.52} = 0.550$$

3
$$P_{N_2} = X_{N_2}P_T = 0.550 \times 580 \text{ torr} = \underline{319 \text{ torr}}$$

4
$$P_T = P_{N_2} + P_{O_2} \qquad P_{O_2} = P_T - P_{N_2}$$
$$P_{O_2} = 580 - 319 = \underline{261 \text{ torr}}$$

Example 10-12

On a humid day the normal atmosphere ($P = 760$ torr) is composed of 2.10% water molecules. Of the remaining gas, 78.0% is N_2 and 21.0% is O_2 (the other 1.0% is a mixture of many gases). What is the partial pressure of H_2O, N_2, and O_2?

Procedure:

1 Find the partial pressure of H_2O.

2 Find the pressure of the remaining gas.

3 Find P_{N_2} and P_{O_2}.

Solution:

1
$$2.10\% \text{ of } P_T \text{ is } 0.0210 \times 760 = \underline{16.0 \text{ torr } (H_2O)}$$

2
$$P_T = P_{H_2O} + P_{N_2} + P_{O_2} + P_{\text{other}}$$
$$(P_{N_2} + P_{O_2} + P_{\text{other}}) = P_T - P_{H_2O} = 760 - 16 = 744 \text{ torr}$$

See Problems 10-30 through 10-37

3
$$P_{N_2} = 0.780 \times 744 \text{ torr} = \underline{580 \text{ torr } (N_2)}$$
$$P_{O_2} = 0.210 \times 744 \text{ torr} = \underline{156 \text{ torr } (O_2)}$$

10-8 Volume and Quantity

Avogadro's law *states that equal volumes of gases at the same pressure and temperature contain equal numbers of molecules.* Obviously, this observation predates the kinetic theory, since this follows from the same reasoning that explained Dalton's law. If gas molecules do not interact, the total number of molecules is important, not their nature or identity. Notice that both containers in Figure 10-8 have the same total number of molecules although container 1 had a pure gas present and container 2 had a mixture. Otherwise, all measurable properties were identical (e.g., P, V, T, and n_{total}).

A corollary of Avogadro's law is that the volume is proportional to the number of molecules (moles) of gas at constant P and T. This is expressed as follows, where n equals the number of moles:

$$V \propto n \qquad \text{(as a proportion)}$$
$$V = kn \qquad \text{(as an equality)}$$
$$\frac{V_1}{n_1} = \frac{V_2}{n_2} \qquad \text{(for two conditions)}$$

See Figure 10-9 for an illustration.

V = 1.0 liter
n = 0.20 mol

V = 2.0 liters
n = 0.40 mol

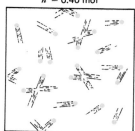

Figure 10-9 AVO-GADRO'S LAW (constant P and T). At constant temperature and pressure, the volume increases as the number of gas molecules increases.

Example 10-13
A balloon that is not inflated but is full of air has a volume of 275 ml and contains 0.0120 mol of air. As shown in Figure 10-10, a piece of Dry Ice (solid CO_2) weighing 1.00 g is placed in the balloon and the neck tied. What is the volume of the balloon after the Dry Ice has evaporated? (Assume constant T and P.)

Solution:

Initial Conditions	*Final Conditions*

$V_1 = 275$ ml $V_2 = ?$

$n_1 = 0.0120$ mol n_2 = mol of air + mol of CO_2

$$= 0.0120 + \left(1.00 \, \cancel{g \, of \, CO_2} \times \frac{1 \, mol}{44.0 \, \cancel{g \, of \, CO_2}}\right)$$

$$= 0.0120 + 0.0227$$
$$= 0.0347 \, mol$$

$$\frac{V_1}{n_1} = \frac{V_2}{n_2}$$

$$V_2 = V_1 \times \frac{n_2}{n_1} = 275 \, ml \times \frac{0.0347 \, \cancel{mol}}{0.0120 \, \cancel{mol}}$$

$$V_2 = \underline{\underline{795 \, ml}}$$

See Problems 10-38 through 10-40

10-9 The Ideal Gas Law
So far, the following proportionalities have been established pertaining to the volume of a gas:

Boyle's law $V \propto \dfrac{1}{P}$

Charles' law $V \propto T$

Avogadro's law $V \propto n$

Figure 10-10 ILLUS-TRATION OF AVO-GADRO'S LAW. The addition of the carbon dioxide to the balloon increases the number of gas molecules, which increases the volume.

These three relations can be combined into one proportionality:

$$V \propto \frac{nT}{P}$$

This can be changed to an equality by introducing a constant of proportionality. The constant used in this case is R and is called the **molar gas constant.** Traditionally, the constant is placed between n and T.

$$V = \frac{nRT}{P} \quad \text{or} \quad PV = nRT$$

The value for the gas constant is

$$R = 0.0821 \frac{\text{liter} \cdot \text{atm}}{\text{K} \cdot \text{mol}} = 62.4 \frac{\text{liter} \cdot \text{torr}}{\text{K} \cdot \text{mol}}$$

Notice that the units of R require specific units for P, V, and T. This law is know as the **ideal gas law.** The ideal gas law allows a variety of calculations involving one sample of gas. From the four possible variables (P, V, n, and T), one can be calculated if the other three are known (assuming that one knows the value of the gas constant, R). The following examples illustrate the use of the ideal gas law.

Example 10-14
What is the pressure of a 1.45-mol sample of a gas if the volume is 20.0 liters and the temperature is 25°C?

Solution:

$$P = ? \quad V = 20.0 \text{ liters} \quad n = 1.45 \text{ mol} \quad T = 25°C = 298 \text{ K}$$

$$PV = nRT$$

$$P = \frac{nRT}{V} = \frac{1.45 \; \cancel{\text{mol}} \times 0.0821 \dfrac{\cancel{\text{liter}} \cdot \text{atm}}{\cancel{\text{K}} \cdot \cancel{\text{mol}}} \times 298 \; \cancel{\text{K}}}{20.0 \; \cancel{\text{liters}}}$$

$$= \underline{\underline{1.77 \text{ atm}}}$$

Example 10-15

What is the temperature (in °C) of a 9.65-g quantity of O_2 in a 4560-ml container if the pressure is 895 torr?

Solution:

$$P = 895 \text{ torr} \quad T = ? \quad V = 4560 \text{ ml} = 4.56 \text{ liters}$$

$$n = 9.65 \text{ g of } O_2 \times \frac{1 \text{ mol}}{32.0 \text{ g of } O_2} = 0.302 \text{ mol}$$

$$PV = nRT \quad T = \frac{PV}{nR}$$

$$T = \frac{895 \text{ torr} \times 4.56 \text{ liters}}{0.302 \text{ mol} \times 62.4 \dfrac{\text{liter} \cdot \text{torr}}{\text{K} \cdot \text{mol}}} = 217 \text{ K}$$

$$T = 217 - 273 = \underline{\underline{-56°C}}$$

Example 10-16

What is the volume of 1.00 mol of a gas at STP?

Solution:

$$P = 1.00 \text{ atm} \quad V = ? \quad T = 273 \text{ K} \quad n = 1.00 \text{ mol}$$

$$PV = nRT \quad V = \frac{nRT}{P}$$

$$V = \frac{1.00 \text{ mol} \times 0.0821 \dfrac{\text{liter} \cdot \text{atm}}{\text{K} \cdot \text{mol}} \times 273 \text{ K}}{1.00 \text{ atm}} = \underline{\underline{22.4 \text{ liters}}}$$

See Problems 10-41 through 10-47

Notice that the ideal gas law is used when *one* set of conditions is given with one missing variable. The combined gas law is used when *two* sets of conditions are given with one missing variable.

This law is called ideal because it follows from the assumptions of the kinetic theory, which describes an ideal gas. The molecules of an ideal gas have no volume and have no attraction for each other. The molecules of a "real" gas obviously have a volume and there is some interaction between molecules, especially at high pressures and low temperatures where the molecules are pressed close together. Fortunately, at normal temperatures and pressures found on the surface of the earth, gases have close to "ideal" behavior. Therefore, the use of the ideal gas law is justified. If we lived on the planet Jupiter, however, where pressure is measured in thousands of earth atmospheres, the ideal gas law would not provide accurate answers. Other relationships would have to be used that take into account the volume of the molecules and the interaction between molecules.

10-10 Molar Volume

The volume calculated in Example 10-16 is known as the **molar volume** of a gas. *One mole of any gas (either pure or a mixture) occupies about 22.4 liters at STP.* This fact can be included in the definition of a mole as described in Chapter 8.

$$1.00 \text{ mol of } CO_2 \begin{cases} \longrightarrow 6.02 \times 10^{23} \text{ molecules} & \text{(molar number)} \\ \longleftrightarrow 44.0 \text{ g} & \text{(molar mass)} \\ \longrightarrow 22.4 \text{ liters at STP} & \text{(molar volume)} \end{cases}$$

Before working sample problems illustrating the use of molar volume, let's summarize the relationships discussed between moles and volume of a gas. In Figure 10-11, notice that moles can be converted to volume at STP using the molar volume relationship (path 1). Moles of a gas can also be converted to volume at some other temperature and pressure using the ideal gas law (path 2). Finally, the volume at STP of a quantity of gas and the volume at some other temperature and pressure relate by the combined gas law (path 3).

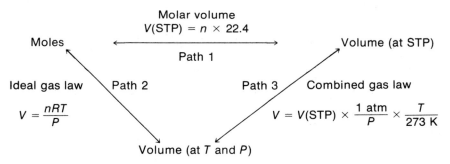

Figure 10-11 MOLES AND THE VOLUME OF A GAS. The number of moles relates to the volume by two paths.

Example 10-17
What is the weight of 4.55 liters of O_2 measured at STP?

Procedure:
Utilizing path 1 in Figure 10-11, convert volume to moles and then moles to the weight. The two possible conversion factors for converting volume to moles are

$$\frac{1 \text{ mol}}{22.4 \text{ liters}} \quad \text{and} \quad \frac{22.4 \text{ liters}}{\text{mol}}$$

Volume (STP) $\longrightarrow$ moles $\longrightarrow$ wt

Solution:

$$4.55 \; \cancel{\text{liters}} \times \frac{1 \; \cancel{\text{mol}}}{22.4 \; \cancel{\text{liters}}} \times \frac{32.0 \; \text{g}}{\cancel{\text{mol}}} = \underline{\underline{6.50 \; \text{g}}}$$

Example 10-18

A sample of gas that weighs 3.20 g occupies 2.00 liters at 17°C and 380 torr. What is the molar mass of the gas?

Procedure:

There are two ways that one can work this problem. The first uses paths 1 and 3 in Figure 10-11:

1 Convert given volume to volume at STP.

2 Convert volume at STP to moles.

3 Convert moles and weight to molar mass as follows: From Chapter 8,

$$n \; (\text{number of moles}) = \text{weight (in g)} \times \frac{1 \; \text{mol}}{\text{molar mass}}$$

or

$$n = \frac{\text{wt}}{MM} \qquad MM = \frac{\text{wt}}{n}$$

Solution:

Initial Conditions	*Final Conditions* (STP)
$V_1 = 2.00$ liters	$V_2 = ?$
$P_1 = 380$ torr	$P_2 = 760$ torr
$T_1 = 17°C = 290$ K	$T_2 = 273$ K

1
$$\frac{P_1 V_1}{T_1} = \frac{P_2 V_2}{T_2} \qquad V_2 = V_1 \times \frac{P_1}{P_2} \times \frac{T_2}{T_1}$$

$$V_2 = 2.00 \; \text{liters} \times \frac{380 \; \cancel{\text{torr}}}{760 \; \cancel{\text{torr}}} \times \frac{273 \; \cancel{K}}{290 \; \cancel{K}} = \underline{0.941 \; \text{liters (STP)}}$$

2
$$0.941 \; \cancel{\text{liters}} \times \frac{1 \; \text{mol}}{22.4 \; \cancel{\text{liters}}} = 0.0420 \; \text{mol of gas}$$

3
$$\text{Molar mass} = \frac{\text{wt}}{n} = \frac{3.20 \; \text{g}}{0.0420 \; \text{mol}} = \underline{\underline{76.2 \; \text{g/mol}}}$$

The alternative procedure is more direct, since it uses the ideal gas law as shown in path 2 in Figure 10-11. To find the molar mass using the ideal gas law, substitute the relationship for n ($n = \text{wt}/MM$) into the gas law.

$$PV = nRT$$

$$n = \frac{\text{wt}}{MM}$$

$$PV = \frac{\text{wt}}{MM} RT$$

Solving for *MM*, we have

$$MM = \frac{(\text{wt})(RT)}{PV}$$

$$= \frac{3.20 \text{ g} \times 62.4 \; \dfrac{\text{liter} \cdot \text{torr}}{\text{K} \cdot \text{mol}} \times 290 \; \text{K}}{380 \; \text{torr} \times 2.00 \; \text{liters}}$$

$$= \underline{\underline{76.2 \text{ g/mol}}}$$

**See Problems 10-48
through 10-50**

10-11 Stoichiometry Involving Gases

Gases are formed or consumed in many chemical reactions. Since the volume of a gas relates to the number of moles by the ideal gas law or the molar volume relationship, we can update the general procedure for stoichiometry problems shown in Figure 9-3. In Figure 10-12 the number of individual molecules has been deleted from the scheme, since problems involving actual numbers of molecules are not often encountered. We have also included in this figure the relationships used to convert from one unit to another. Note that the molar volume relationship ($n = V(\text{STP})/22.4$ liters) can be used to convert from volume to moles when the conditions are at STP.

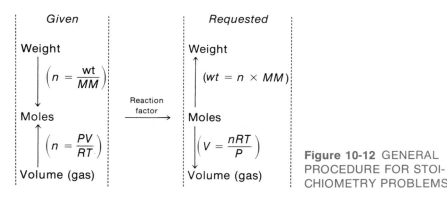

Figure 10-12 GENERAL PROCEDURE FOR STOICHIOMETRY PROBLEMS.

The following examples illustrate the relationship of gas laws to stoichiometry.

Example 10-19
Given the following balanced equation;

$$2\text{Al}(s) + 6\text{HCl}(aq) \longrightarrow 2\text{AlCl}_3(aq) + 3\text{H}_2(g)$$

What weight of Al is needed to produce 50.0 liters of H_2 measured at STP?

Procedure:

The general procedure is shown below. In this case, it is easier to use the molar volume to convert volume directly to moles.

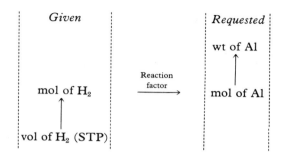

Given
Requested
wt of Al
Reaction factor
mol of H_2
mol of Al
vol of H_2 (STP)

Solution:

$$50.0\ \cancel{\text{liters (STP)}} \times \frac{1\ \cancel{\text{mol of H}_2}}{22.4\ \cancel{\text{liters (STP)}}} \times \frac{2\ \cancel{\text{mol of Al}}}{3\ \cancel{\text{mol of H}_2}} \times \frac{27.0\ \text{g}}{\cancel{\text{mol of Al}}} = \underline{\underline{40.2\ \text{g}}}$$

Example 10-20

Given the following balanced equation:

$$4NH_3(g) + 5O_2(g) \longrightarrow 4NO(g) + 6H_2O(l)$$

What volume of NO gas measured at 550 torr and 25°C will be produced from 19.5 g of O_2?

Procedure:

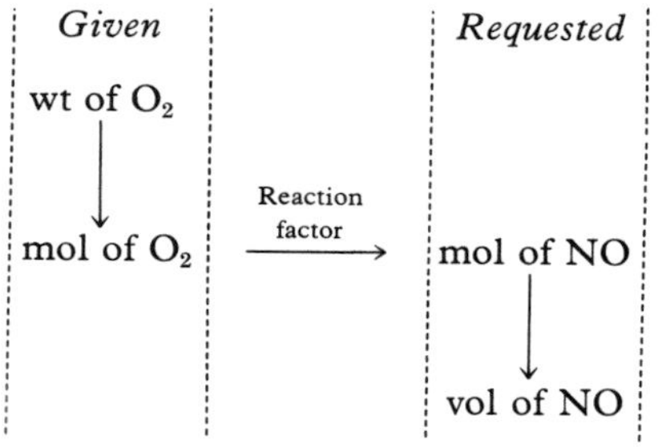

Given
Requested
wt of O_2
Reaction factor
mol of O_2
mol of NO
vol of NO

Solution:

$$19.5\ \cancel{\text{g of O}_2} \times \frac{1\ \cancel{\text{mol of O}_2}}{32.0\ \cancel{\text{g of O}_2}} \times \frac{4\ \text{mol of NO}}{5\ \cancel{\text{mol of O}_2}} = \underline{\underline{0.488\ \text{mol of NO}}}$$

1 Using the ideal gas law,

$$V = \frac{nRT}{P} = \frac{0.488\ \cancel{\text{mol}} \times 62.4\ \dfrac{\text{liter} \cdot \cancel{\text{torr}}}{\cancel{\text{K}} \cdot \cancel{\text{mol}}} \times 298\ \cancel{\text{K}}}{550\ \cancel{\text{torr}}} = \underline{\underline{16.5\ \text{liters}}}$$

2   Convert moles of NO to volume at STP and then volume at STP to volume at 550 torr and 25°C.

$$0.488 \text{ mol} \times 22.4 \text{ liters/mol} = 10.9 \text{ liters (STP)}$$

$$V_2 = V_1 \times \frac{P_1}{P_2} \times \frac{T_2}{T_1}$$

Initial Conditions	Final Conditions
$V_1 = 10.9$ liters	$V_2 = ?$
$P_1 = 760$ torr	$P_2 = 550$ torr
$T_1 = 273$ K	$T_2 = 25°C = 298$ K

See Problems 10-51 through 10-55

$$V_2 = 10.9 \text{ liters} \times \frac{760 \text{ torr}}{550 \text{ torr}} \times \frac{298 \text{ K}}{273 \text{ K}} = \underline{\underline{16.4 \text{ liters}}}$$

Review Questions

1   Describe Torricelli's barometer. What did Torricelli prove?
2   Give the definition of pressure in terms of force and area. How does force differ from pressure?
3   Define the standard of pressure. What are some units of pressure?
4   Describe Charles' observations on the relation of volume of a gas to temperature.
5   How is the Kelvin temperature scale defined? Give the relationship between Kelvin and the Celsius scale.
6   What is the combined gas law?
7   How is Gay-Lussac's law derived from the combined gas law?
8   What is standard temperature and pressure (STP)?
9   Give the assumptions of the kinetic theory of gases.

10   How does kinetic theory explain Boyle's, Charles', and Gay-Lussac's laws?
11   What other observations of the nature of gases are explained by kinetic theory?
12   What is Graham's law, and how is it explained by kinetic theory?
13   What is Dalton's law of partial pressures, and how is it explained by kinetic theory?
14   Define mole fraction. How is it calculated?
15   What is Avogadro's law, and how is it explained by kinetic theory?
16   Give the ideal gas law. Why is it called ideal?
17   What is the molar volume of a gas?
18   Describe two ways to convert moles of a gas to a volume at a certain temperature and pressure.

Problems

Pressure

10-1   Convert the following:
(a) 1650 torr to atm

(b) 3.50×10^{-5} atm to torr
(c) 185 lb/in.² to torr
(d) 5.65 kPa to atm

■ (e) 190 torr to lb/in.^{2}

■ (f) 85 torr to kPa

10-2 The density of water is 1.00 g/ml. If water is substituted for mercury in the barometer, how high (in feet) would be a column of water supported by 1 atm? A water well is 40 ft deep. Can suction be used to raise the water to ground level?

■ **10-3** A tube containing an alcohol (density 0.890 g/ml) is 1.00 m high and has a cross section of 15.0 cm^2. What is the total force at the bottom of the tube? What is the pressure? How high would be an equivalent amount of mercury assuming the same cross section?

Boyle's Law

10-4 A gas has a volume of 6.85 liters at a pressure of 0.650 atm. What is the volume of the gas if the pressure is decreased to 0.435 atm?

■ **10-5** If a gas has a volume of 1560 ml at a pressure of 812 torr, what will be the volume if the pressure is increased to 2.50 atm?

10-6 A gas has a volume of 125 ml at a pressure of 62.5 torr. What is the pressure if the volume is decreased to 115 ml?

*10-7 A gas in a piston engine is compressed by a ratio of 15:1. If the pressure before compression is 0.950 atm, what pressure is required to compress the gas? (Assume constant temperature.)

■ **10-8** A few miles above the surface of the earth the pressure drops to 1.00×10^{-5} atm. What would be the volume of a 1.00-liter sample of gas at sea level pressure (1.00 atm) if it were taken to that altitude? (Assume constant temperature.)

Charles' Law

10-9 A balloon has a volume of 1.55 liters at 25°C. What would be the volume if the balloon is heated to 100°C? (Assume constant P.)

■ **10-10** A sample of gas has a volume of 677 ml at 23°C. What is the volume of the gas if the temperature is increased to 46°C?

10-11 A balloon has a volume of 325 ml at 17°C. What is the temperature if the volume increases to 392 ml?

*10-12 The temperature of a sample of gas is 0°C. When the temperature is increased, the volume increases by a factor of 1.25 (i.e., $V_2 = 1.25V_1$). What is the final temperature in degrees Celsius?

■ **10-13** A quantity of gas has a volume of 3.66×10^4 liters. What will be the volume if the temperature is changed from 455 K to 50°C?

Gay-Lussac's Law

10-14 A confined quantity of gas is at a pressure of 2.50 atm and a temperature of -22°C. What is the pressure if the temperature increases to 22°C?

10-15 A quantity of gas has a volume of 3560 ml at a temperature of 55°C and a pressure of 850 torr. What is the temperature if the volume remains unchanged but the pressure is decreased to 0.652 atm?

10-16 An aerosol spray can has gas under a pressure of 1.25 atm at 25°C. The can explodes when the pressure reaches 2.50 atm. At what temperature will this happen? (Do not throw these cans into a fire!)

■ **10-17** The pressure in an automobile tire is 28.0 lb/in.^{2} on a chilly morning of 17°C. After it is driven for awhile, the temperature of the tire rises to 40.0°C. What is the pressure in the tire if the volume remains constant?

Combined Gas Law

10-18 A 5.50-liter volume of gas has a pressure of 0.950 atm at 0°C. What is the pressure if the volume decreases to 4.75 liters and the temperature increases to 35°C?

10-19 A quantity of gas has a volume of 17.5 liters at a pressure of 6.00 atm and a temperature of 100°C. What is its volume at STP?

10-20 A quantity of gas has a volume of 4.78×10^{-4} ml at a temperature of -50°C and a pressure of 78.0 torr. If the volume changes to 9.55×10^{-5} ml and the pressure to 155 torr, what is the temperature?

■ **10-21** A gas has a volume of 64.2 liters at STP. What is the volume at 77.0°C and 7.55 atm?

■ **10-22** A quantity of gas has a volume of 6.55×10^{-5} liter at STP. What is the pressure if the volume changes to 4.90×10^{-3} liter and the temperature remains at 273°C?

Graham's Law

10-23 A bowling ball weighs 6.00 kg and a bullet weighs 1.50 g. If the bowling ball is rolled down an alley at 20.0 miles/hr, what is the velocity of a bullet having the same kinetic energy?

10-24 Order the following gases in order of increasing average speed (rate of diffusion) at the same temperature.
(a) CO_2 (b) SO_2 (c) N_2 (d) SF_6 (e) N_2O (f) H_2

■ **10-25** A certain gas diffuses twice as fast as SO_3. What is the molar mass of the unknown gas?

*__10-26__ To make enriched uranium for use in nuclear reactors or for weapons, ^{235}U must be separated from ^{238}U. Although ^{235}U is the isotope needed for fission, only 0.7% of U atoms are this isotope. Separation is a difficult and expensive process. Since UF_6 is a gas, Graham's law can be applied to separate the isotopes. How much faster does a $^{235}UF_6$ molecule travel on the average compared to a $^{238}UF_6$ molecule?

Mole Fraction

10-27 What is the mole fraction of each component of a mixture of 1.65 mol of SO_2, 3.42 mol of O_2, and 0.57 mol of SO_3?

■ **10-28** What is the mole fraction of each component of a mixture of 1.86 g of N_2, 2.44 g of O_2, and 3.11 g of CO_2?

10-29 What is the mole fraction of each component of a mixture of 14.5 g of O_2, 1.75 mol of CO_2, 22.0 g of N_2, and 7.50×10^{23} molecules of CO?

Dalton's Law

10-30 A volume of gas is composed of N_2, O_2, and SO_2. If the total pressure is 1050 torr, what is the partial pressure of each gas if the gas is 72.0% N_2 and 8.00% O_2?

■ **10-31** A volume of gas has a total pressure of 2.75 atm. If the gas is composed of 0.250 mol of N_2 and 0.427 mol of CO_2, what is the partial pressure of each gas?

★■ **10-32** A mixture of gases has a total pressure of 725 torr. The gas is composed of 1.50×10^{23} molecules of SF_6, 0.375 mol of CO_2, and 32.5 g of SO_2. What is the partial pressure of each component?

10-33 A volume of gas has a total pressure of 685 torr. If the pressure due to gas A is 215 torr, what is the mole fraction of gas A?

10-34 A container holds two gases, A and B. Gas A has a partial pressure of 0.455 atm and gas B has a partial pressure of 0.175 atm. What is the mole fraction of gas A?

★■ **10-35** A container holds three gases, CO, CO_2, and O_2. There is 0.232 mol of CO, which exerts a partial pressure of 0.115 atm. There are also 0.368 mol of CO_2 and 0.188 mol of O_2 present. What is the total pressure?

10-36 The following gases are all combined into a 2.00-liter container: a 2.00-liter volume of N_2 at 300 torr, a 4.00-liter volume of O_2 at 85 torr, and a 1.00-liter volume of CO_2 at 450 torr. What is the total pressure?

*__10-37__ A mixture of two gases (A and B) has a total pressure of 0.655 atm. How many moles of gas B must be mixed with 2.00 mol of A so that the partial pressure of gas A is 0.355 atm?

Avogadro's Law

10-38 A 0.112-mol quantity of gas has a volume of 2.54 liters at a certain temperature and pressure. What is the volume of 0.0750 mol of gas under the same conditions?

10-39 A balloon has a volume of 75.0 ml and contains 2.50×10^{-3} mol of gas. How many grams of N_2 must be added to the balloon in order for the volume to increase to 164 ml at the same temperature and pressure?

■ **10-40** A 48.0-g quantity of O_2 in a container has a pressure of 0.625 atm. What would be the pressure of 48.0 g of SO_2 in the same container at the same temperature?

Ideal Gas Law

10-41 What is the temperature (in degrees Celsius) of 4.50 liters of a 0.332-mol quantity of gas under a pressure of 2.25 atm?

■ **10-42** A quantity of gas has a volume of 16.5 liters at 325 K and a pressure of 850 torr. How many moles of gas are present?

10-43 What weight of NH_3 gas has a volume of 16,400 ml, a pressure of 0.955 atm, and a temperature of $-23°C$?

■ **10-44** A gas has a density of 8.37 g/liter at a pressure of 1.45 atm and a temperature of 35.0°C. What is the molar mass of the gas?

*****10-45** The Goodyear blimp has a volume of about 2.5×10^7 liters. What is the weight of He (in pounds) in the blimp at 27°C and 780 torr? The average molar mass of air is 29.0 g/mol. What weight of air (in pounds) would the blimp contain? The difference between these two values is the lifting power of the blimp. What weight could the blimp lift? If H_2 is substituted for He, what is the lifting power? Why isn't H_2 used?

10-46 A gaseous compound is 85.7% C and 14.3% H. A 6.58-g quantity of this gas occupies 4500 ml at 77.0°C and a pressure of 1.00 atm. What is the molar mass of the compound, and what is its molecular formula?

*■ **10-47** A good vacuum pump on earth can produce a vacuum with a pressure as low as 1.00×10^{-8} torr. How many molecules are present in each milliliter at a temperature of 27.0°C?

Molar Volume

10-48 A gas has a density of 1.52 g/liter at STP.

What is the molar mass of the gas?

■ **10-49** What would be the volume of 15.0 g of CO at STP?

10-50 What is the weight of 6.78×10^{-4} liter of NO_2 at STP?

Stoichiometry Involving Gases

10-51 Limestone is dissolved by CO_2 according to the following equation:

$$CaCO_3(s) + H_2O(l) + CO_2(g) \longrightarrow Ca(HCO_3)_2(aq)$$

What volume of CO_2 measured at STP would dissolve 115 g of $CaCO_3$?

■ **10-52** Acetylene is produced from calcium carbide as shown by the following reaction:

$$CaC_2(s) + 2H_2O(l) \longrightarrow Ca(OH)_2(s) + C_2H_2(g)$$

What volume of acetylene (C_2H_2) measured at 25.0°C and 745 torr would be produced from 5.00 g of H_2O?

10-53 Butane (C_4H_{10}) burns according to the following equation:

$$2C_4H_{10}(g) + 13O_2(g) \longrightarrow 8CO_2(g) + 10H_2O(l)$$

(a) What volume of CO_2 measured at STP would be produced by 85.0 g of C_4H_{10}?

(b) What volume of O_2 measured at 3.25 atm and a temperature of 127°C would be required to react with 85.0 g of C_4H_{10}?

(c) What volume of CO_2 measured at STP would be produced from 45.0 liters of C_4H_{10} measured at 25°C and 0.750 atm pressure?

10-54 In March 1979 a nuclear reactor overheated, producing a dangerous hydrogen gas bubble at the top of the reactor core. The following reaction occurring at the high temperatures (about 1500°C) accounted for the hydrogen (Zr alloys hold the uranium pellets in long rods):

$$Zr(s) + 2H_2O(g) \longrightarrow ZrO_2(s) + 2H_2(g)$$

If the bubble had a volume of about

2500 liters at 250°C and 65.0 atm pressure, what weight of Zr had reacted?

■ **10-55** Nitric acid is produced according to the equation:

$$3NO_2(g) + H_2O(l) \longrightarrow$$
$$HNO_3(aq) + 2NO(g)$$

What volume of NO_2 measured at $-73°C$ and 1.56×10^{-2} atm would be needed to produce 4.55×10^{-3} mol of HNO_3?

The Nature of Water
and Aqueous Solutions

The large oceans that cover most of the surface of the earth are not composed of pure water. They contain many dissolved substances.

In recent years there has been much interest in the search for evidence of life on other planets. Central to this endeavor is the search for one of the most ordinary compounds on our planet, water. Life, as we know it, depends on this unique chemical substance. For example, water carries nourishment and removes waste in the bloodstream. Water, together with carbon dioxide and minerals, is changed chemically by plants and energy from the sun into great masses of vegetation and then returned to the environment when these substances burn or decay. Water also regulates climate so that life can flourish. The great bodies of water in oceans and lakes store and distribute heat so that much of this planet has a temperate climate.

Water is thought to have played a significant role in the evolution of life on this planet since the vast oceans provided the medium for the many chemical reactions that led to the first living cells. As we will see in this and subsequent chapters, water indeed acts as a medium for a wide variety of chemical reactions. To the chemist, therefore, water is much more than a common and inexpensive substance; it is an intriguing and truely unique chemical.

Before reactions that occur in an aqueous medium are discussed, the nature of water itself and how it interacts with other compounds will be explained. Also in this chapter, the procedure for working stoichiometry problems introduced in Chapter 9 and 10 is expanded to include aqueous reactions.

As background to this chapter, you should be familiar with the following topics:

1 Ionic compounds and how they are identified (Section 6-3).

2 The charges, formulas, and names of ions (Table 7-3).

3 Polarity (Sections 6-11 and 6-12).

4 Stoichiometry (Sections 9-3 and 10-11).

11-1 The Conductivity of Aqueous Solutions

Water supports many chemical reactions because of its extensive ability to act as a solvent. *A* **solvent** *is a medium, usually a liquid, that dissolves or disperses another substance called a* **solute** *to form a homogeneous mixture of solute and solvent called a* **solution** (see Figure 11-1). The solution has the same physical state as the solvent.

One important property of aqueous solutions concerns their ability to act as conductors of electricity. **Electricity** *is simply a flow of negatively charged electrons through a substance called a* **conductor.** Metals such as copper are useful in electrical wires because they allow the flow of electricity along a length of continuous wire. Glass, on the other hand, does not allow the flow of electricity and is called a **nonconductor** or **insulator.** Pure water is also a

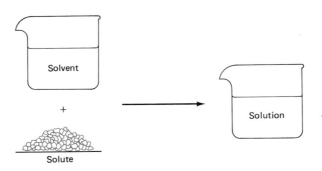

Solvent

+

Solute

Solution

Figure 11-1 A SOLUTION. A solution is a homogeneous mixture of solute and solvent that has the same physical state as the solvent.

nonconductor, as shown in Figure 11-2. In the figure, the wires A and B (called **electrodes**) are separated and immersed in the water. When the battery is connected, the light does not shine. This tells us that the water prohibits the transfer of electrons between the two wires and the circuit is therefore broken. The presence of certain solutes such as sugar and alcohol have no effect on the ability of water to conduct electricity, since the light remains out when the wires are immersed in these solutions. *Compounds whose aqueous solutions do not conduct electricity are called* **nonelectrolytes.**

There is another class of compounds that have a profoundly different effect. When the electrodes are immersed in aqueous solutions of compounds such as sodium chloride or potassium hydroxide, the light shines brightly. Obviously, the presence of these compounds allows the conduction of electricity between wire A and wire B. *Compounds whose aqueous solutions conduct electricity are called* **electrolytes.**

There is a third class of compounds whose aqueous solutions conduct electricity to a very limited amount. These compounds are called **weak electrolytes.** When electrodes are immersed in solutions of weak electrolytes, the light shines but very faintly. Weak electrolytes are discussed in more detail in Chapter 12.

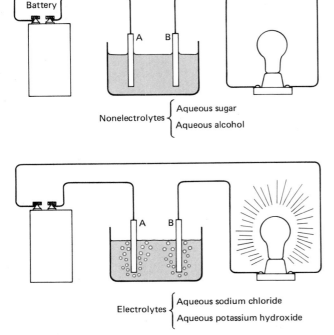

Nonelectrolytes
{ Aqueous sugar
  Aqueous alcohol }

Electrolytes
{ Aqueous sodium chloride
  Aqueous potassium hydroxide }

Figure 11-2 ELECTRO-LYTES AND NON-ELECTROLYTES. Solutions of electrolytes conduct electricity, solutions of nonelectrolytes do not.

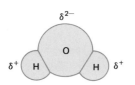

Figure 11-3 THE POLAR NATURE OF WATER. In a water molecules the oxygen is partially negative and the hydrogens are partially positive.

11-2 Water as a Solvent

What is unique about electrolytes that affect the solution in such a way as to allow the conduction of electricity? Before answering this question, we should briefly review the nature of pure water itself. In Chapter 6 we learned that water is composed of polar, covalent molecules. This is a result of the significant difference in electronegativity between the oxygen and the hydrogen atoms and the angular geometry of water (see Figure 11-3).

The presence of dipoles in water molecules has a significant effect on its nature. For example, in both the liquid and solid states there is an electrostatic interaction (attraction of opposite charges) between different water molecules. This is illustrated in Figure 11-4. Notice that each hydrogen is covalently bound to one oxygen, but because it has a positive dipole it may interact with the oxygen of a different molecule which has a negative dipole. *This type of electrostatic interaction is known as a* **dipole–dipole interaction.** If, as is often the case, *the positive dipole is a hydrogen (e.g., water), the dipole–dipole interaction is specifically referred to as* **hydrogen bonding** (although the interaction is not actually a chemical bond). In the liquid state the hydrogen bonding is somewhat random, allowing the groups of molecules to slip past one another. In the solid state the hydrogen bonding is orderly and complete (each oxygen is covalently bound to two hydrogens and interacts electrostatically with two others). This holds the water molecules in a fixed position, giving the solid a rigid shape.

It should be kept in mind that interactions such as those illustrated in Figure 11-4 and several subsequent figures have a three-dimensional, geometric shape. For example, the four hydrogens in ice are arranged around the oxygen in a somewhat distorted tetrahedral shape.

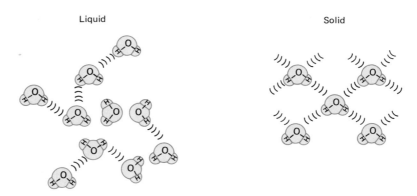

Liquid   Solid

Figure 11-4 THE INTERACTION OF WATER MOLECULES. Hydrogen bonding occurs in both water and ice. In ice the hydrogen bonding is more ordered, causing an open structure.

You can visualize from Figure 11-4 the reason for one unusual property of water. The liquid is a more collapsed, compact structure than the relatively open structure of ice. As a result, ice is less dense than the liquid state. This is the reason that ice floats on water.

We are now ready to appreciate what occurs when solid ionic compounds are added to water. As illustrated in Figure 11-5, there is an electrostatic interaction between the negative dipoles of water molecules and the positive ions on the surface of the crystal. Likewise, the positive dipoles of water are attracted to the negative ions. *The electrostatic interactions between an ion and the dipoles of polar covalent molecules are known as* **ion–dipole forces.**

The ion–dipole forces attempt to lift the ions away from the crystal surface and suspend them in solution. This attempt is countered in what may be considered a tug-of-war by the ion–ion electrostatic interactions (known as the lattice energy) holding the crystal together. The actual quantitative values for these forces are known, but for now we can appreciate that if the forces are comparable the ions can be lifted from the crystal lattice by the water molecules and be held in the aqueous medium surrounded by an "escort" of water molecules. *In solution the ions are said to be* **hydrated.**

By virtue of its polar nature, water can thus dissolve a wide variety of

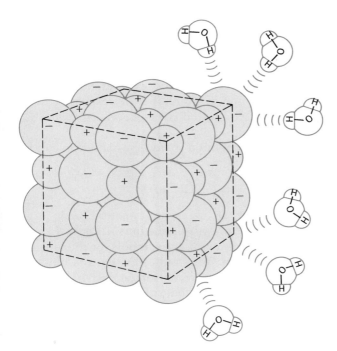

Figure 11-5 THE INTER-ACTION OF WATER AND IONIC COMPOUNDS. There is an electrostatic interaction between the polar water molecules and the ions. This is a dipole–ion force.

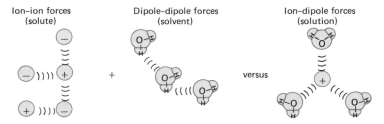

Ion–ion forces (solute)   Dipole–dipole forces (solvent)   Ion–dipole forces (solution)

Figure 11-6 FORCES IN THE SOLUTION PROCESS. When an ionic compound dissolves in water, the ion–dipole forces are comparable to the ion–ion forces holding the crystal together.

polar substances, especially ionic compounds. In some cases, however, the energy that holds a crystal together is strong enough to withstand the forces pulling the crystal apart and little solute dissolves (e.g., AgCl, CaCO₃). *Compounds that dissolve are said to be **soluble**, and those that do not dissolve are said to be **insoluble.***

In summary, the solution process (see Figure 11-6) occurs because the attractive forces between the solute ions and the solvent water molecules (ion–dipole forces) are strong enough to overcome the forces holding the solute together (ion–ion forces) as well as the forces between the water molecules (dipole–dipole forces).*

As an analogy to these interactions, there is the well-known phenomenon of the interaction of a group of young men and a group of young women. The men interact together (boy–boy forces) quite well and are happy, as do the ladies (girl–girl forces). When the two groups come together, however, the boy–girl forces are usually strong enough to overcome the respective boy–boy and girl–girl forces. As a result, a thorough mixing occurs. It is simply a competition of forces: The strongest wins.

The solution process of an ionic compound in aqueous solution can be represented by a chemical equation as follows:

$$NaCl(s) \xrightarrow{xH_2O} Na^+(aq) + Cl^-(aq)$$

The coefficient of H_2O above the arrow is an undetermined, large number of molecules needed to surround or *hydrate* the ions. Its actual value is not important. The abbreviation (*aq*) following the symbols for the ions indicates that the ion is in aqueous solution. This indicates that the ions are hydrated and that positive and negative ions are no longer associated with each other in the solution. The water molecules that surround the ion in several layers tend to diminish the effect of the charge by spreading it out.

---

* There are other factors involved in the solution process that can be important, but generally these forces determine the solubility of a compound in water.

Example 11-1

Write equations illustrating the solution of the following ionic compounds in water: (a) Na_2CO_3, (b) $CaCl_2$, (c) 2 mol of K_3PO_4, and (d) 3 mol of $(NH_4)_2SO_4$.

Solution:

Review the charges and formulas of the ions discussed in Chapter 6 and 7. The ions present in water are the same as those present in the solid ionic compound.

$$\text{(a)} \quad Na_2CO_3(s) \xrightarrow{xH_2O} 2Na^+(aq) + CO_3^{2-}(aq)$$

$$\text{(b)} \quad CaCl_2(s) \xrightarrow{xH_2O} Ca^{2+}(aq) + 2Cl^-(aq)$$

$$\text{(c)} \quad 2K_3PO_4(s) \xrightarrow{xH_2O} 6K^+(aq) + 2PO_4^{3-}(aq)$$

$$\text{(d)} \quad 3(NH_4)_2SO_4(s) \xrightarrow{xH_2O} 6NH_4^+(aq) + 3SO_4^{2-}(aq)$$

See Problem 11-1

The solution of ionic compounds in water form solutions that are electrolytes *because of the presence of ions*. Besides ionic compounds, many polar covalent compounds also dissolve in water. These compounds may form electrolytes, weak electrolytes, or nonelectrolytes depending on the degree of ion formation in the solution. This is illustrated by the following three examples.

1 $\quad HCl(g) \xrightarrow{xH_2O} H^+(aq) + Cl^-(aq)$ HCl is an electrolyte, since all HCl molecules break into ions in solution.

2 $\quad HF(g) \xrightarrow{xH_2O} HF(aq)$

$\quad HF(aq) \xrightarrow[\text{(partial)}]{xH_2O} H^+(aq) + F^-(aq)$ HF is a weak electrolyte, since only a small percentage of the dissolved HF molecules break into ions.

3 $\quad CH_3OH(l) \xrightarrow{xH_2O} CH_3OH(aq)$ CH_3OH (methyl alcohol) is a nonelectrolyte, since no ions are formed when the compound dissolves.

The solution of covalent compounds to form strong and weak electrolytes will be discussed in the next chapter. When covalent compounds dissolve to form nonelectrolytes, it is due to the formation of dipole–dipole forces between solute molecule and water molecules as shown in Figure 11-7.

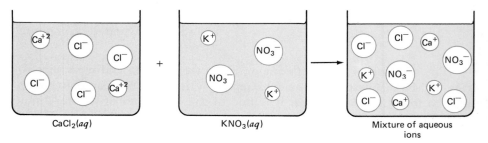

Figure 11-7 METHYL ALCOHOL IN WATER. For a nonelectrolyte in water, there are dipole–dipole interactions between solute and water molecules.

Nonpolar compounds such as gasoline or carbon tetrachloride do not dissolve in water. Since there are no appreciable solute–solvent forces due to the lack of a charge or a dipole on the nonpolar solute molecules, there are no forces strong enough to overcome the dipole–dipole forces of the water molecules themselves. The water molecules prefer to stay together, which means that mixing or solution does not occur.

To understand this, we can use the analogy of boy–girl forces versus boy–boy and girl–girl forces. This time a group of young women pass a group of chickens instead of men. Since girl–chicken forces are almost nil, mixing is not likely and the girls prefer to hang around together. No solution occurs because there is no reason to mix, and actually a better reason not to.

11-3 Ionic Equations and Double Displacement Reactions

When an ionic compound dissolves in water, the cation and the anion behave independently. What happens if two different ionic compounds, such as $CaCl_2$ and KNO_3, are dissolved in water in separate beakers and then mixed? In this particular case, nothing happens. We simply have a solution containing four different ions as illustrated in Figure 11-8. No cation is associated with a particular anion.

In other cases when solutions of ionic compounds are mixed, a chemical reaction occurs. Although there are several types of reactions that may occur, we are now concerned with a reaction that leads to the formation of a **precipitate** (*an insoluble compound*). To understand this, remember that certain ionic compounds *do not* dissolve to any appreciable extent in water because of the strong interaction between the cation and the anion. Such a compound

Figure 11-8 A MIXTURE OF $CaCl_2$ KNO_3 SOLUTIONS. No reaction occurs when these solutions are mixed.

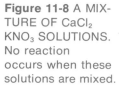

Figure 11-9 A MIXTURE OF NaCl AND $AgNO_3$ SOLUTIONS. When these solutions are mixed, a precipitate forms.

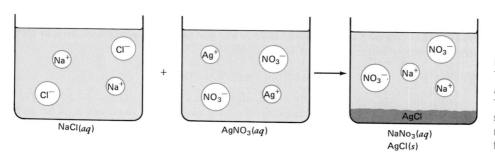

is AgCl (silver chloride). In fact, *any time* Ag^+ *ions and* Cl^- *ions are present in the same solution, they are strongly attracted to each other and combine to form solid AgCl.* Thus if a solution of NaCl and $AgNO_3$ are mixed as shown in Figure 11-9, a white, solid precipitate of AgCl is formed, leaving the soluble compound, $NaNO_3$, in solution.

This type of reaction is an example of a **double displacement** (or metathesis) reaction. (Another type of double displacement reaction is more commonly referred to as an acid–base reaction. These reactions are discussed in Chapters 12 and 15.) *The type of double displacement reaction discussed in this chapter occurs because certain cations form insoluble precipitates with certain anions.* In the above example, AgCl can be removed by filtration and the $NaNO_3$ recovered by boiling away the solvent water.

The following equation represents the chemical reaction illustrated in Figure 11-9:

$$NaCl(aq) + AgNO_3(aq) \longrightarrow AgCl(s) + NaNO_3(aq)$$

This is the molecular* form of the equation. *In the* **molecular equation,** *all reactants and products are shown as compounds.*

When all cations and anions in solution are written separately, the resulting equation is known as the **total ionic equation:**

$$Na^+(aq) + Cl^-(aq) + Ag^+(aq) + NO_3^-(aq) \longrightarrow AgCl(s) + Na^+(aq) + NO_3^-(aq)$$

In the total ionic equation, notice that the $Na^+(aq)$ and the $NO_3^-(aq)$ ions are written in an identical state on both sides of the equation. From algebra you are aware that when identical numbers or variables appear on both sides of an equation, they can be subtracted from both sides of the equation as follows:

$$\begin{array}{rcl} x + y & = & x + 17 + z \\ -x & = & -x \\ \hline y & = & 17 + z \end{array}$$

* "Molecular" is a misnomer in this case, since no real molecules exist in the case of ionic compounds.

In the same manner, identical substances that appear on both sides of a chemical equation can be subtracted out to produce the **net ionic equation.** In our example, the net ionic equation is

$$Ag^+(aq) + Cl^-(aq) \longrightarrow AgCl(s)$$

Ions that are dropped from both sides of an equation are called **spectator ions,** *since they are not involved directly with the chemical reaction.* The net ionic equation is important, since it focuses on the "main action" of what really occurs in solution.

11-4 Solubility of Ionic Compounds

Each compound has a limit as to *the maximum amount of that compound that can dissolve in a given amount of solvent at a certain temperature. This quantity is known as the compound's* **solubility** at a certain temperature. Insoluble compounds such as AgCl have extremely low solubilities in water (but they do dissolve to a very small extent). Soluble compounds such as table sugar (sucrose) have a relatively high degree of solubility. (If you have ever dumped a large amount of sugar in iced tea, you have noticed that even sugar has limits of solubility.) The solubility of a compound in water at a certain temperature can be listed as a definite numerical quantity, but the relative terms "soluble" and "insoluble" will suit our purposes in predicting the occurrence of double displacement reactions.

To predict double displacement reactions, some means of identifying those ionic compounds considered insoluble is needed. As it turns out, there are some relatively straightforward rules that can be memorized or at least referred to in order to predict the formation of insoluble precipitates from the

Table 11-1   Solubility Rules for Some Ionic Compounds

Anion	Cations Forming Insoluble Compounds	Slightly Soluble Compounds
Cl^-, Br^-, I^-	Ag^+, Hg_2^{2+}, Pb^{2+}	$PbCl_2$, $PbBr_2$
NO_3^-, ClO_3^-, ClO_4^-, $C_2H_3O_2^-$	None	$AgC_2H_3O_2$
SO_4^{2-}	Pb^{2+}, Ba^{2+}, Sr^{2+}	$CaSO_4$, Ag_2SO_4
CO_3^{2-}, SO_3^{2-}, PO_4^{3-}	All except IA metals and NH_4^+	
S^{2-}	All except IA metals, IIA metals, and NH_4^+	
OH^-	All except IA metals, Ba^{2+}, Sr^{2+}, and NH_4^+	$Ca(OH)_2$
O^{2-}	All except IA metals, Ca^{2+}, Sr^{2+}, and Ba^{2+}	

more common ions. The rules are tabulated in Table 11-1. If a common ionic compound is not mentioned in the table, it is assumed to be soluble.

As you can see from Table 11-1, all alkali metal (group IA) and ammonium compounds are soluble. The following examples illustrate the use of Table 11-1 to predict the occurrence of double displacement reactions in aqueous solution.

Write the balanced molecular equation, the total ionic equation, and the net ionic equation for any reaction that occurs when solutions of the following are mixed.

Example 11-2

A solution of Na_2CO_3 is mixed with a solution of $CaCl_2$.

Solution:

To begin, write out the formulas of the possible products resulting from a double displacement reaction. In this case the possible products from the exchange of ions are NaCl and $CaCO_3$.

If both of these compounds are soluble, no reaction occurs. In this case, however, Table 11-1 tells us that $CaCO_3$ is insoluble. Thus a reaction occurs as illustrated by the following balanced equation written in molecular form:

$$Na_2CO_3(aq) + CaCl_2(aq) \longrightarrow CaCO_3(s) + 2NaCl(aq)$$

The equation written in total ionic form is

$$2Na^+(aq) + CO_3^{2-}(aq) + Ca^{2+}(aq) + 2Cl^-(aq) \longrightarrow$$
$$CaCO_3(s) + 2Na^+(aq) + 2Cl^-(aq)$$

Notice that the Na^+ and the Cl^- ions are spectator ions. Elimination of the spectator ions on both sides of the equation leaves the net ionic equation

$$Ca^{2+}(aq) + CO_3^{2-}(aq) \longrightarrow CaCO_3(s)$$

Example 11-3

A solution of KOH is mixed with a solution of MgI_2.

Solution:

The possible double displacement reaction products are KI and $Mg(OH)_2$. The information in Table 11-1 indicates that $Mg(OH)_2$ is insoluble. The balanced molecular equation for this reaction is

$$2KOH(aq) + MgI_2(aq) \longrightarrow Mg(OH)_2(s) + 2KI(aq)$$

The total ionic equation is

$$2K^+(aq) + 2OH^-(aq) + Mg^{2+}(aq) + 2I^-(aq) \longrightarrow$$
$$Mg(OH)_2(s) + 2K^+(aq) + 2I^-(aq)$$

Elimination of spectator ions gives the following net ionic equation:

$$Mg^{2+}(aq) + 2OH^-(aq) \longrightarrow Mg(OH)_2(s)$$

**See Problems 11-2
through 11-7**

Example 11-4
A solution of KNO_3 is mixed with a solution of $CaBr_2$.

Solution:
The possible double displacement reaction products are KBr and $Ca(NO_3)_2$. Since these compounds are both soluble, no precipitate forms and a reaction does not occur.

11-5 Concentration: Percent by Weight

It is necessary at this point to introduce a method of expressing *the amount of a solute that is present in a given amount of solution. This is known as the* **concentration** *of the solution.* The chemist has another convenient but vague way of expressing concentration by terming the solution **dilute** *when not very much solute is present or* **concentrated** *when a lot of solute is present.* Obviously, a more quantitative method is necessary in order to work stoichiometry problems involving solutions. This is fundamental to any chemistry laboratory experience.

There are several ways of expressing concentration quantitatively: percent by weight of solute to solution, mole fraction, molarity, molality, and normality. All of these have their special use, but percent by weight and molarity are the most common and are discussed here.

Percent by weight, *a rather straightforward method of expressing concentration, simply relates the weight of the solute as a percent of the total weight of the solution.* Therefore, in 100 g of a solution that is 25% by weight HCl there are 25 g of HCl and 75 g of H_2O. The formula for percent by weight is

$$\% \text{ by weight (solute)} = \frac{\text{weight of solute}}{\text{weight of solution}} \times 100$$

Example 11-5
What is the percent by weight of NaCl if 1.75 g of NaCl is dissolved in 5.85 g of H_2O?

Procedure:
Find the total weight of the solution and then the percent of NaCl.

Solution:

Total weight
    1.75 g of NaCl (solute)
    <u>5.85 g of H_2O (solvent)</u>
    7.60 g of solution

$$\frac{1.75 \text{ g of NaCl}}{7.60 \text{ g of Solution}} \times 100 = \underline{23.0\% \text{ by weight NaCl}}$$

Example 11-6
A solution is 14.0% by weight H_2SO_4. How many moles of H_2SO_4 are in 155 g of solution?

Procedure:

1 Find the weight of H_2SO_4 in the solution.

2 Convert weight to moles.

Solution:

1 Weight of H_2SO_4 is 14.0% of 155 g or

$$0.140 \times 155 \text{ g} = 21.7 \text{ g of } H_2SO_4$$

2 The molar mass of H_2SO_4 is

$$2.0 \text{ g (H)} + 32.1 \text{ g (S)} + 64.0 \text{ g (O)} = 98.1 \text{ g}$$

$$21.7 \text{ g} \times \frac{1 \text{ mol}}{98.1 \text{ g}} = \underline{0.221 \text{ mol of } H_2SO_4}$$

See Problems 11-8 through 11-11

11-6 Concentration: Molarity

Molarity (M) is defined as the number of moles of solute (n) per liter of solution (V):

$$M = \frac{n \text{ (moles of solute)}}{V \text{ (liters of solution)}}$$

Thus a 1.00 M solution of HCl contains 1.00 mole (36.5 g) of HCl dissolved in enough H_2O to make 1.00 liter of solution.

Example 11-7

What is the molarity of a solution made by dissolving 49.0 g of pure H_2SO_4 in enough water to make 250 ml of solution?

Procedure:

Write down the formula for molarity, what you have been given, and then solve for what's requested.

Solution:

$$M = \frac{n}{V}$$

(n) $\quad 49.0 \text{ g of } H_2SO_4 \times \dfrac{1 \text{ mol}}{98.1 \text{ g of } H_2SO_4} = 0.499 \text{ mol}$

(V) $\quad 250 \text{ ml} \times \dfrac{1 \text{ liter}}{10^3 \text{ ml}} = 0.250 \text{ liter}$

$$M = \frac{0.499 \text{ mol}}{0.250 \text{ liter}} = \underline{\underline{2.00}}$$

Example 11-8
What weight of HCl is present in 155 ml of a 0.540 M solution?

$$M = \frac{n}{V} \qquad n = M \times V$$

$$M = 0.540 \qquad V = 155 \text{ ml} = 0.155 \text{ liter}$$

$$n = 0.540 \text{ mol/liter} \times 0.155 \text{ liter} = 0.0837 \text{ mol of HCl}$$

$$0.0837 \text{ mol of HCl} \times \frac{36.5 \text{ g}}{\text{mol of HCl}} = \underline{\underline{3.06 \text{ g of HCl}}}$$

Example 11-9
Concentrated laboratory acid is 35.0% by weight HCl and has a density of 1.18 g/ml. What is its molarity?

Procedure:
Since a volume was not given, you can start with any volume you wish. The molarity will be the *same* for 1 ml as for 25 liters. To make the problem as simple as possible, assume that you have exactly 1 liter of solution ($V = 1.00$ liter) and go from there. The number of moles of HCl (n) in 1 liter can be obtained as follows:

1   Find the weight of 1 liter from the density.

2   Find the weight of HCl in 1 liter using the percent by weight and the weight of 1 liter.

3   Convert the weight of HCl to moles of HCl.

Solution:
Assume that $V = 1.00$ liter.

1   The weight of 1.00 liter (10^3 ml) is

$$10^3 \text{ ml} \times 1.18 \text{ g/ml} = 1180 \text{ g}$$

2   The weight of HCl in 1.00 liter is

$$1180 \text{ g} \times 0.350 = 413 \text{ g of HCl}$$

3   The number of moles of HCl in 1.00 liter is

$$413 \text{ g} \times \frac{1 \text{ mo.}}{36.5 \text{ g}} = 11.3 \text{ mol of HCl}$$

See Problems 11-12 through 11-18

$$M = \frac{n}{V} = \frac{11.3 \text{ mol}}{1.00 \text{ liter}} = \underline{\underline{11.3}}$$

One of the common laboratory exercises for the experienced as well as the beginning chemist is to make a certain volume of a *dilute* solution from a *concentrated* solution. In this case a certain amount (number of moles) of solute is withdrawn from the concentrated solution and diluted with water to form the dilute solution. The calculations involved are best illustrated by the following examples.

Example 11-10
What volume of 11.3 M HCl solution is needed to mix with water to make 1.00 liter of 5.55 M HCl solution?

Procedure:
In this type of problem (known as a dilution problem), there are two solutions, a *concentrated* (M_c = 11.3) and eventually a *dilute* solution (M_d = 5.55, V_d = 1.00 liter). The problem requires that you calculate the volume of concentrated solution needed to provide the number of moles of HCl present in the dilute solution.

1 Calculate the number of moles of HCl needed for the dilute solution.

2 Calculate the volume of concentrated solution that would contain this amount of HCl.

3 Notice that there is a short cut to working this problem.

Solution:

1 Calculate the number of moles of HCl needed for the dilute solution.

$$M_d = \frac{n_d}{V_d} \qquad n_d = M_d \times V_d$$

$$n_d = 5.55 \text{ mol/liter} \times 1.00 \text{ liter} = 5.55 \text{ mol of HCl}$$

2 Calculate the volume of concentrated solution that would contain 5.55 mol of HCl.

$$M_c = \frac{n_c}{V_c} \qquad V_c = \frac{n_c}{M_c}$$

$$V_c = \frac{5.55 \text{ mol}}{11.3 \text{ mol/liter}} = 0.491 \text{ liter}$$

Therefore, to make 1.00 liter of dilute solution, 0.491 liter (491 ml) of concentrated HCl (11.3 M) is diluted to 1.00 liter with water in a manner illustrated in Figure 11-10.

3 Notice that in this problem the 0.491 liter of concentrated acid contains the same number of moles of HCl as the 1.00 liter of dilute solution:

$$n_c = n_d$$

Therefore, since quantities equal to the same quantity are equal to each other, a short cut is possible:

$$M_c \times V_c = n_c \qquad M_d \times V_d = n_d$$

$$M_c \times V_c = M_d \times V_d$$

$$V_c = \frac{M_d \times V_d}{M_c}$$

$$= \frac{5.55 \ \text{mol/liter} \times 1.00 \ \text{liter}}{11.3 \ \text{mol/liter}} = \underline{\underline{0.491 \ \text{liter}}}$$

Example 11-11

What is the molarity of a solution of KCl that is prepared by dilution of 855 ml of a 0.475 M solution to a volume of 1.25 liters?

Procedure:

$$M_c \times V_c = M_d \times V_d$$

$$M_d = \frac{M_c \times V_c}{V_d}$$

Solution:

$$M_c = 0.475 \qquad\qquad M_d = ?$$

$$V_c = 855 \ \text{ml} = 0.855 \ \text{liter} \qquad V_d = 1.25 \ \text{liters}$$

$$M_d = \frac{0.475 \times 0.855 \ \text{liter}}{1.25 \ \text{liters}} = \underline{\underline{0.325}}$$

See Problems 11-19 through 11-24

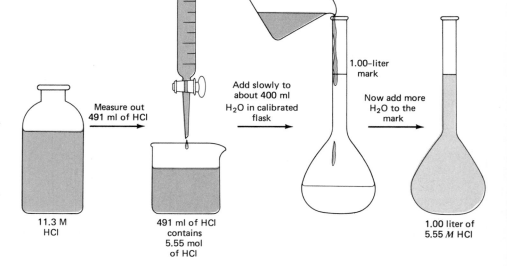

Figure 11-10 DILUTION OF CONCENTRATED HCl. (Note: Water is never added directly to concentrated acid, because it may splatter and cause severe burns.)

11.3 M HCl

Measure out 491 ml of HCl

491 ml of HCl contains 5.55 mol of HCl

Add slowly to about 400 ml H₂O in calibrated flask

1.00–liter mark

Now add more H₂O to the mark

1.00 liter of 5.55 M HCl

11-7 Stoichiometry Involving Solutions

In Chapter 9 a general procedure was presented for working stoichiometry problems. This was expanded in Chapter 10 to include gases. We can now further expand the general procedure to include stoichiometry in aqueous solution, since moles relate to volume of solution by molarity (see Figure 11-11).

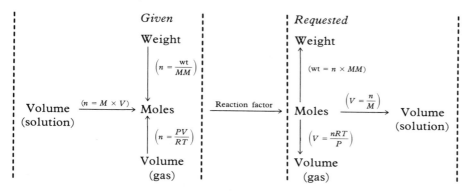

Figure 11-11 GENERAL PROCEDURE FOR STOICHIOMETRY.

The following examples illustrate the role of solutions in stoichiometry.

Example 11-12

Given the following balanced equation:

$$3NaOH(aq) + H_3PO_4(aq) \longrightarrow Na_3PO_4(aq) + 3H_2O$$

What volume of $0.250\,M$ NaOH is required to react completely with 4.90 g of H_3PO_4?

Procedure:

Given		Requested	
wt of H_3PO_4			
↓			
mol of H_3PO_4	Reaction factor →	mol of NaOH ⟶ vol of NaOH (solution)	

Solution:

$$4.90 \text{ g of } H_3PO_4 \times \frac{1 \text{ mol of } H_3PO_4}{98.0 \text{ g of } H_3PO_4} \times \frac{3 \text{ mol of NaOH}}{1 \text{ mol of } H_3PO_4}$$

$$= 0.150 \text{ mol of NaOH}$$

$$\text{Vol of NaOH} = \frac{\text{mol of NaOH}}{M} = \frac{0.150 \text{ mol}}{0.250 \text{ mol/liter}} = \underline{\underline{0.600 \text{ liter}}}$$

Example 11-13
Given the following balanced equation:

$$Cd(NO_3)_2(aq) + K_2S(aq) \longrightarrow CdS(s) + 2KNO_3(aq)$$

What weight of CdS would be produced from 15.8 ml of a 0.122 M $Cd(NO_3)_2$ solution with excess K_2S present?

Procedure:

Given		Requested
		wt of CdS
		↑
Vol of $Cd(NO_3)_2$ ⟶ mol of $Cd(NO_3)_2$ (solution)	Reaction factor ⟶	mol of CdS

Solution:

$$\text{mol of } Cd(NO_3)_2 = \text{vol} \times M$$
$$= 0.0158 \text{ liter} \times 0.122 \text{ mol/liter}$$
$$= 0.00193 \text{ mol of } Cd(NO_3)_2$$

$$0.00193 \text{ mol of } Cd(NO_3)_2 \times \frac{1 \text{ mol of CdS}}{1 \text{ mol of } Cd(NO_3)_2} \times \frac{144 \text{ g of CdS}}{\text{mol of CdS}}$$
$$= \underline{0.278 \text{ g of CdS}}$$

Example 11-14
Given the following balanced equation:

$$2KCl(aq) + Pb(NO_3)_2 \longrightarrow PbCl_2(s) + 2KCl(aq)$$

What volume of 0.200 M KCl is needed to react completely with 185 ml of 0.245 M $Pb(NO_3)_2$ solution?

Procedure:

Given			Requested	
Vol of $Pb(NO_3)_2$ (solution)	⟶ mol of $Pb(NO_3)_2$	Reaction factor ⟶	mol of KCl	⟶ vol of KCl (solution)

Solution:

$$\text{mol of } Pb(NO_3)_2 = V[Pb(NO_3)_2] \times M[Pb(NO_3)_2]$$

$$0.185 \text{ liter} \times 0.245 \text{ mol/liter} = 0.0453 \text{ mol of } Pb(NO_3)_2$$

$$0.0453 \text{ mol of } Pb(NO_3)_2 \times \frac{2 \text{ mol of KCl}}{1 \text{ mol of } Pb(NO_3)_2} = 0.0906 \text{ mol of KCl}$$

$$V \text{ (KCl)} = \frac{n(\text{KCl})}{M(\text{KCl})} = \frac{0.0906 \text{ mol}}{0.200 \text{ mol/liter}} = 0.453 \text{ liter}$$

$$0.453 \text{ liter} \times \frac{10^3 \text{ ml}}{\text{liter}} = \underline{\underline{453 \text{ ml}}}$$

Example 11-15

Given the following balanced equation:

$$2HCl(aq) + K_2S(aq) \longrightarrow H_2S(g) + 2KCl(aq)$$

What volume of H_2S measured at STP would be evolved from 1.65 liter of a 0.552 M HCl solution with excess K_2S present?

Procedure:

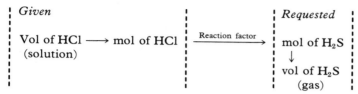

Solution:

$$V(HCl) \times M(HCl) = n(HCl)$$

$$1.65 \; \text{liters} \times 0.552 \; \text{mol/liter} = 0.911 \; \text{mol of HCl}$$

(Since the volume of the gas is at STP, the molar volume relationship can be used rather than the ideal gas law.)

$$0.911 \; \text{mol of HCl} \times \frac{1 \; \text{mol of } H_2S}{2 \; \text{mol of HCl}} \times \frac{22.4 \; \text{liter (STP)}}{\text{mol of } H_2S} = \underline{\underline{10.2 \; \text{liter (STP)}}}$$

See Problems 11-25 through 11-28

Review Questions

1 Describe what is meant by a solvent, solute, and solution.
2 What is electricity?
3 How does a conductor of electricity differ from a nonconductor?
4 What are an electrolyte, a nonelectrolyte, and a weak electrolyte?
5 Why does ice float on water?
6 How does water dissolve ionic compounds? Describe the forces between solute ions; between solvent molecules; between water and solute ions.
7 Define what is meant by the terms soluble and insoluble.
8 What is present in solution that distinguishes an electrolyte and a nonelectrolyte?
9 How does a polar nonelectrolyte dissolve in water?

10 Explain why nonpolar compounds do not dissolve in water.
11 How does an equation written in total ionic form differ from an equation written in molecular form?
12 What is a spectator ion?
13 How does a net ionic equation differ from a total ionic equation?
14 Describe what happens in a double displacement reaction.
15 What is meant by solubility?
16 Define what is meant by the terms concentrated and dilute.
17 What is meant by concentration of a solution? Define percent by weight.
18 Define molarity.

Problems

Solutions of Ionic Compounds

11-1 Write the reaction illustrating the solution of each of the following ionic compounds in water.
(a) $K_2Cr_2O_7$
(b) Li_2SO_4
■ (c) Cs_2SO_3
■ (d) $Ca(ClO_3)_2$
(e) Two moles of $(NH_4)_2S$
(f) Four moles of $Ba(OH)_2$
■ (g) Three moles of MgI_2
(h) Two moles of $Sr(C_2H_3O_2)_2$

Formation of Insoluble Ionic Compounds

11-2 Write the total ionic and the net ionic equation for each of the following reactions.
(a) $K_2S(aq) + Pb(NO_3)_2(aq) \rightarrow$
$$PbS(s) + 2KNO_3(aq)$$
(b) $(NH_4)_2CO_3(aq) + CaCl_2(aq) \rightarrow$
$$CaCO_3(s) + 2NH_4Cl(aq)$$
(c) $2AgClO_4(aq) + K_2CrO_4(aq) \rightarrow$
$$Ag_2CrO_4(s) + 2KClO_4(aq)$$
■ (d) $3Sr(OH)_2(aq) + 2Fe(NO_3)_3(aq) \rightarrow$
$$2Fe(OH)_3(s) + 3Sr(NO_3)_2(aq)$$

11-3 Referring to Table 11-1, determine which of the following compounds are insoluble in water.
(a) Na_2S (f) Ag_2O
(b) $PbSO_4$ (g) $(NH_4)_2S$
(c) $MgSO_3$ ■ (h) HgI_2
(d) NiS ■ (i) $Mg(OH)_2$
(e) Hg_2Br_2 ■ (j) Rb_2SO_4

11-4 Write the balanced molecular reaction that occurs (if any) when solutions of the following are mixed.
(a) KI and $Pb(C_2H_3O_2)_2$
(b) $AgClO_3$ and KNO_3
(c) $Sr(ClO_3)_2$ and $Ba(OH)_2$
(d) MgS and $Hg_2(NO_3)_2$
■ (e) $AlCl_3$ and KOH
■ (f) $Ba(C_2H_3O_2)_2$ and Na_2SO_4

11-5 Write the total ionic equation for any reactions that occurred in Problem 11-4.

11-6 Write the net ionic equation for any reactions that occurred in Problem 11-4.

11-7 Write balanced molecular equations indicating how the following ionic compounds could be prepared by a double displacement reaction using any other ionic compounds. In some cases the desired compound may be soluble and must be recovered by boiling off the solvent water after removal of a precipitate.
(a) $CuCO_3$ (d) NH_4NO_3
(b) $PbSO_4$ ■ (e) $Al_2(SO_3)_3$
(c) Hg_2I_2 ■ (f) $KC_2H_3O_2$

Percent by Weight

11-8 What is the percent by weight of solute in a solution made by dissolving 9.85 g of $Ca(NO_3)_2$ in 650 g of water?

■ **11-9** What is the percent by weight of solute if 14.15 g of NaI is present in 75.55 g of solution?

11-10 A solution is 10.0% by weight $NaOH$. How many moles of $NaOH$ are dissolved in 150 g of solution?

■ **11-11** A solution is 8.5% by weight NH_4Cl. What weight of NH_4Cl is present for each 100 g of water?

Molarity

11-12 How many moles of Epsom salts $(MgSO_4 \cdot 7H_2O)$ are present in 15.6 liters of a 0.0542 M solution?

11-13 Fill in the blanks in the table at the top of the next page.

11-14 What is the molarity of a solution made by dissolving 2.50×10^{-4} g of baking soda $(NaHCO_3)$ in enough water to make 2.54 ml of solution?

11-15 What is the molarity of the hydroxide ion and the barium ion if 13.5 g of $Ba(OH)_2$ is dissolved in enough water to make 475 ml of solution?

11-16 A solution is 25.0% by weight calcium nitrate and has a density of 1.21 g/ml. What is its molarity?

■ **11-17** A solution of concentrated $NaOH$ is 16.4 M. If the density of the solution is 1.43 g/ml, what is the percent by weight $NaOH$?

Solute	M	Amount of Solute	Volume of Solution
(a) KI	_____	2.40 mol	2.75 liters
(b) C_2H_5OH	_____	26.5 g	410 ml
(c) $NaC_2H_3O_2$	0.255	3.15 mol	_____ liters
(d) $LiNO_2$	0.625	_____ g	1.25 liters
(e) $BaCl_2$	_____	0.250 mol	850 ml
(f) Na_2SO_3	0.054	_____ mol	0.45 liter
■(g) K_2CO_3	0.345	14.7 g	_____ ml
■(h) LiOH	1.24	_____ g	1650 ml
■(i) H_2SO_4	0.905	0.178g	_____ ml

★■ **11-18** Concentrated nitric acid is 70.0% HNO_3 and is 14.7 M. What is the density of the solution?

Dilution

11-19 What volume of 4.50 M H_2SO_4 should be diluted with water to form 2.50 liters of 1.50 M acid?

■ **11-20** If 450 ml of a certain solution is diluted to 950 ml with water to form a 0.600 M solution, what was the molarity of the original solution?

11-21 What volume of water in milliliters should be added to 1.25 liters of 0.860 M HCl so that its molarity will be 0.545?

★**11-22** What volume in milliliters of *pure* acetic acid should be used to make 250 ml of 0.200 M $HC_2H_3O_2$? The density of the pure acid is 1.05 g/ml.

■ **11-23** What volume of water in milliliters should be *added* to 400 ml of a solution containing 35.0 g of KBr to make a 0.100 M KBr solution?

★**11-24** What would be the molarity of a solution made by mixing 150 ml of 0.250 M HCl with 450 ml of 0.375 M HCl

Stoichiometry Involving Solutions

11-25 Given the following reaction:

$$3KOH(aq) + CrCl_3(aq) \longrightarrow Cr(OH)_3(s) + 3KCl(aq)$$

What weight of $Cr(OH)_3$ would be produced if 500 ml of 0.250 M KOH were added to a solution containing excess $CrCl_3$?

★**11-26** Given the following reaction:

$$2NaOH(aq) + MgCl_2(aq) \longrightarrow Mg(OH)_2(s) + 2NaCl$$

What weight of $Mg(OH)_2$ would be produced by mixing 250 ml of 0.240 M NaOH with 400 ml of 0.100 M $MgCl_2$?

★■ **11-27** Given the following reaction:

$$CO_2(g) + Ca(OH)_2(aq) \longrightarrow CaCO_3(s) + H_2O$$

What is the molarity of a 1.00-liter solution of $Ca(OH)_2$ that would completely react with 10.0 liters of CO_2 measured at 25.0°C and a pressure of 0.950 atm?

11-28 Given the reaction:

$$2HNO_3(aq) + 3H_2S(aq) \longrightarrow 2NO(g) + 3S(s) + 4H_2O$$

■ (a) What volume of 0.350 M HNO_3 will completely react with 275 ml of 0.100 M H_2S?

■ (b) What weight of sulfur will be produced by 850 ml of 2.45 M HNO_3?

★(c) What volume of NO gas measured at 27°C and 720 torr will be produced from 650 ml of 0.100 M H_2S solution?

12 Acids, Bases, and Salts

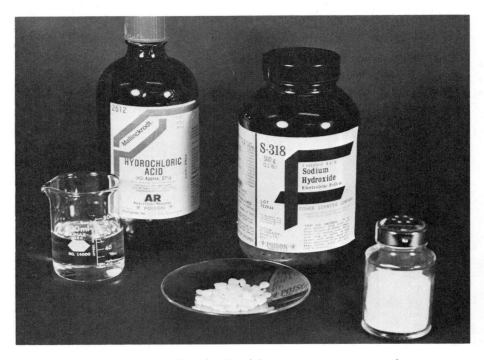

Hydrochloric acid and sodium hydroxide are two very corrosive compounds. When mixed in stoichiometric amounts, ordinary table salt is formed.

A certain type of chemical reaction supplies the English language with some very harsh adjectives. For example, we hear expressions such as "he gave a *vitriolic* criticism," "she has a *caustic* sense of humor," or "they gave the idea the *acid* test." All of these words originate from the corrosive chemical reactions typical of a class of compounds called acids and bases.

The subject of acids and bases is one of the more extensive and complicated topics in chemistry even for the experienced chemist. If the subject is developed one step at a time, however, it becomes quite reasonable and certainly understandable. To this end we will limit ourselves to the discussion of the most familiar yet most important concept of acids and bases. This concerns their behavior in aqueous solution. The discussion will be divided into two parts.

1 The identification of acids and bases and the distinctive reactions they undergo in aqueous solution is discussed in this chapter. This provides a solid working understanding of these concepts for future courses as well as Chapter 15.

2 In Chapter 14, the concept of dynamic equilibrium will be discussed. This will provide a basis for a deeper look into the more subtle action of acids and bases in aqueous solution as discussed in Chapter 15. The concept of pH is introduced at that time.

In more advanced courses, acid and base behavior is broadened by more general definitions that include other solvents and systems.

The background for this chapter includes:

1 Ionic and covalent compounds (Sections 6-4 and 6-5)

2 Nomenclature of acids (Section 7-3)

3 Ionic and net ionic equations (Section 11-3)

4 Molarity (Section 11-6)

12-1 Early Criteria for Acids and Bases

Historically, acids were known by what they did and not by some unique elemental composition. For example, the following facts have been known about the nature of acids since the time of the alchemists in the middle ages.

Acids:

1 Taste sour (don't try this—it's OK for vinegar or carbonated water, but battery acid could ruin your tongue)

2 Cause certain organic dyes to change color (an example is litmus, which turns from blue to red in acids)

3 Dissolve certain metals, such as zinc, with the liberation of a gas (see Figure 12-1)

4 Dissolve limestone ($CaCO_3$), with the liberation of a gas (see Figure 12-1)

5 React with bases to form salts and water

Bases, on the other hand:

1 Taste bitter (again, don't try it)

2 Are slippery or soapy feeling

Figure 12-1 ZINC AND LIME-STONE IN ACID. Both zinc (right) and limestone ($CaCO_3$) (left) react with acid to liberate a gas.

3   Cause certain organic dyes to change color (red litmus turns blue in basic solution)

4   React with acids to form salts and water

It was easy to classify substances as acid, base, or neither from a few simple laboratory tests. It wasn't until less than 100 years ago, however, that the foundation for a currently accepted model was suggested. *In 1884, Svante Arrhenius suggested that* **acids** *are substances that produce H^+ ions and* **bases** *are substances that produce OH^- ions in aqueous solution.*

12-2 The Nature of Acids and Bases

Our first definition of acids is that of Arrhenius. The following reactions illustrate how some well-known acids produce H^+ in solution.

Muriatic or hydrochloric acid	$HCl \longrightarrow H^+ + Cl^-$
Oil of vitriol or sulfuric acid	$H_2SO_4 \longrightarrow H^+ + HSO_4^-$
Vinegar or acetic acid	$HC_2H_3O_2 \longrightarrow H^+ + C_2H_3O_2^-$
Carbonated water or carbonic acid	$H_2CO_3 \longrightarrow H^+ + HCO_3^-$

See Problem 12-1

The acids listed above are not ionic compounds when pure. They form ions only when mixed with water. In fact, a chemical reaction occurs between the acid molecules and the H_2O molecules which results in the removal of a H^+ ion from the rest of the molecule. As discussed in Chapter 6 and 11, acids (such as HCl) and H_2O are polar covalent compounds. When

HCl dissolves in water, there is a dipole–dipole electrostatic attraction between the negative dipole of one molecule and the positive dipole of the other as illustrated in Figure 12-2. (Actually, there are many more H_2O molecules involved with each HCl than the four shown in the figure.)

A tug-of-war has developed. On the one side the H_2O's are pulling on the HCl molecule, attempting to split the molecule into a H^+ and a Cl^- ion. On the other side, the HCl covalent bond is trying mightily to hold the molecule together. In the case of HCl the H_2O molecules are clearly the winners, since the interaction is strong enough to break the bonds of all dissolved HCl molecules. Therefore, HCl is an acid since an H^+ is produced in the solution as a result of the **dissociation** (breaking apart) of HCl. (Since dissociation produces ions in this case, the process is also called **ionization.**)

To illustrate the importance of water in the ionization process, the reaction of an acid with water can be represented as

$$HCl + H_2O \longrightarrow H_3O^+(aq) + Cl^-(aq)$$

Instead of H^+, the acid species is often represented as H_3O^+, which is known as the **hydronium ion.** The hydronium ion is simply a representation of the H^+ ion in a hydrated form. The acid species is represented as H_3O^+ rather than H^+ because it is somewhat closer to what is believed to be the actual situation. In fact, the nature of H^+ in aqueous solution is even more complex than H_3O^+ (i.e., $H_5O_2^+$, $H_7O_3^+$, etc.) In any case, the acid species is represented as H^+, $H^+(aq)$, or $H_3O^+(aq)$ depending on the convenience of the particular situation. *Just remember that all refer to the same species in aqueous solution.* If $H^+(aq)$ is used, it should be understood that it is not just a bare proton in aqueous solution but is associated with water molecules (it is hydrated).

The common property of acids in water is the formation of $H^+(aq)$ by this definition. We can now see by the following net ionic reactions how the $H^+(aq)$ ion accounts for the reactions long known to indicate acid behavior mentioned in the previous section.

1 Acids react with zinc and give off a gas (H_2).

$$Zn(s) + \underline{2H^+(aq)} \longrightarrow Zn^{2+}(aq) + H_2(g)$$

2 Acids react with limestone ($CaCO_3$) and give off a gas (CO_2).

$$CaCO_3(s) + \underline{2H^+(aq)} \longrightarrow Ca^{2+}(aq) + H_2O + CO_2(g)$$

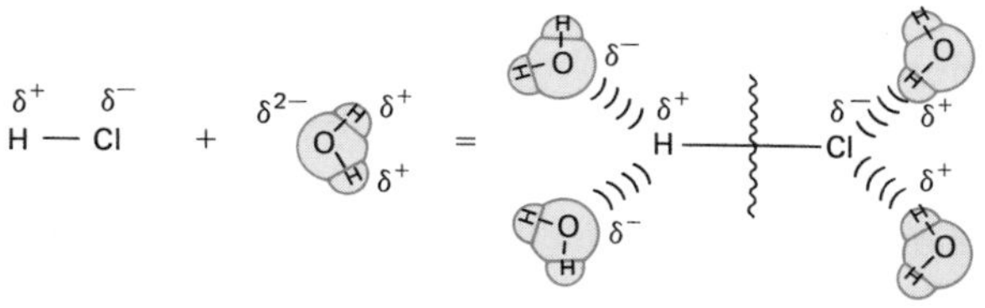

Figure 12-2 THE INTERACTION OF HCl and H_2O. The dipole–dipole interaction between H_2O and HCl leads to the breaking of the HCl bond.

3   Acids react with bases.

$$H^+(aq) + OH^-(aq) \longrightarrow H_2O$$

Other molecules containing hydrogen may not be acidic in H_2O because the hydrogens are (1) too strongly attached to the rest of the molecule to be ionized or (2) not polar enough to create a strong interaction with H_2O. Sometimes both reasons are important. The hydrogens in CH_4, NH_3, and the three hydrogens on the carbon in acetic acid are examples of hydrogens that do not ionize in aqueous solution to form H^+ ion.

Three H's on the C in acetic acid do not ionize (the C—H bond is essentially nonpolar).

The H attached to the O is polar and can be ionized.

Unlike the acids, which are covalent compounds before being ionized by the water molecules, many common bases consist of ionic metal hydroxides in the solid state. The solution process simply breaks down the solid ionic crystals to the same ions in solution as discussed in Section 11-2. For example,

$$Na^+OH^-(s) \xrightarrow{xH_2O} Na^+(aq) + OH^-(aq)$$

See Problem 12-2

Some commonly known bases are caustic soda or lye (NaOH), caustic potash (KOH), lime [$Ca(OH)_2$], and ammonia (NH_3).

12-3 Neutralization and Salts

One of the characteristics of acids and bases described earlier in this chapter is that the $H^+(aq)$ from an acid reacts with the $OH^-(aq)$ from a base to form water:

$$H^+(aq) + OH^-(aq) \longrightarrow H_2O$$

This is the net ionic equation of the process known as neutralization. **Neutralization** *is the reaction between an acid and a base to form a salt and water. A* **salt** *is an ionic compound composed of the cation from a base and the anion from an acid.* It is amazing that acids such as hydrochloric acid and bases such as sodium hydroxide are both corrosive and dangerous compounds. When mixed in exactly stoichiometric amounts, however, the product is simply a solution of common table salt (NaCl).* The corrosive agents [$H^+(aq)$ and

---

* Don't try this without supervision: Much heat evolution, with boiling and splattering, can occur.

OH$^-$(aq)] annihilate each other, leaving in solution the cation from the base (Na$^+$) and the anion from the acid (Cl$^-$). The salt can be recovered by boiling off the water. The following molecular, total ionic, and net ionic equations illustrate the neutralization discussed:

$$NaOH(aq) + HCl(aq) \longrightarrow NaCl(aq) + H_2O$$

$$Na^+(aq) + OH^-(aq) + H^+(aq) + Cl^-(aq) \longrightarrow Na^+(aq) + Cl^-(aq) + H_2O$$

$$H^+(aq) + OH^-(aq) \longrightarrow H_2O$$

The word "salt" is often thought of as just one substance, sodium chloride, as formed in the previous reaction. Actually, a salt can result from many different combinations of anions and cations from a variety of neutralizations.* The following neutralization reactions illustrate the formation of some other salts:

Acid	+	Base	$\longrightarrow$	Salt	+	Water
1 $HNO_3(aq)$	+	$KOH(aq)$	$\longrightarrow$	$KNO_3(aq)$	+	H_2O
2 $H_2SO_4(aq)$	+	$2NaOH(aq)$	$\longrightarrow$	$Na_2SO_4(aq)$	+	$2H_2O$
3 $2HClO_4(aq)$	+	$Ca(OH)_2(aq)$	$\longrightarrow$	$Ca(ClO_4)_2(aq)$	+	$2H_2O$
4 $2H_3PO_4(aq)$	+	$3Ba(OH)_2(aq)$	$\longrightarrow$	$Ba_3(PO_4)_2(aq)$	+	$6H_2O$

Notice in the preceeding reactions that one H$^+$ must be available for each OH$^-$ (and vice versa) for *complete* neutralization. For example, in reaction 2 above, 2 mol of NaOH is required to react with 1 mol of H$_2$SO$_4$, since the latter compound can produce 2 mol of H$^+$(aq) ions as follows.

$$H_2SO_4(aq) \longrightarrow 2H^+(aq) + SO_4{}^{2-}(aq)$$

For a similar reason, 2 mol of HClO$_4$ is required for each mole of Ca(OH)$_2$ as shown in reaction 3. To illustrate these two reactions in more detail, the total ionic and net ionic equations are shown in the following equations:

2 $2H^+(aq) + \cancel{SO_4{}^{2-}(aq)} + \cancel{2Na^+(aq)} + 2OH^-(aq) \longrightarrow \cancel{2Na^+(aq)} + \cancel{SO_4{}^{2-}} + 2H_2O$

$$2H^+(aq) + 2OH^-(aq) \longrightarrow 2H_2O$$

3 $2H^+(aq) + \cancel{2ClO_4{}^-(aq)} + \cancel{Ca^{2+}(aq)} + 2OH^-(aq) \longrightarrow \cancel{Ca^{2+}(aq)} + \cancel{2ClO_4{}^-(aq)} + 2H_2O$

$$2H^+(aq) + 2OH^-(aq) \longrightarrow 2H_2O$$

What happens if *1 mol* of H$_2$SO$_4$ is mixed with only *1 mol* of NaOH, which is only half of the amount of NaOH needed for complete neutralization? In this case, notice that H$_2$SO$_4$ ionizes in two steps, with the first ionization as follows:

$$H_2SO_4 \longrightarrow H^+(aq) + HSO_4{}^-(aq) \qquad \text{First ionization of } H_2SO_4$$

* Salts are also produced by double displacement reactions as discussed in Chapter 11 or directly from the elements (Chapter 9).

Thus the reaction of *1 mol* of NaOH with *1 mol* of H_2SO_4 is illustrated by the following equations:

$$H_2SO_4(aq) + NaOH(aq) \longrightarrow NaHSO_4(aq) + H_2O$$

$$\underline{H^+(aq)} + \cancel{HSO_4^-(aq)} + \cancel{Na^+(aq)} + \underline{OH^-(aq)} \longrightarrow$$
$$\cancel{Na^+(aq)} + \cancel{HSO_4^-(aq)} + \underline{H_2O}$$

$$H^+(aq) + OH^-(aq) \longrightarrow H_2O$$

Notice that the salt formed in this neutralization ($NaHSO_4$) has one acidic hydrogen remaining on the anion, since the original acid (H_2SO_4) was not completely neutralized. The hydrogen sulfate anion can produce $H^+(aq)$ as follows:

$$HSO_4^-(aq) \longrightarrow H^+(aq) + SO_4^{2-}(aq) \qquad \text{Second ionization of } H_2SO_4$$

This salt is known as an acid salt. *An **acid salt** is an ionic compound that contains acidic hydrogens on the anion. Acid salts result from the partial neutralization of* **polyprotic acids** *(acids with more than one acidic hydrogen.)* The stepwise neutralization of a polyprotic acid is further illustrated by the neutralization of H_3PO_4 with KOH. If *1 mol* of the KOH is added to *1 mol* of H_3PO_4, reaction (a) occurs:

(a) 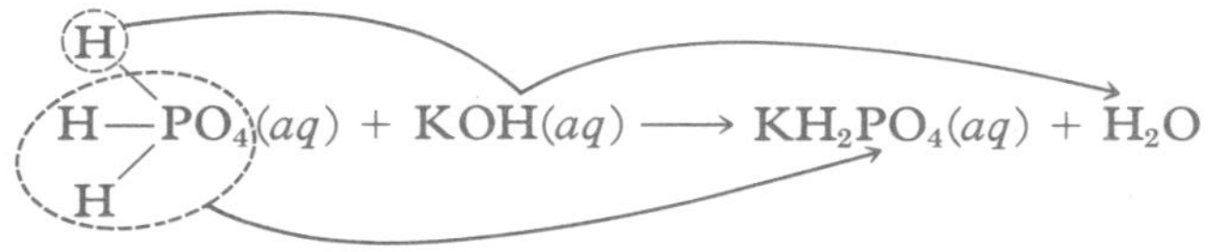
$$H-PO_4(aq) + KOH(aq) \longrightarrow KH_2PO_4(aq) + H_2O$$

If 1 mol of KOH is then added to the KH_2PO_4 solution, reaction (b) occurs:

(b) 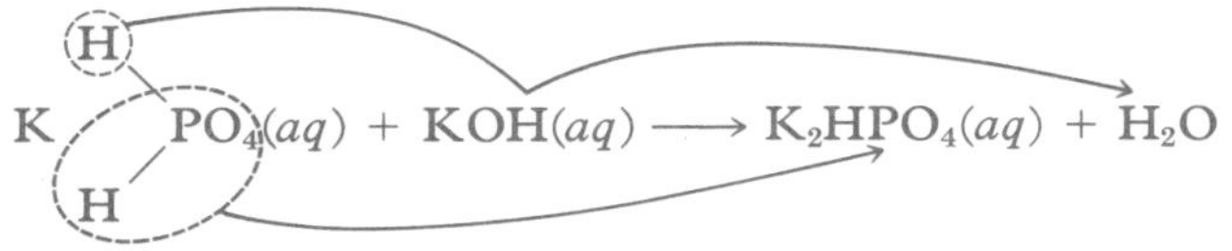
$$K\ PO_4(aq) + KOH(aq) \longrightarrow K_2HPO_4(aq) + H_2O$$

Finally, by adding 1 mol of KOH to the K_2HPO_4 solution, reaction (c) occurs:

(c) 
$$K_2(H)-(PO_4)(aq) + KOH \longrightarrow K_3PO_4(aq) + H_2O$$

Notice that each additional OH^- removes one H^+ from the acid or acid salt.

The total reaction, which is the algebraic sum of all three reactions, is

$$H_3PO_4(aq) + 3KOH(aq) \longrightarrow K_3PO_4(aq) + 3H_2O$$

This is the reaction that occurs regardless of whether the KOH is added 1 mol at a time or all 3 mol at once. In these reactions the H_3PO_4 is identified as an acid, KH_2PO_4 and K_2HPO_4 as acid salts, and K_3PO_4 as a salt. (See Figure 12-3.)

Figure 12-3 THE NEUTRALIZATION OF H_3PO_4. The hydrogens on H_3PO_4 can be neutralized one at a time.

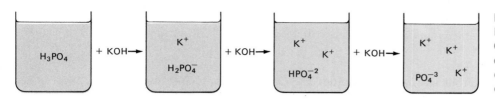

Example 12-1
Identify each of the following as either an acid, a base, a salt, or an acid salt: (a) $KClO_4$, (b) H_2S, (c) $Ba(HSO_3)_2$, (d) $Al(OH)_3$.

Answer:
(a) $KClO_4$: salt (b) H_2S: acid (c) $Ba(HSO_3)_2$: acid salt (d) $Al(OH)_3$: base

Example 12-2
Write the balanced equation in molecular form illustrating the complete neutralization of $Al(OH)_3$ with H_2SO_4.

Procedure:
Each mole of base produces *3 mol* of OH^-; each mole of acid produces *2 mol* of H^+. Therefore, 2 mol of base (producing 6 mol of OH^-) reacts with 3 mol of acid (producing 6 mol of H^+) for complete neutralization.

Answer:

$$2Al(OH)_3 + 3H_2SO_4 \longrightarrow Al_2(SO_4)_3 + 6H_2O$$

Example 12-3
Write the balanced molecular equation illustrating the reaction of 1 mol of H_3PO_4 with 1 mol of $Ca(OH)_2$.

Procedure:
One mole of base produces 2 mol of OH^-, which reacts with 2 mol of H^+. The removal of 2 mol of H^+ from 1 mol of H_3PO_4 leaves the HPO_4^{2-} ion in solution.

Answer:

$$H_3PO_4 + Ca(OH)_2 \longrightarrow CaHPO_4 + 2H_2O$$

See Problems 12-3 through 12-8

12-4 The Strengths of Acids

A 0.10 *M* solution of hydrochloric acid burns holes in your clothes, and you too if not washed off immediately. A 0.10 *M* solution of acetic acid is not particularly corrosive, and when present in vinegar adds a little tang to your salad. Hydrochloric acid and acetic acid at the same concentrations are, at best, distant cousins. The difference lies in the *strength* of their acid behavior, which relates directly to the concentration of $H^+(aq)$ (or H_3O^+) produced in the respective solutions.

Hydrochloric acid is a strong acid. *A **strong acid** is completely ionized in aqueous solution.* This is illustrated by the following equation:

$$HCl + H_2O \longrightarrow H_3O^+(aq) + Cl^-(aq) \quad \text{(complete)}$$

Other strong acids include HNO_3, HBr, HI, $HClO_4$, and H_2SO_4. Most other common acids are considered weak acids in water. *A **weak acid** is only partially ionized in water* (usually less than 5%), which means that the reaction below is incomplete. An example of a reaction of a weak acid is

$$HF(aq) + H_2O \longrightarrow H_3O^+(aq) + F^-(aq) \quad \text{(partial)}$$

In this case the vast majority of acid molecules are present in solution in the molecular form, as shown on the left side of the equation. The few HF molecules that ionize or dissociate create a small concentration of H_3O^+ which accounts for the acidity of the solution. Because of the low concentration of ions, a weak acid is also classified as a weak electrolyte (see Section 11-1).

The partial ionization of a weak acid is one example of a chemical reaction that reaches a state of equilibrium. *In a reaction at **equilibrium**, two reactions are occurring simultaneously.* In the ionization of HF, for example, a forward reaction occurs (to the right), producing ions (H_3O^+ and F^-), and a reverse reaction occurs (to the left), producing covalent compounds (HF and H_2O).

Forward: $$HF(aq) + H_2O \longrightarrow H_3O^+(aq) + F^-(aq)$$

Reverse: $$H_3O^+(aq) + F^-(aq) \longrightarrow HF(aq) + H_2O$$

At equilibrium, the forward and the reverse reactions occur at the same rate. Equilibrium is illustrated by a double arrow ($\rightleftharpoons$) rather than a single arrow, which represents a complete reaction.

$$HF(aq) + H_2O \rightleftharpoons H_3O^+(aq) + F^-(aq)$$

When a system is at equilibrium, the concentration of all species (reactants and products) remains the same but the identity of the individual molecules changes. The reaction thus *appears* to have gone so far to the right and then stopped. In fact, at equilibrium, a **dynamic** (*constantly changing*) situation exists where two reactions going in opposite directions at the same rate keep the concentrations of all species constant.

An analogy to a situation where the amount remains constant but the identity changes is found in the water of a lake confined by a dam. As shown in Figure 12-4, if the amount of water entering the lake equals the amount leaving the lake over the spillway, the level of the lake is at a point of equilibrium. A dynamic situation exists, however, since the amount of water in the lake is constant but the water itself is always changing.

In the chemical system, whenever one HF molecule ionizes, one H_3O^+ and one F^- react to reform one HF and H_2O. In this case the equilibrium lies

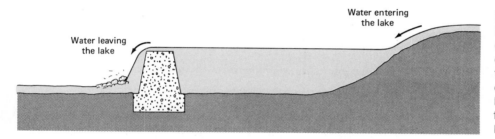

Figure 12-4 EQUI-LIBRIUM. The level of the lake is at equilibrium when the flow of water entering the lake equals the flow of water leaving the lake.

far to the left, since most of the HF is not ionized. We will have much more to say about equilibrium in Chapter 14. The subject is introduced now so that we may appreciate the reason for incomplete reactions such as the ionization of a weak acid or base. (In Chapter 15 the equilibrium involved in the ionization of weak acids and bases is examined in more detail.)

The following examples illustrate the difference in acidity (the difference in H_3O^+ concentration) between a strong and a weak acid. In these examples the appearance of an ion in brackets (e.g., $[H_3O]^+$) stands for the concentration of that ion in *molarity*.

Example 12-4

What is $[H_3O^+]$ in a 0.10 *M* HNO_3 solution?

Solution:

HNO_3 is a strong acid, so the following reaction goes 100% to the right:

$$HNO_3 + H_2O \longrightarrow H_3O^+(aq) + NO_3^-(aq)$$

The initial concentration of HNO_3 is 0.10 *M*, and all of the HNO_3 ionizes to form H_3O^+. The reaction stoichiometry tells us that 1 mol of H_3O^+ is formed for every mole of HNO_3 initially present. Therefore, 0.10 mol of HNO_3 forms 0.10 mol of H_3O^+ in 1 liter. Thus,

$$[H_3O^+] = \underline{\underline{0.10\ M}}$$

Example 12-5

What is $[H_3O^+]$ in a 0.100 *M* $HC_2H_3O_2$ solution that is 1.34% ionized?

Solution:

Since $HC_2H_3O_2$ is a weak acid, the following ionization reaches equilibrium when 1.34% of the initial $HC_2H_3O_2$ ionizes:

$$HC_2H_3O_2(aq) + H_2O \rightleftharpoons H_3O^+(aq) + C_2H_3O_2^-(aq)$$

The $[H_3O^+]$ is calculated by multiplying the original concentration of acid by the percent ionization in decimal form:

$$[H_3O^+] = [\text{original conc. of acid}] \times \frac{\%\ \text{ionization}}{100}$$

**Review Appendix
B on percent
problems**

In this case,

$$[H_3O^+] = [0.100] \times \frac{1.34}{100} = \underline{\underline{1.34 \times 10^{-3} \ M}}$$

(Review Appendix B for similar problems in the use of percent.)

Notice in the above examples that the $[H_3O^+]$ is almost 100 times greater in the strong acid solution compared to the weak acid solution at the same initial concentration of acid. The much depressed $[H_3O^+]$ in an acetic acid ($HC_2H_3O_2$) solution accounts for its comparatively mild acid behavior.

12-5 The Strengths of Bases

Caustic soda (NaOH) is a good agent to use to clean drains but not floors. It would ruin the floors and probably the hand that applied it. (Many tragedies could be avoided by keeping caustic soda or lye out of the reach of children.) If you want to clean the floor, a milder caustic agent is needed. Household ammonia is used for this purpose, because it produces a much lower OH^- concentration than lye.

Sodium hydroxide is a **strong base** because 100% of the NaOH that dissolves forms Na^+ and OH^- ions:

$$NaOH(s) \longrightarrow Na^+(aq) + OH^-(aq)$$

The alkali metal hydroxides and the alkaline earth hydroxides [except $Be(OH)_2$] are all strong bases, although some of the alkaline earth hydroxides dissolve to only a limited extent. These compounds are all ionic in the solid state. As we mentioned before, the solution process simply breaks down the solid lattice of ions into hydrated ions.

Ammonia is a **weak base,** as shown by the following equation:

$$NH_3(aq) + H_2O \rightleftharpoons NH_4^+(aq) + OH^-(aq)$$

**See Problems 12-9
through 12-12**

Since this equilibrium reaction lies far to the left, only a comparatively small concentration of OH^- ions is produced. Many other neutral nitrogen compounds, such as methylamine (CH_3NH_2) and pyridine (C_5H_5N), behave as weak bases.

12-6 Brønsted-Lowry Acids and Bases

The reaction of NH_3 as a base is an interesting phenomenon, since pure NH_3 obviously does not contain an OH^- ion to donate directly to the water as do metal hydroxides. Ammonia is a base by virtue of the lone pair of electrons on the nitrogen, which can add a H^+ ion removed from a H_2O molecule. The loss of a H^+ from a H_2O leaves a OH^- in solution:

$$H-\overset{..}{\underset{|}{N}}-H \; + \; \textcircled{H}-\overset{..}{\underset{H}{O}}: \; \rightleftharpoons \; H-\overset{\overset{H}{|}}{\underset{|}{N}}-H \; \overset{+}{} \; + \; :\overset{..}{\underset{..}{O}}-H^-$$

Although the production of OH^- classifies NH_3 as a legitimate Arrhenius base, the **Brønsted-Lowry definition** of acid-base behavior is also useful in describing this behavior. In this approach an **acid** *is defined as a proton* (H^+) *donor* and a **base** *as a proton acceptor.* Notice in the reaction between NH_3 and H_2O that the NH_3 accepts a H^+ from a water molecule. Thus, by this definition, NH_3 is a base and H_2O is an acid, since the H_2O donates an H^+.

As mentioned in the previous section NH_3 is a *weak* base, because the reaction reaches a point of equilibrium that lies far to the left. This means that the reverse reaction also takes place to a significant extent:

$$NH_3 \; + \; H_2O \; \rightleftharpoons \; H-\overset{\textcircled{H}}{\underset{\underset{H}{|}}{N}}-H \; \overset{+}{} \; + \; OH^-$$

In the reverse reaction, NH_4^+ is an acid, since it donates a H^+ to the base, OH^-.

Notice in the reaction that when a base (NH_3) reacts, it forms an acid (NH_4^+). When an acid (H_2O) reacts, it forms a base (OH^-). The NH_3–NH_4^+ and the H_2O–OH^- pairs are known as **conjugate acid – base pairs.** Thus a reaction in the Brønsted-Lowry sense constitutes two conjugate pairs as illustrated in the following equation (A = Brønsted-Lowry acid, B = Brønsted-Lowry base):

$$NH_3 \; + \; H_2O \; \rightleftharpoons \; NH_4^+ \; + \; OH^-$$

$$B_1 \qquad A_2 \qquad A_1 \qquad B_2$$

The Brønsted-Lowry definition transforms the concept of acids and bases from just the production of certain ions (H_3O^+ or OH^-) to the dynamics of a chemical reaction—in particular, a proton exchange reaction. Acids and bases, by this definition, are so defined by the role they play in the proton exchange and not necessarily by the fact that they produce H_3O^+ or OH^- ions in water as in the Arrhenius definition. Since there are two acids and two bases present, the total reaction can be considered as simply a competition for the H^+ ion. Neutralization between an acid and a base in this case creates a new acid and base. It is worthwhile to be comfortable with the meaning of conjugate acids and bases, since these concepts are used in Chapter 15. (The Brønsted-Lowry definition is also quite useful in discussions of proton exchange reactions in solvents other than water, such as pure NH_3 and H_2SO_4.)

Although this approach is more general than the Arrhenius definition, all substances that act as acids and bases in water by this previous definition maintain their particular status as acids or bases in the Brønsted-Lowry definition. For example, HCl is an acid, since it donates a H^+ to H_2O:

$$HCl + H_2O \longrightarrow H_3O + Cl^-$$
$$A_1 \qquad B_2 \qquad\qquad A_2 \qquad B_1$$

In this reaction H_2O is a base, since it accepts a H^+ to form H_3O^+. (Remember that H_2O is an acid when NH_3 is present.) *A compound that can act as either an acid or base, depending on what other substance is present, is called* **amphoteric.** One must determine whether an amphoteric substance is an acid or a base from the particular reaction. For example, we cannot determine whether H_2O is a Brønsted-Lowry acid or base unless we know what other substance is present (HCl or NH_3).

As mentioned previously, the reaction of HCl with H_2O is essentially an irreversible reaction. Since Cl^- does not react with H^+ (the reverse reaction), it does not show base properties in water. It is important to note, then, that a substance may be termed a "conjugate base" because it results from the loss of a H^+ from an acid. However, a conjugate base does not necessarily *react* as a base (combine with H^+) in water. On the other hand, the conjugate base (e.g., F^-) of a weak acid (e.g., HF) does act as a base, which means that the reverse reaction from the ionization occurs to an appreciable extent. These two examples suggest an inverse relation between the strength of an acid and the strength of its conjugate base. That is, the stronger the acid in water, the weaker is its conjugate base and vice versa.

This relationship is illustrated in Figure 12-5 by a "seesaw." Notice that when one side is up, indicating a strong acid or base, the other side is down, indicating a weak base or acid, respectively.

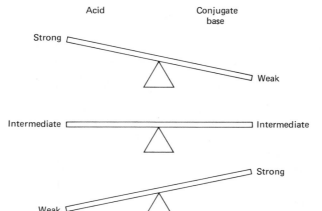

Figure 12-5 AN ACID AND ITS CONJUGATE BASE. There is an inverse relationship between the strength of an acid and the strength of its conjugate base.

Table 12-1 Relative Strengths of Some Acids and Bases

	Acid		Base	
		$\xrightarrow{\;-H^+\;}$		
		$\xleftarrow{\;+H^+\;}$		
1 Very strong	$HClO_4$		ClO_4^-	Very weak
	HBr		Br^-	(negligible)
	HCl		Cl^-	
	HNO_3		NO_3^-	
2 Intermediate	H_2SO_3		HSO_3^-	
	HNO_2		NO_2^-	
	HF		F^-	
	$HC_2H_3O_2$		$C_2H_3O_2^-$	Intermediate
	H_2S		HS^-	
	H_2CO_3		HCO_3^-	
	$HOCl$		OCl^-	
	HCN		CN^-	
3 Very weak (negligible)	NH_3		NH_2^-	Very strong
	H_2		H^-	
	CH_4		CH_3^-	

(left margin: Increasing acid strength) *(right margin: Increasing base strength)*

In Table 12-1 some neutral acids are listed in order of *decreasing* acid strength on the left. From the inverse relationship between the acid and its conjugate base, notice that the conjugate bases are listed in order of *increasing* strength on the right.

The acids and bases in this table can be divided into three groups.

Group 1   Acid $\longleftrightarrow$ Conjugate base

                Very strong   Negligible
                 in water   in water

In this group all acids are identical, since they are as strong as possible in water. That is they are all 100% ionized. The conjugate bases, on the other hand, show negligible base behavior in water. (They may show base behavior in other solvent media, however.) Examples are

$$HNO_3 + H_2O \xrightarrow{100\%} H_3O^+ + NO_3^-$$

$$HNO_3 + H_2O \xleftarrow{\;\;\;\not\Vert\;\;} H_3O^+ + NO_3^-$$

($\not\rightarrow$ means that the reaction does not occur). *The molecular form of the acid, which is HNO_3 on the right, does not exist in water.*

Group 2   Acid $\longleftrightarrow$ Conjugate base

                Intermediate   Intermediate
                 in water   in water

In this group all acids behave differently depending on their degree of ionization. Although all acids in this group are only partially ionized, those at the top produce a higher $[H_3O^+]$ than those at the bottom. In any case, the conjugate bases in this group all show some base behavior in water. Examples are

$$HNO_2 + H_2O \xrightarrow{\text{Partial}} H_3O^+ + NO_2^-$$

$$HNO_2 + H_2O \xleftarrow{\text{Partial}} H_3O^+ + NO_2^-$$

Total $\quad \overline{HNO_2 + H_2O \rightleftharpoons H_3O^+ + NO_2^-}$

Both the molecular form of the acid (HNO_2 on the right) and the conjugate base (NO_2^- on the left) exist in aqueous solution in equilibrium.

Group 3   Acid $\longleftrightarrow$ Conjugate base

Negligible   Very strong
in water   in water

In this group, the acids show negligible acid behavior in water. (In fact, NH_3 behaves as a base.) Their conjugate bases do exist in certain ionic compounds, however, and behave as very strong bases in water. These bases react completely with either H_3O^+ or H_2O to form the conjugate acid. For example:

$$CH_4 + H_2O \xrightarrow{\quad} \!\!\!\!\!\!\!| \; H_3O^+ + CH_3^-$$

$$CH_4 + H_2O \xleftarrow{100\%} H_3O^+ + CH_3^-$$

Notice that this situation is just the opposite of that for group 1. *In this case, only the molecular form of the acid (CH_4 on the left) exists in water. The conjugate base (CH_3^- on the right) does not exist in water.*

Groups 1 and 3 are straightforward situations, since reactions occur either completely or not at all. The acids in group 2 are partially ionized and require quantitative relationships indicating the extent of ionization. These relationships, called equilibrium constants, are discussed in Chapter 15. In further development of this topic, Table 12-1 can be expanded to include many more compounds, including several ions (e.g., HSO_4^-) that act as acids in water.

Example 12-6
Write the equation illustrating the reaction of H_2S as a Brønsted-Lowry acid. What is the conjugate base of H_2S?

Answer:

$$H_2S(aq) + H_2O \rightleftharpoons H_3O^+(aq) + HS^-(aq)$$

The conjugate base of H_2S is $\underline{\underline{HS^-}}$.

Example 12-7

The acid salt anion, HPO_4^{2-}, is amphoteric. Write one equation illustrating how it behaves as a Brønsted-Lowry base with H_3O^+ and one equation illustrating how it behaves as a Brønsted-Lowry acid with OH^- in water.

Answer:

Base: $HPO_4^{2-}(aq) + H_3O^+ \rightarrow H_2PO_4^-(aq) + H_2O$

Acid: $HPO_4^{2-}(aq) + OH^- \rightarrow PO_4^{3-}(aq) + H_2O$

Example 12-8

Referring to Table 12-1, complete the following equations. When no reaction occurs, write N.R. When an equilibrium exists, indicate it by double arrows.

(a) $HBr + H_2O$

(b) $NH_2^- + H_3O^+$

(c) $OCl^- + H_3O^+$

(d) $ClO_4^- + H_3O^+$

(e) $H_2 + H_2O$

(f) $HCN + H_2O$

Answers:

(a) $HBr + H_2O \rightarrow H_3O^+(aq) + Br^-(aq)$

(b) $NH_2^- + H_3O^+ \rightarrow NH_3(aq) + H_2O$

(c) $OCl^- + H_3O^+ \rightleftharpoons HOCl(aq) + H_2O$

(d) $ClO_4^- + H_3O^+ \rightarrow$ N.R.

(e) $H_2 + H_2O \rightarrow$ N.R.

(f) $HCN + H_2O \rightleftharpoons H_3O^+(aq) + CN^-(aq)$

See Problems 12-13 through 12-22

12-7 Oxides as Acids and Bases

Carbonated water is tangy because the carbon dioxide gas dissolves in water to produce the weak carbonic acid:

$$CO_2(g) + H_2O \rightleftharpoons H_2CO_3(aq) + H_2O \rightleftharpoons H_3O^+(aq) + HCO_3^-(aq)$$

Since CO_2 can be obtained algebraically by subtracting out the elements of water from the formula of carbonic acid, *the oxide is called an* **acid anhydride** (*acid without water*).

$$\begin{array}{r} H_2CO_3 \\ -H_2O \\ \hline CO_2 \end{array}$$

Most (but not all) nonmetal oxides form acids when dissolved in water. Some other examples of the more common acid anhydrides and their reactions with water to produce the acid are the following:

$$SO_2(g) + H_2O \longrightarrow H_2SO_3(aq)$$

$$N_2O_5(l) + H_2O \longrightarrow 2HNO_3(aq)$$

$$B_2O_3(s) + 3H_2O \longrightarrow 2H_3BO_3(aq)$$

$$SO_3(g) + H_2O \longrightarrow H_2SO_4(aq)$$

Notice that the central nonmetal has the *same oxidation state in both the oxide and the acid*. This is a convenient way to tell in some cases which oxide goes with which acid. For example, is N_2O_3 the anhydride for HNO_3 or HNO_2? The answer is HNO_2, since both compounds (HNO_2 and N_2O_3) have N in a $+3$ oxidation state.

The action of oxides as acids in water is very well illustrated by some current problems with the environment and the burning of coal. When high-sulfur coal is burned in industrial plants, a significant amount of SO_2 is released into the atmosphere if it is not first removed (see Example 9-9). By itself SO_2 is an irritant and generally unpleasant, but when it dissolves in rain water it forms sulfurous acid (H_2SO_3). Thus the rain can have an appreciable H_3O^+ concentration.* In the Northeast region of this country this is cause for concern, since acidic rain leaches soil, dissolves limestone ($CaCO_3$) on buildings, and kills fish in lakes. In southern Scandinavia, for example, the fish cannot live in many lakes because of the high acidity. This area is downwind from northern Europe, in which a large amount of industry is concentrated. Carbon dioxide, which is also present in the atmosphere, makes the rain acidic, but H_2CO_3 is a much weaker acid than H_2SO_3 and does not cause such dramatic effects. (See Figure 12-6.)

* On April 10, 1974, a rain fell on Pilochry, Scotland, which had the same acidity as vinegar. This was the most acidic rain recorded. Nitrogen oxides formed in automobile engines also contribute to acid rain by forming nitrous and nitric acids.

Figure 12-6 THE EFFECT OF ACID RAIN. Acid rain caused by air pollution is responsible for the deterioration of this ancient statue.

Another problem is that when there is SO_2 present in the atmosphere, there is also some SO_3 present. This latter oxide forms H_2SO_4 when hydrated, and you'll remember that this is one of the strong acids. It doesn't take much sulfuric acid in the atmosphere to ruin paint jobs on cars and cause synthetic fibers to disintegrate. (When this happens, people in the area of the offending industrial plant get quite annoyed, to say the least.)

Ionic metal oxides dissolve in water to form bases and thus are known as **base anhydrides.** Some examples of these reactions are

$$Na_2O(s) + H_2O \longrightarrow 2NaOH(aq)$$

$$CaO(s) + H_2O \longrightarrow Ca(OH)_2(aq)$$

$$Bi_2O_3(s) + 3H_2O \longrightarrow 2Bi(OH)_3(aq)$$

Salt formation occurs in a reaction between an acid anhydride and a base anhydride. For example, the following reaction occurs, forming the same salt as is formed in the neutralization of H_2SO_3 with $Ca(OH)_2$ in aqueous solution.

$$SO_2(g) + CaO(s) \longrightarrow CaSO_3(s)$$

$$H_2SO_3(aq) + Ca(OH)_2(aq) \longrightarrow CaSO_3(s) + 2H_2O$$

The former reaction is typical of reactions that are being studied as a possible way to remove SO_2 from the combustion products of an industrial plant so that some of our abundant high-sulfur coal can be used without harming the environment.

See Problems 12-23 through 12-26

Review Questions

1 Give the Arrhenius definition of an acid and a base.
2 Describe how water is involved in the ionization of an acid in water.
3 What is a hydronium ion? How is it the same or different from a H^+ ion in aqueous solution?
4 Define neutralization. How is a salt formed by neutralization?
5 How can an acid salt be classified as an acid? As a salt?
6 Describe the difference between a strong and a weak acid.
7 Name the strong acids.
8 Why is a weak acid classified as a weak electrolyte?

9 Describe the concept of equilibrium as found in a weak acid solution.
10 Give the Brønsted-Lowry definition of acids and bases.
11 What is meant by conjugate acid–base pairs?
12 Describe how a substance can be amphoteric. When does H_2O act as a Brønsted-Lowry base? A Brønsted-Lowry acid?
13 Discuss the relationship between the strength of an acid and its conjugate base.
14 What is an acid anhydride? What oxides generally form acid anhydrides?
15 What is a base anhydride? What oxides generally form base anhydrides?

Problems

Acids, Bases, and Salts

12-1 Give the formula and the name of the neutral acids formed from each of the following anions:
(a) NO_3^- (b) NO_2^- (c) ClO_3^- (d) SO_3^{2-}
■ (e) AsO_4^{3-} ■ (f) S^{2-} ■ (g) BrO_4^-

12-2 Give the formula and the name of the neutral bases formed from each of the following cations:
(a) Cs^+ (b) Sr^{2+} (c) Al^{3+} (d) Mn^{3+}
■ (e) Li^+ ■ (f) Fe^{2+}

12-3 Identify each of the following as an acid, base, salt, or acid salt:
(a) H_2S (e) $Ba(HSO_4)_2$
(b) $BaCl_2$ ■ (f) K_2SO_4
(c) H_3AsO_4 ■ (g) $LiOH$
(d) $Ni(OH)_2$ ■ (h) $LiHCO_3$

Neutralization

12-4 Write the balanced equation showing the total neutralization of:
(a) KOH by $HC_2H_3O_2$
(b) $Ca(OH)_2$ by HI
■ (c) H_2SO_4 by $Ca(OH)_2$

12-5 Write the total ionic equations and the net ionic equations for reactions (b) and (c) in Problem 12-4.

12-6 Write balanced acid–base neutralization reactions that would lead to formation of the following salts or acid salts:
(a) $CaBr_2$ ■ (b) $Sr(ClO_2)_2$ ■ (c) $Ba(HS)_2$
(d) $Mg_3(PO_4)_2$

■ **12-7** Write three equations illustrating the stepwise neutralization of H_3AsO_4 with $NaOH$. Write the total reaction.

★**12-8** Write the equation illustrating the reaction between 1 mol of $Ca(OH)_2$ and 2 mol of H_3PO_4.

Strengths of Acids and Bases

12-9 What is $[H_3O^+]$ in a 0.55 M $HClO_4$ solution?

12-10 What is $[H_3O^+]$ in a 0.55 M solution of a weak acid, HX, that is 3.0% ionized?

■ **12-11** What is $[OH^-]$ in a 1.45 M solution of NH_3 if the NH_3 is 0.95% ionized?

★**12-12** What is $[H_3O^+]$ in a 0.354 M solution of H_2SO_4? Assume that the first ionization is complete but that the second is only 25% complete.

Brønsted-Lowry Acids and Bases

12-13 What is the conjugate base of each of the following?
(a) HNO_3 ■ (d) CH_4
(b) H_2SO_4 ■ (e) H_2O
(c) HPO_4^{2-} ■ (f) NH_3

12-14 What is the conjugate acid of each of the following?
(a) NH_2CH_3 ■ (d) O^{2-}
(b) HPO_4^{2-} ■ (e) HCN
(c) NO_3^- ■ (f) H_2O

12-15 Write reactions indicating Brønsted-Lowry acid behavior with H_2O for each of the following. Indicate conjugate acid–base pairs.
(a) H_2SO_3 (b) $HClO$ (c) HBr (d) HSO_3^-
■ (e) H_2S ■ (f) H_2 ■ (g) NH_4^+

12-16 Write reactions indicating Brønsted-Lowry base behavior with H_2O for each of the following. Indicate conjugate acid–base pairs.
(a) NH_3 (b) N_2H_4 (c) HS^-
■ (d) H^- ■ (e) F^-

12-17 Referring to Table 12-1, complete the following equations. When no reaction occurs, write N.R. When an equilibrium exists, indicate it by double arrows.
(a) $HNO_3 + H_2O$ (d) $HOCl + H_2O$
(b) $NO_2^- + H_3O^+$■ (e) $NO_3^- + H_3O^+$
(c) $CH_4 + H_2O$ ■ (f) $ClO_4^- + H_2O$

12-18 Most acid salts are amphoteric, as illustrated in Example 12-7. Write equations showing how HS^- can act as a Brønsted-Lowry base with H_3O^+ and as a Brønsted-Lowry acid with OH^-.

★**12-19** Both C_2^{2-} and its conjugate base HC_2^- are very strong bases in H_2O. In fact, both act as Brønsted-Lowry bases with H_2O as well as with H_3O^+. From this information, complete the following equation:

$$CaC_2(s) + 2H_2O \longrightarrow \underline{\hspace{2cm}}(g)$$
$$+ Ca^{2+}(aq) + 2 \underline{\hspace{1.5cm}}(aq)$$

[The gas formed (acetylene) can be burned as it is produced. This reaction was once important for this purpose as a source of light in old miners' lamps.]

■ **12-20** Sodium hypochlorite ($NaOCl$) is the principle ingredient in commercial bleaches. When dissolved in water it forms a slightly basic solution. Complete the following equation illustrating this behavior.

$$NaOCl(s) + H_2O \rightleftharpoons Na^+(aq)$$
$$+ \underline{\hspace{2cm}} + \underline{\hspace{2cm}}$$

■ **12-21** Aqueous solutions of NaF are slightly basic, but solutions of $NaCl$ are not. Show the equation illustrating how the F^- ion reacts with water to produce OH^- ions.

12-22 Calcium carbonate is the principal ingredient in Rolaids, Tums, and other products used to neutralize the H_3O^+ from excess stomach acid (HCl). Show an equation illustrating how the CO_3^{2-} ion acts as a Brønsted-Lowry base with H_3O^+. Bicarbonate of soda ($NaHCO_3$) also acts as an antacid (base) in water. Show an equation illustrating how the HCO_3^- ion reacts with H_3O^+ to produce $CO_2(g)$.

Oxides as Acids and Bases

12-23 Write the acid or base formed when each of the following anhydrides is dissolved in water:
(a) SrO (b) SeO_3 (c) P_4O_{10} (d) Cs_2O
(e) N_2O_3 (f) Cl_2O_5
■ (g) Bi_2O_3 ■ (h) Br_2O ■ (i) CoO

12-24 Carbon dioxide is removed from manned space capsules by bubbling the air through a LiOH solution. Show the reaction and the product formed.

12-25 Complete the following reaction:

$$Li_2O(s) + N_2O_5(g) \longrightarrow \underline{\hspace{2cm}}(s)$$

■ **12-26** The sulfite ion (SO_3^{2-}) and sulfur trioxide (SO_3) look similar, but one forms a strongly acid solution whereas the other is weakly basic. Write equations illustrating this behavior.

13 Oxidation–Reduction Reactions

The lead-acid battery in your car stores chemical energy that can be converted into electrical energy to start the engine.

Turn the keys in your car and a miracle of chemistry occurs. A chemical reaction begins in the car battery which generates a flow of electrical current which is used to start the engine. After the engine is started, a device called an alternator reverses the flow of current and forces electricity back through the battery, which regenerates the original compounds. Electricity is a form of energy involving a flow of electrons. Obviously, the chemical reaction in the battery involves the conversion of chemical energy directly into electrical energy. This reaction is far from unique, however. Reactions such as combustion, decay, and metabolism also involve a flow or exchange of electrons. This important class of chemical reactions is the subject of this chapter.

As background, review the following:

1 The formation of ions and binary compounds (Sections 6-3 and 6-4)

2 Oxidation states and their calculation (Section 7-1)

3 The names and formulas of ions (Section 7-3)

4 Ionic equations (Section 11-3)

13-1 The Nature of Oxidation and Reduction

Sodium metal burns in the presence of chlorine gas to form table salt, NaCl (see Figure 13-1). This is represented by the equation

$$2Na(s) + Cl_2(g) \longrightarrow 2Na^+Cl^-(s)$$

The active metal and poisonous gas combine to form a compound used for our food.

Let's take a close look at each element involved in this reaction to identify the changes that have occurred. First, the Na has lost an electron:

$$Na \longrightarrow Na^+ + e^-$$

Notice that the oxidation state of the Na has changed from zero for the element to $+1$ for the ion. *A substance whose oxidation state has increased (by losing electrons) in a chemical reaction is said to be* **oxidized.**

The Cl_2 has undergone the following change:

$$2e^- + Cl_2 \longrightarrow 2Cl^-$$

Notice that the oxidation state of each Cl in Cl_2 has decreased from zero to -1. *A substance whose oxidation state has decreased (by gaining electrons) in a chemical reaction is said to be* **reduced.**

Reactions involving an exchange of electrons are known as **oxidation–reduction** *or simply* **redox reactions.**

The substance that accepts the electrons from the substance oxidized is called the **oxidizing agent.** Notice that this is the same as the substance re-

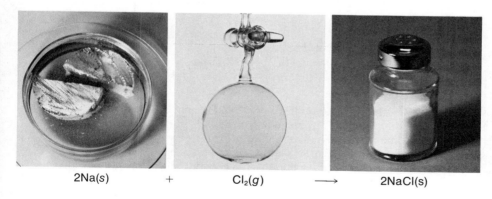

2Na(s)   +   Cl₂(g)   ⟶   2NaCl(s)

Figure 13-1 THE FORMATION OF NaCl FROM ITS ELEMENTS. An active metal reacts with a poisonous gas to form sodium chloride.

duced. *The substance that gives up the electrons to the substance reduced is called the* **reducing agent.** The reducing agent is thus the substance oxidized.

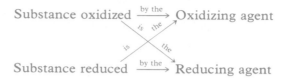

The reaction of Na and Cl_2 is summarized as follows:

Reactant	Change	Product	Agent
Na	Oxidation	Na^+	Reducing
Cl_2	Reduction	Cl^-	Oxidizing

Notice that oxidation–reduction reactions are the sum of two half-reactions. *A* **half-reaction** *illustrates either the oxidation or the reduction reaction separately.* Two half-reactions add to make a total reaction in such a way that *electrons lost* in the oxidation half-reaction equal the *electrons gained* in the reduction half-reaction. Therefore, 2 mol of Na react to produce 2 mol of electrons needed for the 1 mol of Cl_2 as follows:

Oxidation half-reaction	$2Na \longrightarrow 2Na^+ + 2e^-$
Reduction half-reaction	$2e^- + Cl_2 \longrightarrow 2Cl^-$
Total reaction	$2Na + Cl_2 \longrightarrow 2Na^+Cl^-$

We will have more to say about half-reactions and the balancing of equations later. First, let's concentrate on the identification of substances oxidized and reduced and of oxidizing and reducing agents.

Example 13-1

In the following unbalanced equations, indicate the substance oxidized, the substance reduced, the oxidizing agent, and the reducing agent. Indicate the products that contain the elements that were oxidized or reduced.

(a) $Al + HCl \rightarrow AlCl_3 + H_2$
(b) $CH_4 + O_2 \rightarrow CO_2 + H_2O$
(c) $MnO_2 + HCl \rightarrow MnCl_2 + Cl_2 + H_2O$
(d) $K_2Cr_2O_7 + SnCl_2 + HCl \rightarrow CrCl_3 + SnCl_4 + KCl + H_2O$

Procedure:

It is essential that you thoroughly review and memorize the rules for calculating oxidation states as discussed in Section 7-1. In the equations, we wish to identify the element, ion, or molecule that contains an element undergoing a change in oxidation state. At first, it may be necessary to calculate the oxidation state of every element in the equation until you can recognize the changes by inspection.

Solution:

(a) Ox. state of element

$$O \quad +1 \; -1 \qquad +3 \; -1 \qquad 0$$
$$Al + HCl \longrightarrow AlCl_3 \; + H_2$$

Reactant	Change	Product	Agent
Al	Oxidation	$AlCl_3$	Reducing
HCl	Reduction	H_2	Oxidizing

(b) Ox. state of element

$$-4 \; +1 \qquad 0 \qquad +4 \; -2 \qquad +1 \; -2$$
$$CH_4 \; + O_2 \longrightarrow CO_2 \; + H_2O$$

Reactant	Change	Product	Agent
CH_4	Oxidation	CO_2	Reducing
O_2	Reduction	CO_2, H_2O	Oxidizing

(c) Ox. state of element

$$+4 \; -2 \qquad +1 \; -1 \qquad +2 \; -1 \qquad 0 \qquad +1 \; -2$$
$$MnO_2 \; + HCl \longrightarrow MnCl_2 \; + Cl_2 + H_2O$$

Reactant	Change	Product	Agent
HCl	Oxidation	Cl_2	Reducing
MnO_2	Reduction	$MnCl_2$	Oxidizing

(d) Ox. state of element

$$+1 \; +6 \; -2 \qquad +2 \; -1 \qquad +1 \; -1 \qquad +3 \; -1 \qquad +4 \; -1 \qquad +1 \; -2$$
$$K_2Cr_2O_7 \; + SnCl_2 \; + HCl \longrightarrow CrCl_3 \; + SnCl_4 \; + H_2O$$

Reactant	Change	Product	Agent
$SnCl_2$	Oxidation	$SnCl_4$	Reducing
$K_2Cr_2O_7$	Reduction	$CrCl_3$	Oxidizing

See Problems 13-1, 13-2, 13-3

13-2 Balancing Redox Equations: Oxidation State Method

There are two widely used procedures for balancing redox equations. The **bridge** or **oxidation state** *method discussed in this section focuses only on the element undergoing a change in oxidation state.* This method serves as a helpful introduction to the concepts and is effective in balancing uncomplicated redox equations. *The second procedure is the* **ion–electron** *method, which focuses on the entire molecule or ion containing the element undergoing a* *change in oxidation state.* This latter method is generally used in general and subsequent chemistry courses and is discussed in the next section.

The following reaction will be used to illustrate the procedures for balancing equations by the oxidation state method:

$$HNO_3(aq) + H_2S(aq) \longrightarrow NO(g) + S(s) + H_2O$$

1 Identify the elements whose oxidation states have changed.

$$\underset{H\underline{N}O_3}{\overset{+5}{}} + \underset{H_2\underline{S}}{\overset{-2}{}} \longrightarrow \underset{\underline{N}O}{\overset{+2}{}} + \underset{\underline{S}}{\overset{0}{}} + H_2O$$

2 Draw a bridge between the same elements whose oxidation states have changed, indicating the electrons gained or lost. This is the change in oxidation state. *Be sure that the elements in question are balanced on both sides of the equation if they are not the same.*

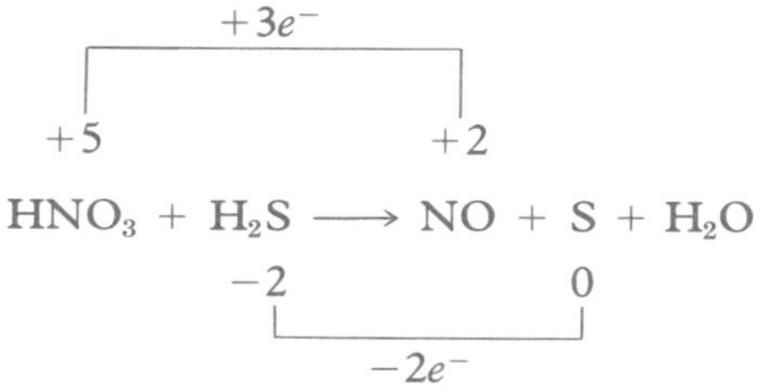

3 Multiply the two numbers ($+3$ and -2) by whole numbers that produce a common number. For 3 and 2 the common number is 6. (e.g., $+3 \times \underline{2} = +6$; $-2 \times \underline{3} = -6$). Use these multipliers as coefficients of the respective compounds or elements.

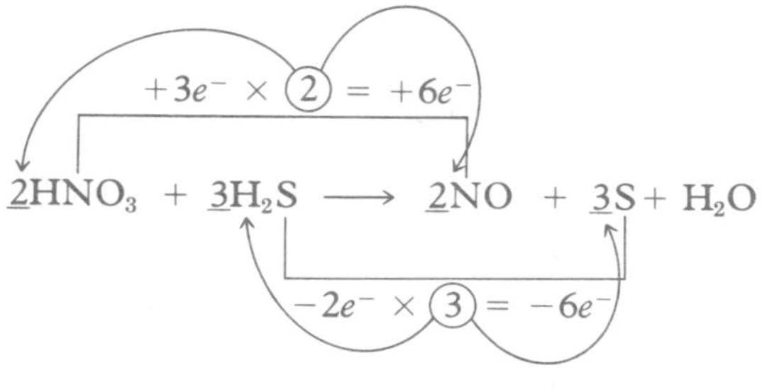

Notice that six electrons are lost (bottom) and six are gained (top).

4 Balance the rest of the equation by inspection. Notice that there are eight H's on the left, so *four* H_2O's are needed on the right. If the equation has been balanced correctly, the O's should balance. Notice that they do.

$$2HNO_3 + 3H_2S \longrightarrow 2NO + 3S + 4H_2O$$

Example 13-2

Balance the following equations by the oxidation state method.

(a) $Zn + AgNO_3 \longrightarrow Zn(NO_3)_2 + Ag$

$$
\overset{\overset{\displaystyle -2e^-}{\overbrace{\hspace{4cm}}}}{\underset{0 \qquad\qquad\qquad +2}{Zn + AgNO_3 \longrightarrow Zn(NO_3)_2 + Ag}}
$$

$$
\underset{\underset{+1e^-}{\underbrace{\hspace{4cm}}}}{\underset{+1 \qquad\qquad\qquad\qquad\qquad 0}{}}
$$

The oxidation process (top) should be multiplied by 1 and the reduction process (bottom) should be multiplied by 2.

$$
\overset{\overset{\displaystyle -2e^- \times 1 = -2e^-}{\overbrace{\hspace{4cm}}}}{Zn + 2AgNO_3 \longrightarrow Zn(NO_3)_2 + 2Ag}
$$

$$
\underset{\underset{+1e \times 2 = +2e}{\underbrace{\hspace{4cm}}}}{}
$$

The final balanced equation is

$$Zn + 2AgNO_3 \longrightarrow Zn(NO_3)_2 + 2Ag$$

(b) $Cu + HNO_3 \longrightarrow Cu(NO_3)_2 + H_2O + NO_2$

$$
\overset{\overset{\displaystyle -2e^- \times 1 = -2e^-}{\overbrace{\hspace{4cm}}}}{\underset{0 \qquad\qquad\qquad +2}{Cu + HNO_3 \longrightarrow Cu(NO_3)_2 + H_2O + NO_2}}
$$

$$
\underset{\underset{+1e^- \times 2 = +2e^-}{\underbrace{\hspace{5cm}}}}{\underset{+5 \qquad\qquad\qquad\qquad\qquad +4}{}}
$$

The equation, so far, appears as follows:

$$Cu + 2HNO_3 \longrightarrow Cu(NO_3)_2 + H_2O + 2NO_2$$

Notice, however, that four N's are present on the right but only two on the left. The addition of two more HNO_3's balances the N's, and the equation is completely balanced with two H_2O's on the right:

$$Cu + 4HNO_3 \longrightarrow Cu(NO_3)_2 + 2H_2O + 2NO_2$$

(In this aqueous reaction, HNO_3 serves two functions. Two HNO_3's are reduced to two NO_2 and the other two HNO_3's provide anions for the Cu^{2+} ion. These later NO_3^- ions are present in the solution as spectator ions. Spectator ions are not oxidized, reduced, or otherwise changed during the reaction.)

(c) $Al + H_2SO_4 \longrightarrow Al_2(SO_4)_3 + H_2$

The elements undergoing a change in oxidation state are Al and H. Before calculating electrons gained or lost, both elements in question must be balanced on both sides of the equation. In this case, two Al's lose a total of six electrons and two H's gain a total of two electrons.

$$2(-3e^-) \times 1 = -6e^-$$

$$\overset{0}{2Al} + H_2SO_4 \longrightarrow \overset{+6}{Al_2(SO_4)_3} + H_2$$

$$\overset{+2}{} \qquad\qquad \overset{0}{}$$

$$2(+1e^-) \times 3 = +6e^-$$

$$2Al + 3H_2SO_4 \longrightarrow Al_2(SO_4)_3 + 3H_2$$

See Problems 13-4, 13-5

13-3 Balancing Redox Equations: The Ion–Electron Method

In the ion–electron method (also known as the half-reaction method), the total reaction is separated into half-reactions which are then balanced separately and added. Although this method is somewhat more involved than the oxidation state method, it is more realistic for redox reactions in aqueous solutions. The ion–electron method recognizes that the entire molecule or ion, not just one element, undergoes a change. This method also provides the proper background for the study of electrochemistry, which involves the applications of balanced half-reactions. This is apparent later in this chapter.

The rules for balancing equations are somewhat different in acidic solution [containing $H^+(aq)$ ion] than in basic solution [containing $OH^-(aq)$ ion]. The two cases are taken up separately, with acid solution reactions discussed first. To simplify the equations, only the net ionic equations are balanced.

The balancing of an equation in aqueous acid solution is illustrated by the following unbalanced equation:

$$Cr_2O_7^{2-}(aq) + Cl^-(aq) + H^+(aq) \longrightarrow Cr^{3+}(aq) + Cl_2(g) + H_2O$$

1 Separate the molecule or ion that contains an element that has been oxidized or reduced and the product containing the element that changed. If necessary, calculate the oxidation states of individual elements until you are able to recognize the species that changed. *It is actually not necessary to know the oxidation state.* The reduction process is

$$Cr_2O_7^{2-} \longrightarrow Cr^{3+}$$

2 If necessary, balance the element undergoing a change in oxidation state. In this case it is Cr.

$$Cr_2O_7^{2-} \longrightarrow 2Cr^{3+}$$

3 Balance the oxygens by adding H_2O on the side needing the oxygens (one H_2O for each O needed).

$$Cr_2O_7^{2-} \longrightarrow 2Cr^{3+} + 7H_2O$$

4 Balance the hydrogens by adding H^+ on the other side of the reaction from the H_2O's ($2H^+$ for each H_2O added). Notice that the H and O have not undergone a change in oxidation state.

$$14H^+ + Cr_2O_7^{2-} \longrightarrow 2Cr^{3+} + 7H_2O$$

5 The elements in the half-reaction are now balanced. Check to make sure. The charge on both sides of the half-reaction must now be balanced. To do this, add the appropriate number of electrons to the *more positive* side. The total charge on the left is $(14 \times +1) + (-2) = +12$. The total charge on the right is $(2 \times +3) = +6$. By adding $6e^-$ on the left, the charges balance on both sides and the half-reaction is balanced:

$$6e^- + 14H^+ + Cr_2O_7^{2-} \longrightarrow 2Cr^{3+} + 7H_2O$$

6 Repeat the same procedure for the other half-reaction:

$$Cl^- \longrightarrow Cl_2$$

$$2Cl^- \longrightarrow Cl_2$$

$$2Cl^- \longrightarrow Cl_2 + 2e^-$$

7 The two half-reactions are added so that the electrons cancel (electrons gained equal electrons lost). Notice the oxidation process is multiplied through by 3, since $6e^-$ are needed for the reduction process.

$$3[2Cl^- \longrightarrow Cl_2 + 2e^-]$$

$$6Cl^- \longrightarrow 3Cl_2 + 6e^-$$

Addition produces the balanced net ionic equation:

$$\cancel{6e^-} + 14H^+ + Cr_2O_7^{2-} \longrightarrow 2Cr^{3+} + 7H_2O$$
$$6Cl^- \longrightarrow 3Cl_2 + \cancel{6e^-}$$
$$\overline{14H^+(aq) + 6Cl^-(aq) + Cr_2O_7^{2-}(aq) \longrightarrow 2Cr^{3+}(aq) + 3Cl_2(g) + 7H_2O}$$

Example 13-3
Balance the following reactions occurring in acid solution by the ion–electron method.

(a) $MnO_4^-(aq) + SO_2(g) + H_2O \rightarrow Mn^{2+}(aq) + SO_4^{2-}(aq) + H^+(aq)$

Reduction: $MnO_4^- \longrightarrow Mn^{2+}$

H_2O: $MnO_4^- \longrightarrow Mn^{2+} + 4H_2O$

H^+: $\qquad\qquad\qquad\qquad 8H^+ + MnO_4^- \longrightarrow Mn^{2+} + 4H_2O$

e^-: $\qquad\qquad\qquad 5e^- + 8H^+ + MnO_4^- \longrightarrow Mn^{2+} + 4H_2O$

Oxidation: $\qquad\qquad\qquad\qquad\qquad SO_2 \longrightarrow SO_4^{2-}$

H_2O: $\qquad\qquad\qquad\qquad 2H_2O + SO_2 \longrightarrow SO_4^{2-}$

H^+: $\qquad\qquad\qquad\qquad 2H_2O + SO_2 \longrightarrow SO_4^{2-} + 4H^+$

e^-: $\qquad\qquad\qquad\qquad 2H_2O + SO_2 \longrightarrow SO_4^{2-} + 4H^+ + 2e^-$

The reduction reaction is multiplied by 2 and the oxidation by 5 to produce 10 electrons for each process as shown below.

$$2(5e^- + 8H^+ + MnO_4^- \longrightarrow Mn^{2+} + 4H_2O)$$
$$5(2H_2O + SO_2 \longrightarrow SO_4^{2-} + 4H^+ + 2e^-)$$
$$\cancel{10e^-} + 16H^+ + 2MnO_4^- \longrightarrow 2Mn^{2+} + 8H_2O$$
$$\underline{10H_2O + 5SO_2 \longrightarrow 5SO_4^{2-} + 20H^+ + \cancel{10e^-}}$$
$$10H_2O + 16H^+ + 5SO_2 + 2MnO_4^- \longrightarrow$$
$$5SO_4^{2-} + 2Mn^{2+} + 8H_2O + 20H^+$$

Notice that H_2O and H^+ are present on both sides of the equation. Therefore, $8H_2O$ and $16H^+$ can be subtracted from *both sides,* leaving the final balanced net ionic equation:

$$2MnO_4^-(aq) + 5SO_2(g) + 2H_2O \longrightarrow$$
$$2Mn^{2+}(aq) + 5SO_4^{2-}(aq) + 4H^+(aq)$$

(b) $Cu(s) + NO_3^-(aq) \longrightarrow Cu^{2+}(aq) + H_2O + NO(g)$

Reduction: $\qquad\qquad\qquad\qquad NO_3^- \longrightarrow NO$

H_2O: $\qquad\qquad\qquad\qquad NO_3^- \longrightarrow NO + 2H_2O$

H^+: $\qquad\qquad\qquad 4H^+ + NO_3^- \longrightarrow NO + 2H_2O$

e^-: $\qquad\qquad 3e^- + 4H^+ + NO_3^- \longrightarrow NO + 2H_2O$

Oxidation: $\qquad\qquad\qquad\qquad Cu \longrightarrow Cu^{2+}$

e^-: $\qquad\qquad\qquad\qquad Cu \longrightarrow Cu^{2+} + 2e^-$

Multiply the reduction half-reaction by 2 and the oxidation half-reaction by 3 and then add the two half-reactions:

See Problems 13-6, 13-7, 13-8

$$\cancel{6e^-} + 8H^+ + 2NO_3^- \longrightarrow 2NO + 4H_2O$$
$$\underline{3Cu \longrightarrow 3Cu^{2+} + \cancel{6e^-}}$$
$$8H^+(aq) + 2NO_3^-(aq) + 3Cu(s) \longrightarrow 3Cu^{2+}(aq) + 2NO(g) + 4H_2O$$

In a basic solution, OH^- ion is present rather than H^+. Therefore, the procedure is adjusted to allow the half-reactions to be balanced with OH^- ions and H_2O molecules in the basic solution.

The balancing of an equation in aqueous base solution is illustrated by the following unbalanced equation:

$$MnO_4^-(aq) + C_2O_4^{2-}(aq) + H_2O \longrightarrow MnO_2(s) + CO_2(g) + OH^-(aq)$$

1 Separate the molecule or ion that contains an element that has been oxidized or reduced and the product containing that element. The reduction process is:

$$MnO_4^- \longrightarrow MnO_2$$

2 For every oxygen needed on the oxygen-deficient side, add *two* OH^- ions. This provides one O and one H_2O:

$$O\boxed{H + OH} = O + H_2O$$
$$MnO_4^- \longrightarrow MnO_2 + 4OH^-$$

3 For every *two* OH^- ions added on the one side, add *one* H_2O to the other side:

$$2H_2O + MnO_4^- \longrightarrow MnO_2 + 4OH^-$$

4 Balance the charge by adding electrons to the more positive side as before:

$$3e^- + 2H_2O + MnO_4^- \longrightarrow MnO_2 + 4OH^-$$

5 Repeat the same procedure for the other half-reaction:

$$C_2O_4^{2-} \longrightarrow CO_2$$
$$C_2O_4^{2-} \longrightarrow 2CO_2$$
$$C_2O_4^{2-} \longrightarrow 2CO_2 + 2e^-$$

6 Multiply the half-reactions so that electrons gained equal electrons lost. In this example the oxidation half-reaction is multiplied by 3 and the reduction half-reaction is multiplied by 2:

$$\cancel{6e^-} + 4H_2O + 2MnO_4^- \longrightarrow 2MnO_2 + 8OH^-$$
$$\underline{3C_2O_4^{2-} \longrightarrow 6CO_2 + \cancel{6e^-}}$$
$$2MnO_4^-(aq) + 3C_2O_4^{2-}(aq) + 4H_2O \longrightarrow$$
$$2MnO_2(s) + 6CO_2(g) + 8OH^-(aq)$$

Example 13-4

Balance the following equation in basic solution by the ion–electron method:

$$Bi_2O_3(s) + OCl^-(aq) + OH^-(aq) \longrightarrow BiO_3^-(aq) + Cl^-(aq) + H_2O$$

Reduction: $\qquad\qquad OCl^- \longrightarrow Cl^-$

OH^-: $\qquad\qquad\qquad OCl^- \longrightarrow Cl^- + 2OH^-$

H_2O: $\qquad\qquad\quad H_2O + OCl^- \longrightarrow Cl^- + 2OH^-$

e^-: $2e^- + H_2O + OCl^- \longrightarrow Cl^- + 2OH^-$

Oxidation: $Bi_2O_3 \longrightarrow BiO_3^-$

Bi: $Bi_2O_3 \longrightarrow 2BiO_3^-$

OH^-: $6OH^- + Bi_2O_3 \longrightarrow 2BiO_3^-$

H_2O: $6OH^- + Bi_2O_3 \longrightarrow 2BiO_3^- + 3H_2O$

e^-: $6OH^- + Bi_2O_3 \longrightarrow 2BiO_3^- + 3H_2O + 4e^-$

Multiply the reduction half-reaction by 2 and add to the oxidation half-reaction:

$$\begin{array}{c} \cancel{4e^-} + 2H_2O + 2OCl^- \longrightarrow 2Cl^- + 4OH^- \\ 6OH^- + Bi_2O_3 \longrightarrow 2BiO_3^- + 3H_2O + \cancel{4e^-} \\ \hline 2H_2O + 6OH^- + Bi_2O_3 + 2OCl^- \longrightarrow 2Cl^- + 4OH^- + 2BiO_3^- + 3H_2O \end{array}$$

By eliminating H_2O and OH^- duplications, we have the balanced net ionic equation:

$$Bi_2O_3(s) + 2OCl^-(aq) + 2OH^-(aq) \longrightarrow 2BiO_3^-(aq) + 2Cl^-(aq) + H_2O$$

See Problems 13-9, 13-10

13-4 Spontaneous Redox Reactions

Scientists are ambitious people. They are rarely satisfied with simply *observing* phenomena, since they also try to *explain* what is happening. That's not all. The real test of their explanation or theory is to use it to *predict* the results of other experiments relating to the observed phenomena. This sequence is known as the **scientific method.** We will apply the scientific method to some simple redox reactions that were also classified as single replacement reactions in Chapter 9. By observing the results of some experiments, we can propose a generalization or theory and from this predict the results of other experiments.

When a strip of nickel is placed in an aqueous $ZnCl_2$ solution, no reaction is apparent (experiment 1 in Figure 13-2). Conversely, if a strip of zinc is

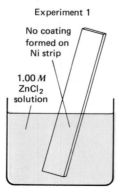

Experiment 1

No coating formed on Ni strip

1.00 M ZnCl₂ solution

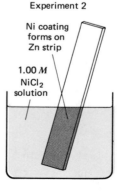

Experiment 2

Ni coating forms on Zn strip

1.00 M NiCl₂ solution

Figure 13-2 THE REACTION OF Zn AND Ni^{2+}. Zinc replaces the Ni^{2+} in a $NiCl_2$ solution.

placed in an aqueous $NiCl_2$ solution, a reaction takes place as shown by the dull coating of Ni formed on the Zn strip (experiment 2).

These two experiments have proved that the following reaction takes place only in the direction written:

$$Zn(s) + NiCl_2(aq) \longrightarrow ZnCl_2(aq) + Ni(s)$$

Just as an acid–base reaction can be considered a competition for a H^+ (by the Brønsted-Lowry definition), an oxidation–reduction reaction can be considered a competition for electrons. In this case we have shown that Ni^{2+} has a stronger tendency to pick up electrons (to act as an oxidizing agent or be reduced) to form Ni than does Zn^{2+} to form Zn.

$$Zn^{2+} + 2e^- \longrightarrow Zn$$

occurs less readily than

$$Ni^{2+} + 2e^- \longrightarrow Ni$$

(In this discussion, we are focusing on the comparative tendency of an ion to be reduced. Shortly, we will show that there is an inverse tendency of the corresponding metal to be oxidized.)

Let's expand our investigation of the tendency of some other ions to be reduced. In experiment 3 (see Figure 13-3), a strip of nickel is immersed in a $CuCl_2$ solution. A reaction takes place as evidenced by the brown coating of Cu on the Ni. As expected, no reaction occurs when a strip of Cu is immersed in a $NiCl_2$ solution (experiment 4, Figure 13-3).

We have now shown that the following reaction also takes place:

$$Ni(s) + CuCl_2(aq) \longrightarrow NiCl_2(aq) + Cu(s)$$

The strength of the two ions as oxidizing agents can be compared as before.

$$Ni^{2+} + 2e^- \longrightarrow Ni$$

occurs less readily than

$$Cu^{2+} + 2e^- \longrightarrow Cu$$

Experiment 3

Cu coating forms on Ni strip

1.00 M $CuCl_2$ solution

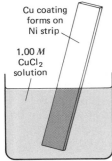

Experiment 4

No coating formed on Cu strip

1.00 M $NiCl_2$ solution

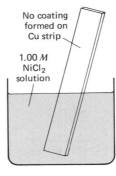

Figure 13-3 THE REACTION OF Ni AND Cu^{2+}. Nickel replaces the Cu^{2+} in a $CuCl_2$ solution.

The results of all four experiments can now be put together, listing the metal ions in order of increasing strength as oxidizing agents.

$$Zn^{2+} + 2e^- \longrightarrow Zn$$

occurs less readily than

$$Ni^{2+} + 2e^- \longrightarrow Ni$$

which occurs less readily than

$$Cu^{2+} + 2e^- \longrightarrow Cu$$

Thus Cu^{2+} is the strongest oxidizing agent of the three ions.

When an ion that is a *strong* oxidizing agent (e.g., Cu^{2+}) is reduced, the metal that is formed (e.g., Cu) is a *weak* reducing agent. This phenomenon is analogous to the acid–conjugate base relationship discussed in Section 12-6. There we found that a strong acid ionizes to produce a weak conjugate base and vice versa. Here we find that a strong oxidizing agent reacts to produce a weak reducing agent and vice versa. This inverse relationship is illustrated by the "seesaw" in Figure 13-4.

Just as the ions were arranged in order of increasing strength as oxidizing agents, the metals can also be arranged in order of increasing strength as reducing agents.

$$Cu \longrightarrow Cu^{2+} + 2e^-$$

occurs less readily than

$$Ni \longrightarrow Ni^{2+} + 2e^-$$

which occurs less readily than

$$Zn \longrightarrow Zn^{2+} + 2e^-$$

Thus Zn is the strongest reducing agent of the three metals.

We are now in a position to summarize all of this information into one table of half-reactions. In Table 13-1, the ions on the left are listed in order of *increasing* strength as oxidizing agents as you go down. This reaction is indicated by the reaction arrow to the right ($\rightarrow$). The metals on the right are thus

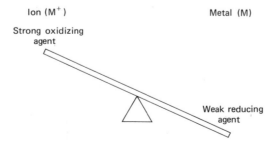

Ion (M^+)   Metal (M)

Strong oxidizing agent

Weak reducing agent

Figure 13-4 AN ION AND ITS METAL. There is an inverse relationship between the oxidizing strength of an ion and the reducing strength of the metal.

Table 13-1   Zn, Ni, and Cu

Weakest oxidizing agent	$Zn^{2+} + 2e^- \rightleftharpoons Zn$	Strongest reducing agent
↓	$Ni^{2+} + 2e^- \rightleftharpoons Ni$	↑
Strongest oxidizing agent	$Cu^{2+} + 3e^- \rightleftharpoons Cu$	Weakest reducing agent

listed in order of *decreasing* strength as reducing agents as you go down. This reaction is indicated by the reaction arrow to the left ($\leftarrow$).

The results of these experiments suggest a generalization (or theory). *A* **spontaneous redox reaction** *occurs whereby the stronger oxidizing agent reacts with the stronger reducing agent to produce a weaker oxidizing agent and a weaker reducing agent. A* **spontaneous reaction** *is one that occurs naturally without external help or stimulus.*

We can now use the generalization and Table 13-1 to predict the results of an experiment before it is run. Which of the following two reactions occurs spontaneously?

(a)
$$Cu(s) + ZnCl_2(aq) \longrightarrow CuCl_2(aq) + Zn(s)$$

or the reverse reaction

(b)
$$Zn(s) + CuCl_2(aq) \longrightarrow ZnCl_2(aq) + Cu(s)$$

In Table 13-1, notice that Cu^{2+} is a stronger oxidizing agent than Zn^{2+} and that Zn is a stronger reducing agent than Cu. Since the stronger react, reaction (b) above is predicted to be the spontaneous reaction. In Figure 13-5, the spontaneous reaction is further illustrated. The *oxidizing agent* that appears *lower* in Table 13-1 reacts with a *reducing agent* that appears *higher* in the Table. Notice that the predicted spontaneous reaction can be visu-

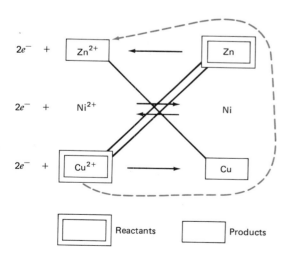

Figure 13-5 Cu AND Zn REACTION. The stronger oxidizing agent reacts with the stronger reducing agent.

Table 13-2   Reducing and Oxidizing Agents

Weakest oxidizing agent	$Na^+ + e^- \rightleftarrows Na$	Strongest reducing agent
	$Al^{3+} + 3e^- \rightleftarrows Al$	
	$Mg^{2+} + 2e^- \rightleftarrows Mg$	
	$2H_2O + 2e^- \rightleftarrows H_2 + 2OH^-$	
	$Zn^{2+} + 2e^- \rightleftarrows Zn$	
	$Fe^{2+} + 2e^- \rightleftarrows Fe$	
Increasing strength of oxidizing agent	$Ni^{2+} + 2e^- \rightleftarrows Ni$	Increasing strength of reducing agent
	$Sn^{2+} + 2e^- \rightleftarrows Sn$	
	$Pb^{2+} + 2e^- \rightleftarrows Pb$	
	$2H^+ + 2e^- \rightleftarrows H_2$	
	$Cu^{2+} + 2e^- \rightleftarrows Cu$	
	$Ag^+ + e^- \rightleftarrows Ag$	
	$Br_2 + 2e^- \rightleftarrows 2Br^-$	
Strongest oxidizing agent	$Cl_2 + 2e^- \rightleftarrows 2Cl^-$	Weakest reducing agent
	$F_2 + 2e^- \rightleftarrows 2F^-$	

alized as occurring in a counterclockwise direction (the dashed line). An experiment quickly proves our prediction. A Zn strip immersed in a $CuCl_2$ solution forms a Cu coating.

We can now expand Table 13-1 by including more ions and some molecular oxidizing agents in their proper positions in the table. This has been accomplished in Table 13-2.

In some cases more elaborate experiments than we have illustrated here are required to locate precisely certain half-reactions in Table 13-2. Positions in the table are subject to certain restrictions on the concentration of ions present and the pressure of any gas present. Specifically, it is assumed that all ions are present at $1.00\ M$ and gas at 1.00 atm pressure.

This table can now be put to effective use in predicting the occurrence of many redox reactions.

In which direction will the following reactions be spontaneous?

Example 13-5

$$Pb^{2+}(aq) + 2Cl^-(aq) \xleftrightarrow{?} Pb(s) + Cl_2(g)$$

In Table 13-2, notice that Cl_2 is a stronger oxidizing agent than Pb^{2+} and that Pb is a stronger reducing agent than Cl^-. Therefore, the reaction is spontaneous to the left. Metallic lead will dissolve in an aqueous solution of Cl_2. (Cl_2 in swimming pools is hard on many metals, as you can see.)

$$Pb(s) + Cl_2(g) \longrightarrow PbCl_2(aq)$$

Example 13-6

$$Mg(s) + 2H_2O \xleftrightarrow{?} Mg^{2+}(aq) + 2OH^-(aq) + H_2(g)$$

H_2O is a stronger oxidizing agent than Mg^{2+}, so the reaction proceeds to the <u>right</u>. All metals above H_2O react with water. These are considered the *active* metals because of their high chemical reactivity.

Example 13-7

$$Ni(s) + 2H^+(aq) \xleftrightarrow{?} Ni^{2+}(aq) + H_2(g)$$

H^+ is a stronger oxidizing agent than Ni^{2+}. This reaction also proceeds to the <u>right</u>. Notice that all metals above H^+ react with aqueous acid. Those <u>between</u> H^+ and H_2O react in acid solution but not in pure H_2O. On the other hand, Cu and Ag do not react with H^+, since Cu^{2+} is a better oxidizing agent than H^+. This is one good reason why copper and silver can be used in coins and jewelry. They are not generally affected by aqueous acid solutions.

See Problems 13-11 through 13-16

13-5 Voltaic Cells

If a car is halfway down the side of a hill and it begins to roll, does it spontaneously roll up or down hill? A silly question perhaps, but the concept is very similar to why chemical reactions proceed in one direction and not the other. The car rolls down the hill because the bottom of the hill represents a lower energy state than the top of the hill. As the car rolls down the hill, it releases the difference in energy between the two states, mostly in the form of heat from friction. Likewise, the redox reactions we have discussed are spontaneous in one direction primarily because the reactants are at a higher energy state than the products (see Figure 13-6). When the reaction proceeds, the difference in energy between the reactants and the products can be released as heat energy (exothermic reaction) or converted directly to electrical energy. *The* **voltaic cell** *(also called the* **galvanic cell***) uses a spontaneous redox reaction to generate electrical energy through an external circuit.* The following spontaneous reaction is a redox reaction that can be used to form a voltaic cell:

$$Zn(s) + CuSO_4(aq) \longrightarrow ZnSO_4(aq) + Cu(s)$$

This is called the **Daniell cell;** it was used to generate electricity for the early telegraph.

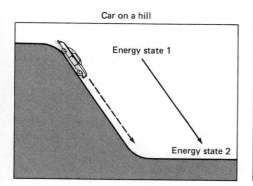

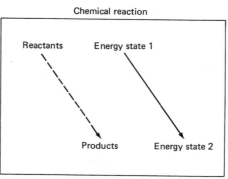

Figure 13-6 ENERGY STATES. A chemical reaction proceeds in a certain direction for the same reason that a car rolls *down* a hill.

In order to detour the electrons through a wire, the two half-reactions must be separated into compartments. This is illustrated in Figure 13-7. *The surfaces in a cell at which the reactions take place are called the* **electrodes.** *The electrode at which oxidation takes place is called the* **anode.** *Reduction takes place at the* **cathode.**

In the compartment on the left, the strip of Zn serves as the anode, since the following reaction occurs when the circuit is connected:

$$Zn \longrightarrow Zn^{2+} + 2e^-$$

When the circuit is complete, the two electrons travel in the external wire to the Cu electrode, which serves as the cathode. The reduction reaction occurs at the cathode:

$$2e^- + Cu^{2+} \longrightarrow Cu$$

To maintain neutrality in the solution, some means must be provided for the movement of a SO_4^{2-} ion (or some other negative ion) from the right compartment where a Cu^{2+} has been removed to the left compartment where a Zn^{2+} has been produced. The *salt bridge* is an aqueous gel that allows ions to migrate between compartments but does not allow the mixing of solutions. (If Cu^{2+} ions wandered into the left compartment, they would form a coating of Cu on the Zn electrode, thus short circuiting the cell.)

As the cell discharges (generates electrical energy), the Zn electrode gets smaller and the Cu electrode gets larger. If the external circuit is open (switch off), the reaction stops. Thus the cell holds its charge until needed and the electrical energy is stored.

Two of the most common voltaic cells in use today are the dry cell (flashlight battery) and the lead–acid cell (car battery). (*The word* **battery** *means a collection of one or more separate cells joined together in one unit.*)

One cell of a lead–acid storage battery is illustrated in Figure 13-8. Each cell is composed of two grids separated by an inert spacer. One grid of a fully charged battery contains metallic lead. The other contains PbO_2, which is

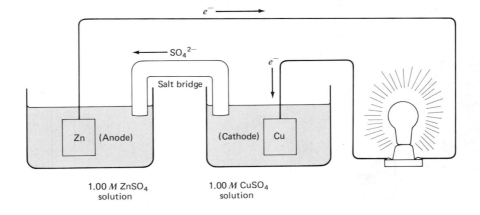

Figure 13-7 THE DANIELL CELL. This chemical reaction produced electricity for the first telegraphs.

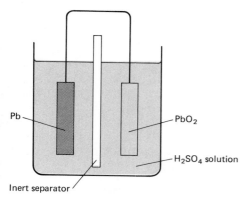

Pb

PbO$_2$

H$_2$SO$_4$ solution

Inert separator

Figure 13-8 THE LEAD STORAGE BATTERY. The lead storage battery produces electricity to start your car.

insoluble in H$_2$O. Both grids are immersed in a sulfuric acid solution (battery acid). When the battery is discharged by connecting the electrodes, the following half-reactions take place spontaneously:

Anode:
$$Pb(s) + H_2SO_4(aq) \longrightarrow$$
$$PbSO_4(s) + 2H^+(aq) + 2e^-$$

Cathode:
$$2e^- + 2H^+(aq) + PbO_2(s) + H_2SO_4(aq) \longrightarrow$$
$$PbSO_4(s) + 2H_2O$$

Total reaction:
$$Pb(s) + PbO_2(s) + 2H_2SO_4(aq) \longrightarrow$$
$$2PbSO_4(s) + 2H_2O$$

The electrons released at the Pb anode travel through the external circuit to run lights, starters, radios, or whatever is needed. The electrons return to the PbO$_2$ cathode to complete the circuit. As the reaction proceeds, both electrodes are converted to PbSO$_4$ and the H$_2$SO$_4$ is depleted. Since PbSO$_4$ is also insoluble, it remains attached to the grids as it forms. The degree of discharge of a battery can be determined by the density of the battery acid. Since the density of a fully discharged battery is 1.05 g/ml, the difference in density between this value and the density of a fully charged battery (1.35 g/ml) gives the amount of charge remaining in the battery. The hydrometer discussed in Chapter 2 is used to determine the density of the acid. As the electrodes convert to PbSO$_4$, the battery loses power and eventually becomes "dead."

The convenience of a car battery is that it can be recharged. After the engine starts, an alternator or generator is engaged to push electrons back into the cell in the direction they came during discharge. This forces the reverse, nonspontaneous reaction to proceed:

$$2PbSO_4(s) + 2H_2O \longrightarrow Pb(s) + PbO_2(s) + 2H_2SO_4(aq)$$

When the battery is fully recharged the alternator shuts off, the circuit is open, and the battery is ready for the next start.

The dry cell (invented by Lelanché in 1866) is not rechargeable to any extent but is comparatively inexpensive and easily portable. (In contrast, the

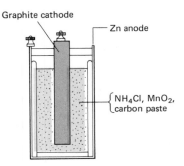

Graphite cathode

Zn anode

NH_4Cl, MnO_2, carbon paste

Figure 13-9 THE DRY CELL. The dry cell is comparatively inexpensive, light, and portable.

lead–acid battery is heavy, expensive, and must be kept upright.) The dry cell illustrated in Figure 13-9 consists of a zinc anode which is the outer rim and an inert graphite electrode. (An inert electrode provides a reaction surface but does not itself react.) In between is an aqueous paste containing NH_4Cl, MnO_2, and carbon. The reactions are as follows:

Anode: $$Zn(s) \longrightarrow Zn^{2+}(aq) + 2e^-$$

Cathode: $$2NH_4^+(aq) + 2MnO_2(s) + 2e^- \longrightarrow$$
$$Mn_2O_3(s) + 2NH_3(aq) + H_2O$$

Many other batteries have their special advantages. Nickel–cadmium batteries are popular as replacements for dry cells. Although initially more expensive, they are rechargeable and so will last longer. For space travel the fuel cell, which converts some of the energy from the reaction between H_2 and O_2 (to form water) directly to electrical energy, is used. Fuel cells are very expensive but generate a continual flow of electricity without recharging. Currently, government and industry are conducting a great amount of research for a replacement for the lead–acid battery for use in electric cars. A small electric car needs at least 18 lead–acid batteries, which have to be completely replaced every few months depending on use. Also, much of the power of the lead–acid batteries must be used just to move the batteries around, let alone the passengers. Some encouraging results have been announced with zinc–nickel oxide batteries that are lighter, last longer, and carry more charge than lead–acid batteries. Production of automobiles using these batteries is planned for the mid-1980s.

See Problems 13-17, 13-18

13-6 Electrolytic Cells

In the previous section we mentioned that a car spontaneously rolls down a hill, releasing energy as it goes. The car can roll up the hill, but obviously energy must be supplied from some source before this can happen. Redox reactions can be forced in the nonspontaneous direction but, like the car on the hill, energy must be supplied. This is what happened in the previous section when the lead–acid battery was recharged with energy supplied from the engine. In this case the electrical energy is converted back into chemical en-

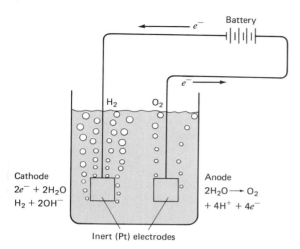

Cathode

$2e^- + 2H_2O$

$H_2 + 2OH^-$

Anode

$2H_2O \longrightarrow O_2$

$+ 4H^+ + 4e^-$

Inert (Pt) electrodes

Figure 13-10 THE ELECTROLYSIS OF WATER. With an input of electrical energy, water can be decomposed to its elements.

ergy. *Cells that convert electrical energy into chemical energy are called* **electrolytic cells.**

An example of an electrolytic cell is shown in Figure 13-10. When sufficient electrical energy is supplied to the electrodes from an outside source, the following nonspontaneous reaction occurs:

$$2H_2O \longrightarrow 2H_2(g) + O_2(g)$$

In order for this electrolysis to occur, an electrolyte such as Na_2SO_4 must be present in solution. Pure water alone does not have a sufficient concentration of ions to allow conduction of electricity.

Electrolysis has many useful applications. For example, silver or gold can be electroplated onto cheaper metals. In Figure 13-11 the metal spoon is the cathode, with the silver bar serving as the anode. When electricity is sup-

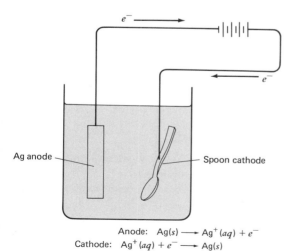

Ag anode

Spoon cathode

Anode: $Ag(s) \longrightarrow Ag^+(aq) + e^-$

Cathode: $Ag^+(aq) + e^- \longrightarrow Ag(s)$

Figure 13-11 ELECTROPLATING. With an input of energy, a spoon can be coated wih silver.

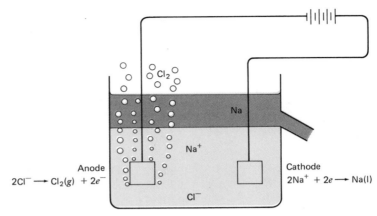

Figure 13-12 THE ELECTROLYSIS O▊ NaCl. Elemental sodium and chlorine are prepared by the electrolytes of ordinary table salt in the molten state.

plied, the Ag anode produces Ag^+ ions and the spoon cathode reduces Ag^+ ions to give a layer of Ag. The silver-plated spoon can be polished and made to look as good as sterling silver.

Electrolytic cells are used commercially to produce many metals such as sodium, magnesium, and aluminum. The electrolysis of molten (melted) salts, where no water is present, serves this purpose. For example, electrolysis of molten NaCl is the prime source of commercial quantities of elemental chlorine and sodium. As shown in Figure 13-12, at the high temperature of the process, the Na forms in the molten state and is subsequently drained from the top of the cell.

See Problems 13-19, 13-20

Review Questions

1   Define oxidation. Define reduction.
2   What is an oxidizing agent? A reducing agent?
3   What occurs in an oxidation–reduction reaction?
4   Discuss the two methods for balancing oxidation–reduction reactions.
5   What is the difference in the procedures for balancing ionic half-reactions in acidic and basic solutions?
6   Ni^{2+} is a stronger oxidizing agent than Zn^{2+}. What does this mean?
7   Discuss the relationship between the strength of an ion as an oxidizing agent and the strength of the corresponding metal as a reducing agent.

8   Describe how Table 13-2 can be used to predict spontaneous oxidation–reduction reactions.
9   How does a voltaic cell apply spontaneous redox reactions to supply energy?
10   What is an electrode? An anode? A cathode?
11   How does the Daniell cell generate electricity?
12   How does the lead–acid storage battery work? The dry cell?
13   What is an electrolytic cell, and how does it differ from a voltaic cell?
14   Describe the process of electroplating.

Problems

Oxidation-Reduction

13-1 Which of the following reactions are oxidation–reduction reactions?
(a) $2H_2 + O_2 \rightarrow 2H_2O$
(b) $CaCO_3 \rightarrow CaO + CO_2$
(c) $2Na + 2H_2O \rightarrow 2NaOH + H_2$
(d) $2HNO_3 + Ca(OH)_2 \rightarrow$
$$Ca(NO_3)_2 + 2H_2O$$
■ (e) $AgNO_3 + KCl \rightarrow AgCl + KNO_3$
■ (f) $Zn + CuCl_2 \rightarrow ZnCl_2 + Cu$

13-2 Identify each of the following half-reactions as either oxidation or reduction:
(a) $Na \rightarrow Na^+ + e^-$
(b) $Zn^{2+} + 2e^- \rightarrow Zn$
■ (c) $Fe^{2+} \rightarrow Fe^{3+} + e^-$
■ (d) $O_2 + 4H^+ + 4e^- \rightarrow 2H_2O$
(e) $S_2O_8^{2-} + 2e^- \rightarrow 2SO_4^{2-}$

13-3 For each of the following unbalanced equations, complete the table below.
(a) $MnO_2 + H^+ + Br^- \rightarrow$
$$Mn^{2+} + Br_2 + H_2O$$
(b) $CH_4 + O_2 \rightarrow CO_2 + H_2O$
(c) $Fe^{2+} + MnO_4^- + H^+ \rightarrow$
$$Fe^{3+} + Mn^{2+} + H_2O$$
■ (d) $Al + H_2O \rightarrow AlO_2^- + H_2 + H^+$
■ (e) $Mn^{2+} + Cr_2O_7^{2-} + H^+ \rightarrow$
$$MnO_4^- + Cr^{3+} + H_2O$$

Balancing Equations by the Oxidation State Method

13-4 Balance each of the following equations by the oxidation state method:
(a) $NH_3 + O_2 \rightarrow NO + H_2O$
(b) $Sn + HNO_3 \rightarrow SnO_2 + NO_2 + H_2O$
■ (c) $Cr_2O_3 + Na_2CO_3 + KNO_3 \rightarrow$
$$CO_2 + Na_2CrO_4 + KNO_2$$
■ (d) $Se + BrO_3^- + H_2O \rightarrow H_2SeO_3 + Br^-$

13-5 Balance the equations in Problem 13-3 by the oxidation state method.

Balancing Equations by the Ion-Electron Method

13-6 Balance each of the following half-reactions in acid solution:
(a) $Sn^{2+} \rightarrow SnO_2$
(b) $CH_4 \rightarrow CO_2$
(c) $Fe^{3+} \rightarrow Fe^{2+}$
(d) $I_2 \rightarrow IO_3^-$
(e) $NO_3^- \rightarrow NO_2$
■ (f) $P_4 \rightarrow H_3PO_4$
■ (g) $NO_3^- \rightarrow NH_4^+$
■ (h) $Fe_2O_3 \rightarrow Fe^{3+}$

13-7 Balance each of the following by the ion–electron method. All are in acid solution.
(a) $S^{2-} + NO_3^- + H^+ \rightarrow S + NO + H_2O$
(b) $I_2 + S_2O_3^{2-} \rightarrow S_4O_6^{2-} + I^-$
(c) $SO_3^{2-} + ClO_3^- \rightarrow Cl^- + SO_4^{2-}$
■ (d) $Fe^{2+} + H_2O_2 + H^+ \rightarrow Fe^{3+} + H_2O$
■ (e) $AsO_4^{3-} + I^- + H^+ \rightarrow$
$$I_2 + AsO_3^{2-} + H_2O$$

Reaction	Species Oxidized*	Product of Oxidation	Species Reduced	Product of Reduction	Oxidizing Agent	Reducing Agent
(a)	Br^-	Br_2	MnO_2	Mn^{2+}	MnO_2	Br^-
(b)						
(c)						
(d)						
(e)						

* Element, molecule, or ion.

13-8 Balance each of the following by the ion–electron method. All are in acid solution.
(a) $Mn^{2+} + BiO_3^- + H^+ \rightarrow$
$$MnO_4^- + Bi^{3+} + H_2O$$
(b) $IO_3^- + SO_2 + H_2O \rightarrow$
$$I_2 + SO_4^{2-} + H^+$$
■ (c) $Se + BrO_3^- + H_2O \rightarrow H_2SeO_3 + Br^-$
(d) $P_4 + HOCl + H_2O \rightarrow$
$$H_3PO_4 + Cl^- + H^+$$
■ (e) $Al + Cr_2O_7^{2-} + H^+ \rightarrow$
$$Al^{3+} + Cr^{3+} + H_2O$$
(f) $Zn + H^+ + NO_3^- \rightarrow$
$$Zn^{2+} + NH_4^+ + H_2O$$

13-9 Balance each of the following half-reactions in basic solution:
(a) $SnO_2^{2-} \rightarrow SnO_3^{2-}$
(b) $ClO_2^- \rightarrow Cl_2$
(c) $Si \rightarrow SiO_3^{2-}$
(d) $Al \rightarrow Al(OH)_4^-$
★■ (e) $NO_3^- \rightarrow NH_3$
■ (f) $S^{2-} \rightarrow SO_4^{2-}$

13-10 Balance each of the following by the ion–electron method. All are in basic solution.
(a) $S^{2-} + OH^- + I_2 \rightarrow SO_4^{2-} + I^- + H_2O$
(b) $MnO_4^- + OH^- + I^- \rightarrow$
$$MnO_4^{2-} + IO_4^- + H_2O$$
(c) $BiO_3^- + SnO_2^{2-} + H_2O \rightarrow$
$$Bi^{3+} + SnO_3^{2-} + OH^-$$
■ (d) $ClO_2 + OH^- \rightarrow$
$$ClO_2^- + ClO_3^- + H_2O$$
■ (e) $OH^- + Cr_2O_3 + NO_3^- \rightarrow$
$$CrO_4^{2-} + NO_2^- + H_2O$$
★(f) $CrI_3 + OH^- + Cl_2 \rightarrow$
$$CrO_4^{2-} + IO_4^- + Cl^- + H_2O$$
(Hint: two ions are oxidized; include both in one half-reaction.)

Spontaneous Redox Reactions

13-11 Given the following information concerning metal strips immersed in certain solutions:

Metal Strip	Solution	Reaction
Cd	$NiCl_2$	Ni coating formed
Cd	$FeCl_2$	N.R. (no reaction)
Zn	$CdCl_2$	Cd coating formed
Fe	$CdCl_2$	N.R.

Where does Cd^{2+} rank in Table 13-2 as a reducing agent?

13-12 Gold does not dissolve in water or acid. Only F_2 will dissolve gold to form aqueous Au^{3+} solutions. Where does Au^{3+} rank as an oxidizing agent in Table 13-2?

13-13 Using Table 13-2, predict whether the following reactions occur in aqueous solution. If not, write N.R. (no reaction).
(a) $2Na + 2H_2O \rightarrow H_2 + 2NaOH$
(b) $Pb + Zn^{2+} \rightarrow Pb^{2+} + Zn$
(c) $Fe + 2H^+ \rightarrow Fe^{2+} + H_2$
(d) $Fe + 2H_2O \rightarrow Fe^{2+} + 2OH^- + H_2$
(e) $Cu + 2Ag^+ \rightarrow 2Ag + Cu^{2+}$
■ (f) $Sn^{2+} + Pb \rightarrow Pb^{2+} + Sn$
■ (g) $Ni^{2+} + H_2 \rightarrow 2H^+ + Ni$
■ (h) $Cu + F_2 \rightarrow CuF_2$
■ (i) $Ni^{2+} + 2Br^- \rightarrow Ni + Br_2$

13-14 Tell whether a reaction occurs in each of the following cases. Write a balanced reaction indicating the spontaneous reaction if one occurs.
(a) Some iron nails are placed in a $CuCl_2$ solution.
(b) Silver coins are dropped into an acid solution [$H^+(aq)$].
■ (c) A copper penny is placed in a $Pb(NO_3)_2$ solution.
■ (d) Ni metal is placed in a $Pb(NO_3)_2$ solution.
(e) Aluminum metal is placed in water. (Aluminum is actually coated with a protective layer of Al_2O_3, which keeps the metallic Al from coming in contact with water.)
(f) Iron nails are placed in a $ZnBr_2$ solution.
(g) H_2 gas is bubbled into an acid solution of Cu^{2+}.

■ **13-15** Br_2 can be prepared from the reaction of Cl_2 with NaBr dissolved in sea water. Explain. Write the reaction. Can Cl_2 be used to prepare F_2 from NaF solutions?

13-16 In Chapter 12 we mentioned the corrosiveness of acid rain. Why does rain containing a high $H^+(aq)$ concentration cause more damage to iron exposed in bridges and buildings than pure H_2O? Write the reaction between Fe and $H^+(aq)$.

Voltaic Cells

■ **13-17** Describe how a voltaic cell could be constructed from a strip of iron, a strip of lead, a $Fe(NO_3)_2$ solution, and a $Pb(NO_3)_2$ solution. Write the anode reaction, the cathode reaction, and the total reaction.

****13-18** Judging from the relative difference in the strength of the oxidizing agents (Fe^{2+} vs. Pb^{2+}) and (Zn^{2+} vs. Cu^{2+}), which do you think would be the more powerful cell, the one in Problem 13-17 or the Daniell cell? Why?

Electrolytic Cells

■ **13-19** Why can't elemental Na be formed in the electrolysis of aqueous NaCl? Write the reaction that occurs at the cathode.

13-20 Certain metals can be purified by electrolysis. If a mixture of Ag, Zn, and Fe was dissolved in acid to form the metal ions and then the aqueous solution was electrolyzed, which metal ion would be reduced to the metal the easiest?

14 Reaction Kinetics and Equilibrium

Like all chemical reactions the spoiling of food is much slower at low temperatures. Thus we store food for days in a refrigerator that would otherwise spoil in a matter of hours.

All chemical reactions slow as the temperature drops. This seems like a broad statement, but there is evidence of this all around us. For example, we store food in a refrigerator because the spoiling or decay of food is a chemical reaction that occurs very slowly at the temperature in a refrigerator but comparatively quickly at the temperature of a warm room. (In a freezer the rate of spoiling can be stopped almost entirely.) Snakes and turtles are animals that become lethargic on chilly mornings. This is because these reptiles are cold-blooded animals whose body temperatures fluctuate with the temperature of the air. Thus, as the temperature drops, their rate of metabolism slows. Metabolism is also a chemical reaction that generates energy from food. As a result, cold-blooded animals must seek shelter in holes as winter approaches or they won't be able to move at all. Warm-blooded animals such as humans continuously generate enough heat internally to keep the body temperature regulated at around 98°F (37°C). This constant internal temperature allows metabolism to continue at a constant rate even in cold weather.

Before we can appreciate the factors (such as temperature) that affect the rate of a chemical reaction, we will first examine *how* reactant molecules are transformed into product molecules. This leads us to a discussion of reaction rates in terms of chemical equilibrium, which was introduced in Chapter 12. Finally, chemical equilibrium is discussed in quantitative terms.

As background for this chapter, review the nature of gases (the kinetic theory of gases) discussed in Section 10-5.

14-1 The Mechanism of a Reversible Reaction

The path or method whereby reactant molecules transform into product molecules is known as the **mechanism** *of the reaction. The study of reaction rates and their relation to the mechanism of a reaction is known as* **chemical kinetics.** To illustrate the relation of mechanism to reaction rates, the following straightforward example is employed.

At moderately high temperatures (e.g., 400°C), elemental iodine is a gas.

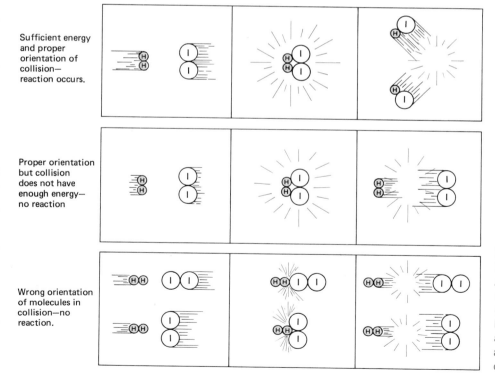

Sufficient energy and proper orientation of collision— reaction occurs.

Proper orientation but collision does not have enough energy— no reaction

Wrong orientation of molecules in collision—no reaction.

Figure 14-1 THE REACTION OF H_2 AND I_2. Reactions occur only when colliding molecules have the minimum amount of energy and the right orientation.

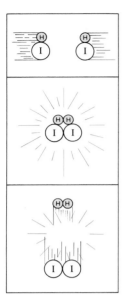

Figure 14-2 THE REACTION OF 2HI. Collisions between HI molecules having the minimum energy and correct orientation lead to the formation of reactants.

At this temperature the following reaction takes place entirely in the gas phase:

$$H_2(g) + I_2(g) \longrightarrow 2HI(g)$$

Recall that one of the assumptions of the kinetic theory of gases is that gas molecules are in constant motion undergoing collisions with each other and the sides of the container. *The **collision theory** of chemical reactions assumes that reactions take place through the collisions of molecules.* If the collision between two reacting molecules takes place (1) in exactly the right orientation and (2) with enough energy, a reshuffling of electrons takes place, with new bonds forming and old bonds breaking. If these conditions are not met, the molecules simply recoil from each other unchanged. This is illustrated in Figure 14-1.

In this reaction, if H_2 and I_2 are mixed at a high temperature, the concentration of HI begins to build as proper collisions lead to products. In many reactions, products continue to form until at least one of the reactants (the limiting reactant) is completely consumed. *Such a reaction that continues completely to the right is called an **irreversible reaction**.* In the reaction chosen as an example, however, this does not happen. To understand this, let's now look at a different experimental situation. At the same temperature as before, we start with pure HI instead of H_2 and I_2. In this case, the HI reacts to form H_2 and I_2, which is exactly the reverse of the previous reaction:

$$2HI(g) \longrightarrow H_2(g) + I_2(g)$$

This reaction also results from a collision (between two HI molecules) that has the proper orientation and energy as shown in Figure 14-2.

*This reaction is an example of a **reversible reaction**, since it proceeds in either direction under the same conditions.* Such a reaction is indicated by double arrows:

$$H_2(g) + I_2(g) \rightleftharpoons 2HI(g)$$

Let's now go back to the first experiment, where we started with H_2 and I_2. At first only one reaction occurs: the formation of HI. However, as the concentration of HI begins to build, the reverse reaction becomes important. Eventually, a point is reached *where the rate of the forward reaction equals the rate of the reverse reaction.* This is known as the **point of equilibrium.** At this point the concentration of all reactants and products remains constant. This phenomenon is sometimes referred to as a dynamic equilibrium, which means that reactions are occurring but there is no change in the amounts of reactants and products. (See Figure 14-3.)

---

* Theoretically, no reaction is completely irreversible unless one of the products is removed, such as would happen when a gas is formed and then escapes. As we saw in Chapter 13, reactions that proceed in one direction spontaneously can be forced in the other direction in an electrolytic cell.

(At the start)   (At equilibrium)

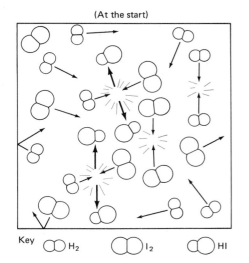

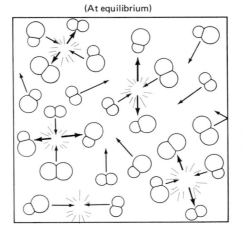

Key H_2 I_2 HI

Figure 14-3 EQUILIBRIUM. At the start, only the forward reaction occurs. At equilibrium, the forward and reverse reactions occur at the same rate.

If we could shrink ourselves to the molecular level, we would immediately become aware of this dynamic (changing) situation. If one I atom in an I_2 molecule was marked to that it could traced, we would notice that at one moment it is present in an I_2 molecule, later it is part of a HI molecule, and still later part of another I_2 molecule:

$$\overset{*I}{\underset{I}{|}} \overset{+ H_2}{\longrightarrow} \overset{H}{\underset{*I}{|}} \overset{+ HI}{\longrightarrow} \overset{I}{\underset{*I}{|}} \overset{+ H_2}{\longrightarrow} \text{ etc.}$$

An analogy to equilibrium is a party where there are initially 100 people in a house. Eventually, about 40 people find their way outside where there is more room. Throughout the evening people are going in and out, but there always seems to be 60 inside and 40 outside. The party has reached a dynamic equilibrium at the point when, for every person moving inside, one goes out. The numbers inside and out stay the same, although the individuals change.

Likewise, when a reaction is at equilibrium, the identities of reactant and product molecules are constantly changing, but the *amount* of each does not change.

14-2 Examples of Reactions at Equilibrium

We have previously discussed several equilibrium reactions. In Chapter 9, the following reaction was used to illustrate the calculation of percent yield of an incomplete reaction:

$$N_2(g) + 3H_2(g) \rightleftharpoons 2NH_3(g)$$

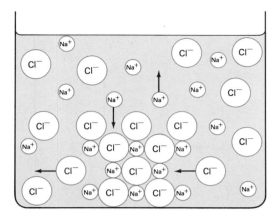

Figure 14-4 SOLUBILITY AND EQUILIBRIUM. At equilibrium the rate of solution of ions equals the rate at which ions are forming the solid.

The reason a 100% yield was not obtained was because the reaction reached a state of equilibrium. When gaseous N_2 and H_2 are mixed, gaseous NH_3 is formed. Eventually, the reaction reaches a state of equilibrium, when the NH_3 decomposes to N_2 and H_2 as fast as it is being formed.

In Chapter 11, the solubility of ionic compounds was mentioned. When the solution is saturated (the limit of solubility reached), an equilibrium is established between the undissolved solid phase and the dissolved ions:

$$NaCl(s) \rightleftharpoons Na^+(aq) + Cl^-(aq)$$

In this case a dynamic situation also exists, as the identity of the ions in solution is changing although the concentration is not (see Figure 14-4).

This phenomenon can be demonstrated by the art of growing large crystals from small ones. If a comparatively large crystal of a water-soluble substance such as $CuSO_4$ is suspended in a saturated solution of that substance, the small crystals in the bottom of the beaker get smaller and the large crystal gets larger. The concentration of the substance in solution does not change, nor does the total weight of crystals. It is obvious, however, by the changing size of the crystals, that a dynamic situation exists (see Figure 14-5).

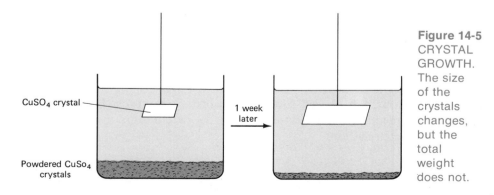

CuSO$_4$ crystal

Powdered CuSo$_4$ crystals

1 week later

Figure 14-5 CRYSTAL GROWTH. The size of the crystals changes, but the total weight does not.

In Chapter 12, weak acids and bases were found to react with water to a limited extent to form ions in an equilibrium reaction such as

$$HF(aq) + H_2O \rightleftharpoons H_3O^+(aq) + F^-(aq)$$

Only a few percent of the HF molecules were ionized at any one time, with most acid molecules existing in solution as unionized HF molecules. We will return to these equilibrium reactions in Chapter 15 for a detailed discussion.

14-3 Reaction Rates and the Equilibrium Constant

In the previous section it was shown that colliding molecules must have a minimum amount of energy in order for a reaction to proceed. Obviously, the more frequent the collisions having the required energy, the faster the reaction rate:

$$r \text{ (rate)} \propto \text{(energy of colliding molecules)}$$

The kinetic energy (K.E.) of a moving object is

$$\text{K.E.} = \tfrac{1}{2}mv^2 \qquad (m = \text{mass}, v = \text{velocity})$$

Thus, the faster an object is moving, the more energy it has. As an analogy, we have all probably witnessed the fact that automobiles have more energy at higher speeds as evidenced by the results of collisions at various speeds. Newer cars, however, are built to sustain a collision at 5 miles/hr (front and rear) without damage. Just as a car must be moving at a minimum velocity in order to be damaged in a collision, gas molecules must also be moving at a minimum velocity in order to have enough energy to undergo a reaction in a collision.

It was mentioned in Chapter 10 that the average velocity of molecules is proportional to the temperature. Thus as the temperature of a reaction mixture is raised, the molecules move faster, resulting in more collisions having the minimum energy for a reaction. This is the reason why the rates of all reactions increase as the temperature increases.

All reactions differ, however, as to the minimum energy needed to initiate a chemical reaction. For example, molecules of white phosphorous (an allotrope of the element) have enough energy to undergo a rapid reaction with oxygen even at room temperature (25°C). (As a result, white phosphorus is stored under water to prevent ignition.) On the other hand, hydrogen reacts with oxygen to form water, but a mixture of hydrogen and oxygen does not react until the temperature is at least 500°C unless ignited by a spark. (When the mixture does begin to react, the hydrogen−oxygen mixture is said to *detonate*. This is when the reaction occurs so rapidly that the quick release of the chemical energy in the form of heat occurs as an explosion.)

In addition to the energy of the collisions, the rate of a reaction depends on the frequency of collisions:

$$r \text{ (rate)} \propto \text{(frequency of collisions)}$$

 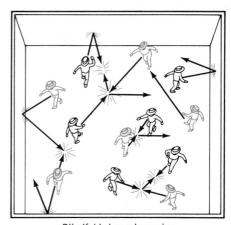

Figure 14-6 THE
EFFECT OF
VELOCITY ON
COLLISIONS. Col-
lisions occur more
frequently (and
with more force) as
the velocity
increases.

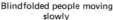

Blindfolded people moving
slowly

Blindfolded people moving
quickly

As the temperature of the gas increases, the velocity increases, which means that the frequency of collisions increases. As an analogy, imagine a room containing some blindfolded prople. As they move around they collide randomly with each other and with the walls of the room. If, on a signal, they began moving faster, there will be more frequent collisions. (Since they are moving faster, those collisions would also hurt more, because they have more energy; see Figure 14-6.)

In summary, a rise in temperature increases the rate of a reaction for two reasons: the collisions have more energy, and the collisions are more frequent.

Let's continue the analogy of the room containing blindfolded, colliding people. The frequency of collisions depends not only on the velocity of movement but, naturally, on the number of people in the room. Now let's say that half of the people are wearing blue hats and the other half are wearing gray hats. Obviously, there are blue–blue, gray–gray, and blue–gray collisions, but we will focus on the blue–gray collisions as analogous to two different reacting molecules. If the number of people wearing blue hats is suddenly doubled, the frequency of blue–gray collisions is also doubled. Conversely, if the number of people with gray hats is doubled (and the number of blue hats is held constant), there is also a doubling of the rate of blue-gray collisions. See Figure 14-7.

In the case of a chemical reaction, the frequency of collisions (and thus the rate of the reaction) is also dependent on the number of reactant molecules in a fixed volume (which is the concentration). In the reaction of H_2 and I_2, the rate of formation of HI doubles if the concentration of H_2 doubles (and $[I_2]$* is held constant). Mathematically this can be expressed as

$$r_f \propto [H_2] \qquad [I_2] \text{ constant} \qquad r_f = \text{rate of forward reaction}$$

* $[I_2]$ symbolizes the concentration of I_2 in moles per liter.

More people—more collisions

Figure 14-7 THE EFFECT OF CONCEN-TRATION ON COLLISIONS. Collisions occur more frequently when there are more people in the room.

The result is similar if $[I_2]$ doubles and $[H_2]$ is held constant:

$$r_f \propto [I_2] \qquad [H_2] \text{ constant}$$

These two relations can be combined into one relation known as the rate law for the forward reaction:

$$r_f \propto [H_2][I_2]$$

or, as an equality,

$$r_f = k_f[H_2][I_2]$$

See Problem 14-1

where k_f is called the **rate constant** for the forward reaction. *The* **rate law** *for a reaction expresses the effect that changes in concentration of reactants have on the rate of the reaction.*

Notice in this rate law that if *both* $[H_2]$ and $[I_2]$ are doubled, the rate doubles twice or *fourfold*. In the reverse reaction, both colliding molecules are identical. Therefore, the doubling of $[HI]$ in effect doubles the concentration of *both* colliding molecules, and the rate increases *fourfold*. The rate law for the reverse reaction (r_r) is thus

$$r_r = k_r[HI][HI] = k_r[HI]^2$$

where k_r is the rate constant for the reverse reaction.

If H_2 and I_2 are mixed in a container at 400°C, initially only HI is formed. Notice, however, that as this forward reaction proceeds, $[H_2]$ and $[I_2]$ must naturally decrease since they are being consumed. *From the rate law for the forward reaction, this means that the rate of this reaction also decreases (top curve in Figure 14-8).* Conversely, as the reaction proceeds to the right, $[HI]$ increases, which means that the rate of the reverse reaction increases (bottom curve in Figure 14-8). Eventually a point is reached (point C in Figure 14-8)

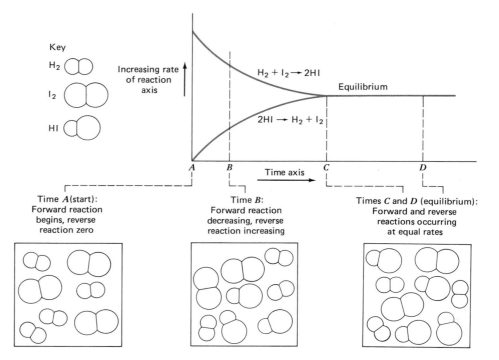

Figure 14-8 H_2, I_2, HI EQUILIBRIUM. Equilibrium is achieved when the rates of the forward and reverse reactions are equal.

where the rates of the two reactions become the same; this is the point of equilibrium.

At time C in Figure 14-8, we have the following relation from the definition of equilibrium:

$$r_f = r_r$$

$$k_f[\text{H}_2][\text{I}_2] = k_r[\text{HI}]^2$$

$$\frac{k_f}{k_r} = \frac{[\text{HI}]^2}{[\text{H}_2][\text{I}_2]}$$

Since the ratio of two constants is another constant, the expression can be simplified as follows:

$$\frac{k_f}{k_r} = K_{\text{eq}}$$

$$K_{\text{eq}} = \frac{[\text{HI}]^2}{[\text{H}_2][\text{I}_2]}$$

K_{eq} is called the **equilibrium constant,** and the fraction to the right of the equality is called the **equilibrium constant expression** for this particular reaction.

In our example, we assumed a mechanism for the reaction and then rationalized a rate law from this proposed mechanism. In fact, the sequence is just the opposite. Chemists first determine a rate law from experiments that study the effect of concentrations on the rate of reaction. From this rate law they then propose a mechanism that is consistent with the experiments. In the case of H_2 and I_2, products form from a simple one-step collision of reactant molecules. Other reactions have rate laws and the corresponding proposed mechanism more complex, where products form only after two or more steps and each step involves a collision. Regardless of the rate law, however, the equilibrium constant expression for every reaction at equilibrium is constructed in the same manner as follows:

$$a\text{A} + b\text{B} + \text{etc.} \rightleftharpoons d\text{D} + e\text{E} + \text{etc.}$$

$$K_{eq} = \frac{[\text{D}]^a[\text{E}]^e \text{ etc.}}{[\text{A}]^a[\text{B}]^b \text{ etc.}}$$

Notice that the *coefficients* in the balanced equation become *exponents* in the equilibrium constant expression for products in the *numerator* and reactants in the *denominator*.

Example 14-1

Write equilibrium constant expressions for each of the following reactions:

(a) $N_2(g) + 3H_2(g) \rightleftharpoons 2NH_3(g)$
(b) $4NH_3(g) + 5O_2(g) \rightleftharpoons 4NO(g) + 6H_2O(g)$
(c) $PCl_3(g) + Cl_2(g) \rightleftharpoons PCl_5(g)$

Solutions:

(a)
$$K_{eq} = \frac{[NH_3]^2}{[N_2][H_2]^3}$$

(b)
$$K_{eq} = \frac{[NO]^4[H_2O]^6}{[NH_3]^4[O_2]^5}$$

(c)
$$K_{eq} = \frac{[PCl_5]}{[PCl_3][Cl_2]}$$

See Problems 14-2, 14-3

Table 14-1   The Value of K_{eq}

For the reaction $H_2(g) + I_2(g) \rightleftharpoons 2HI(g)$,

$$K_{eq} = \frac{[HI]^2}{[H_2][I_2]} \qquad [X] = \text{mol/liter}$$

Exp. No.	Initial Concentration			Equilibrium Concentration			K_{eq}
	$[H_2]$	$[I_2]$	$[HI]$	$[H_2]$	$[I_2]$	$[HI]$	
1	2.000	2.000	0	0.428	0.428	3.144	53.9
2	0	0	2.000	0.214	0.214	1.572	54.0
3	1.000	1.000	1.000	0.321	0.321	2.358	54.0

For a particular reaction, K_{eq} is always constant at a specific temperature. Table 14-1 shows that this relation is true regardless of whether we start with H_2 and I_2 (Experiment 1), HI (Experiment 2), or a mixture of all three (Experiment 3).

Example 14-2

What is the value for K_{eq} for the following system at equilibrium?

$$2NO(g) + O_2(g) \rightleftharpoons 2NO_2(g)$$

At a certain temperature, the equilibrium concentrations of the gases are $[NO] = 0.890$; $[O_2] = 0.250$; $[NO_2] = 0.0320$.

Solution:

For this reaction,

$$K_{eq} = \frac{[NO_2]^2}{[NO]^2[O_2]}$$

$$= \frac{[0.0320]^2}{[0.890]^2[0.250]} = \frac{1.024 \times 10^{-3}}{0.198}$$

$$= \underline{5.17 \times 10^{-3}}$$

Example 14-3

For the equilibrium

$$N_2(g) + 3H_2(g) \rightleftharpoons 2NH_3(g)$$

complete the table and compute the value of the equilibrium constant.

Initial Concentration			Equilibrium Concentration		
$[H_2]$	$[N_2]$	$[NH_3]$	$[H_2]$	$[N_2]$	$[NH_3]$
0.200	0.200	0	_____	_____	0.0450

Procedure:

1 As in other stoichiometry problems, find the $[H_2]$ and $[N_2]$ that reacted to form the 0.0450 mol/liter of NH_3 (Section 9-3).

2 Find the $[H_2]$ and $[N_2]$ remaining at equilibrium by subtracting the concentration that reacted from the initial concentration; that is,

$$[N_2]_{eq} = [N_2]_{initial} - [N_2]_{reacted}$$

3 Substitute the concentrations of all compounds present at equilibrium into the equilibrium constant expression and solve to find the value of K_{eq}.

Solution:

1 If 0.0450 mol of NH_3 is formed, calculate the number of moles of N_2 that reacted (per liter).

$$0.0450 \; \cancel{\text{mol of NH}_3} \times \frac{1 \text{ mol of N}_2}{2 \; \cancel{\text{mol of NH}_3}} = 0.0225 \text{ mol of N}_2 \text{ reacted}$$

The number of moles of H_2 that reacted are

$$0.0450 \; \cancel{\text{mol of NH}_3} \times \frac{3 \text{ mol of H}_2}{2 \; \cancel{\text{mol of NH}_3}} = 0.0675 \text{ mol of H}_2 \text{ reacted}$$

2   At equilibrium,

$$[N_2]_{eq} = 0.200 - 0.0225 = 0.178 \text{ mol/liter}$$

$$[H_2]_{eq} = 0.200 - 0.0675 = 0.132 \text{ mol/liter}$$

3
$$K_{eq} = \frac{[NH_3]^2}{[N_2][H_2]^3} = \frac{(0.0450)^2}{(0.178)(0.132)^3} = \underline{\underline{4.95}}$$

Example 14-4
In the preceding equilibrium, what would be the concentration of NH_3 at equilibrium if the concentration of $N_2 = 0.22$ mol/liter and $H_2 = 0.14$ mol/liter at equilibrium?

Procedure:
In this problem, use the value of K_{eq} found in the previous problem. The $[NH_3]$ can be found by substituting the concentrations of the species given and solving for the one unknown.

Solution:

$$K_{eq} = \frac{[NH_3]^2}{[N_2][H_2]^3} \qquad [N_2] = 0.22 \text{ mol/liter} \qquad K_{eq} = 4.95$$
$$[H_2] = 0.14 \text{ mol/liter}$$

$$= \frac{[NH_3]^2}{[0.22][0.14]^3} = 4.95$$

$$[NH_3]^2 = 2.99 \times 10^{-3}$$

$$[NH_3] = \underline{\underline{5.5 \times 10^{-2}}} = \underline{\underline{0.055 \text{ mol/liter}}}$$

See Problems 14-4 through 14-13

The magnitude of K_{eq} for a reaction can tell us about the *position* of the equilibrium. A large value for K_{eq} (e.g., $>10^3$) mathematically means a large numerator compared to the denominator. In terms of reactants and products, this means that the concentrations of products in the numerator must be considerably larger than the concentrations of reactants in the denominator. Thus the equilibrium lies to the right. Conversely, a small value for K_{eq} (e.g., $<10^{-3}$) indicates that the reactants are favored and the equilibrium lies significantly to the left. Values for K_{eq} in between indicate appreciable concentrations of reactants and products present at equilibrium.

14-4 The Effect of Stress and Catalysts on Equilibrium

When a chemical system is at equilibrium, any change in conditions may affect the point of equilibrium. Such changes include a change in temperature, pressure, or concentration of a reactant or product. How a system at equilibrium reacts to a change in conditions is summarized by **Le Châtelier's principle,** which states: *When stress is applied to a system at equilibrium, the system reacts in such a way to counteract the stress.*

This principle can be illustrated by a long-distance jogger. When jogging it is desireable to maintain a steady pace, which is analogous to equilibrium. At first the body maintains the pace rather easily, with little change in respiration or heartbeat. After a couple of miles the body begins to tire, which is analogous to stress on the system. If nothing changed, the jogger would slow and finally stop. The body counteracts the stress, however, by faster breathing, faster heartbeat, and increased perspiration. Thus the human body counteracts the stress so that the steady pace (equilibrium) is maintained (see Figure 14-9).

We will illustrate Le Châtelier's principle with the following important industrial process that reaches an equilibrium:

$$2SO_2(g) + O_2(g) \rightleftharpoons 2SO_3(g)$$

The SO_2 is produced by the combustion of sulfur or sulfur compounds. The SO_3 produced from this reaction is then hydrated to produce sulfuric acid (H_2SO_4):

$$SO_3(g) + H_2O \longrightarrow H_2SO_4(aq)$$

Currently, about *40 million tons* of H_2SO_4 is produced in the United States each year by this process (see Figure 14-10). The value of this one chemical was $3.2 billion in this country alone in 1980.

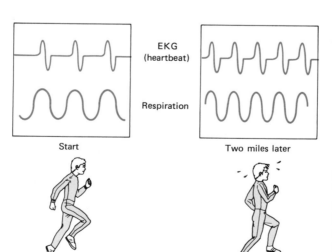

Figure 14-9 STRESS AND THE JOGGER. To compensate for the increased rate of metabolism of a jogger, the heartbeat and respiration increase.

Figure 14-10 A SULFURIC ACID PLANT. This large plant converts sulfur into some of the millions of tons of sulfuric acid manufactured in this country each year.

It is important to convert the maximum amount of SO_2 to SO_3 by forcing the point of equilibrium to the right as far as possible.

1 If the system is at equilibrium and additional O_2 is introduced, the system counteracts that stress by removing some of the added O_2. It can do this by reaction with SO_2. An increase in the concentration of compounds on one side of a reaction at equilibrium ultimately leads to an increase on the other side as the system shifts to remove the stress of the added compound.

The selective removal of a compound from a reaction mixture, such as formation of a precipitate or evolution of a gas, has a similar effect. The system shifts in such a way so as to replace the loss. If the loss is complete, the reaction in the direction of that loss is complete (an irreversible reaction).

The changes in the point of equilibrium can be illustrated by a simple example. In the following hypothetical reaction, equilibrium is established when [A] = 1.00 mol/liter and [B] = 2.00 mol/liter.

$$A \rightleftharpoons B \qquad K_{eq} = \frac{[B]}{[A]} = \frac{2.00}{1.00} = 2.00$$

If an additional 1.00 mol/liter of A is now introduced into the reaction container, notice that the system is no longer at equilibrium.

$$\frac{[B]}{[A]} = \frac{2.00}{(1.00 + 1.00)} = 1.00 \neq K_{eq}$$

For the system to return to equilibrium, the numerator becomes larger (more B is formed) and the denominator smaller (some A must react). Thus we see mathematically that introducing additional components present on one side of an equation leads to a shift to the other side. In the example above, equilibrium is reestablished when

$$[A] = 1.33 \text{ mol/liter}$$

$$[B] = 2.67 \text{ mol/liter}$$

$$\frac{2.67}{1.33} = 2.00 = K_{eq}$$

This indicates that the added A increases the concentration of B by 0.67 mol/liter.

In our industrial example, the conversion of SO_2 to SO_3 is aided by a large excess of O_2 in the reaction container.

2 A change in pressure also affects the equilibrium. Notice that the reaction produces a volume reduction when it goes to the right. Three moles of gas ($2SO_2 + 1O_2$ or 67.2 liters at STP) react to form 2 mol of gas ($2SO_3$ or 44.8 liters at STP). Therefore, the system can counteract an increase in pressure by contracting or going to a lower volume. The reaction responds to an increase in pressure by reaction to the right, which means formation of additional SO_3.

3 The reaction of SO_2 with O_2 not only produces SO_3 as a product but also heat energy since this is an exothermic reaction.

$$2SO_2(g) + O_2(g) \rightleftharpoons 2SO_3(g) + \text{heat}$$

The heat is considered a component of the reaction and is produced in amounts directly proportional to the amount of SO_3 produced. If the reaction mixture at equilibrium is cooled, this removes heat from the system. According to Le Châtelier, the system shifts to the right to replace the lost heat. Since SO_3 is formed along with the heat, cooling the mixture produces more SO_3. On the other hand, heating the mixture forces the equilibrium to shift to the left. By shifting to the left, the system attempts to remove the added heat. This phenomenon is reflected in the value for K_{eq}. *For exothermic reactions, K_{eq} becomes larger as the temperature decreases.* This means that, for exothermic reactions, formation of products (on the right) becomes more favorable at lower temperatures.

In all reactions, however, the rate of reaction decreases as the temperature decreases (i.e., the refrigerator effect). Since time is money in industry, the process of maximizing the yield of SO_3 must be compromised to an extent. The reaction must be run at a high enough temperature so that it proceeds to equilibrium in a reasonable time. This reaction is actually carried out at about 400°C.

There is one other factor that can affect the rates at which reactions occur, and this is the presence of a catalyst. *A* **catalyst** *is a substance that is not* *consumed by a reaction but whose presence increases the rate of both the for-ward and reverse reactions.* Thus a catalyst affects the rate at which equilibrium is reached but does not change the point of equilibrium (i.e., the distribution of reactants and products at equilibrium.) Catalysts work in different ways in different reactions. They may be intimately mixed with the reaction mixture or they may simply provide a surface on which the reactions take place. Their main function is to provide an alternative reaction mechanism for the forward and reverse reactions. This alternative mechanism apparently requires less energy in the collisions, which means that more collisions lead to a reaction and equilibrium is achieved more quickly.

In the reaction of SO_2 and O_2, a surface catalyst such as finely divided platinum or palladium is used to achieve equilibrium rapidly. This allows the reaction to be run at a lower temperature than would otherwise be possible. Without a catalyst the reaction would have to be run at a considerably higher temperature, with a corresponding lower yield of SO_3.

In summary, the production of SO_3 from SO_2 is favored by:

1   A large excess concentration of O_2

2   A high pressure in the reaction container

3   A low temperature (but this slows the reaction)

4   The presence of a catalyst (which speeds up the reaction).

See Problems 14-14
through 14-17

While on the topic of catalysts, we should mention their use in helping to reduce air pollution. Automobile exhaust contains poisonous CO and unburned gasoline (mainly C_8H_{18}) which contribute to smog. The catalytic converter on the automobile contains finely divided platinum and palladium in an attachment on the exhaust pipe. These metals provide a surface for the following reactions, which occur only at very high temperature without a catalyst:

$$2CO(g) + O_2(g) \longrightarrow 2CO_2(g)$$

$$2C_8H_{18}(g) + 25O_2(g) \longrightarrow 16CO_2(g) + 18H_2O(g)$$

Both of these reactions are exothermic, which explains why the catalytic converter becomes quite hot when the engine is running.

One drawback of the catalytic converter in an automobile is that it also converts SO_2 to SO_3 as mentioned earlier. The combustion of sulfur impurities in gasoline produces SO_2, which is then converted to SO_3 in the converter. The SO_3 reacts with H_2O to form H_2SO_4, which is a strong acid. The effect of H_2SO_4 from this source on the environment has not yet been determined.

Review Questions

1 Describe the mechanism of the H_2-I_2 reaction. What is the "collision theory"?
2 Give the two conditions necessary before colliding molecules can react.
3 What is the difference between a reversible and an irreversible reaction?
4 Describe what is meant by a *dynamic* equilibrium.
5 Describe how a dynamic equilibrium is involved in a saturated solution with undissolved solid present.
6 Give two reasons why reaction rates increase as the temperature increases.
7 Why do reaction rates increase as the concentration of colliding molecules increases?

8 What is meant by the rate law for a reaction?
9 What is an equilibrium constant?
10 How is the equilibrium constant expression set up for a reaction?
11 What is LeChâtelier's principle?
12 Describe the effect of changes in concentration of reactants or products on a system at equilibrium.
13 Will a change in pressure (or volume) always affect a system at equilibrium?
14 What effect does a change in temperature have on a system at equilibrium?
15 How does a catalyst affect a reversible reaction?

Problems

The Collision Theory
14-1 Explain the following facts from your understanding of "collision theory" of chemical reactions.
   (a) The rates of chemical reactions approximately double for each 10°C rise in temperature.
   (b) Eggs cook slower at higher altitudes, where water boils at temperatures less than 100°C.
   (c) H_2 and O_2 do not react to form H_2O except at very high temperatures (unless initiated by a spark).
   (d) Wood burns explosively in pure O_2 but slowly in air which is about 20% O_2.
   (e) A wasp is lethargic at temperatures below 65°F.
   ■ (f) Charcoal burns faster if you blow on the coals.
   ■ (g) A 0.10 M boric acid solution ($[H_3O^+] = 7.8 \times 10^{-6}$) can be used for eye wash, but a 0.10 M hydrochloric acid solution ($[H_3O^+] = 0.10$) would cause severe damage.

   ■ (h) To keep apple juice from fermenting to apple cider, the apple juice must be kept cold.

The Equilibrium Constant Expression
14-2 Write the equilibrium constant expression for each of the following equilibria:
   (a) $CO(g) + Cl_2(g) \rightleftharpoons COCl_2(g)$
   (b) $CH_4(g) + 2H_2O(g) \rightleftharpoons CO_2(g) + 4H_2(g)$
   ■ (c) $4HCl(g) + O_2(g) \rightleftharpoons 2Cl_2(g) + 2H_2O(g)$
14-3 Write the equilibrium constant expression for each of the following equilibria:
   (a) $3O_2(g) \rightleftharpoons 2O_3(g)$
   (b) $N_2(g) + 2O_2(g) \rightleftharpoons 2NO_2(g)$
   ■ (c) $C_2H_2(g) + 2H_2(g) \rightleftharpoons C_2H_6(g)$

The Value of K_{eq}
14-4 Given the following system:

$$3O_2(g) \rightleftharpoons 2O_3(g)$$

At equilibrium, $[O_2] = 0.35$ and $[O_3] = 0.12$. What is K_{eq} for the reaction at this temperature?

■ **14-5** Given the following system:

$$N_2(g) + 2O_2(g) \rightleftharpoons 2NO_2(g)$$

At a certain temperature there are 1.25×10^{-3} mol of N_2, 2.50×10^{-3} mol of O_2, and 6.20×10^{-4} mol of NO_2 in a 1.00-liter container. What is K_{eq} for this reaction at this temperature?

14-6 Given the following system:

$$CH_4(g) + 2H_2O(g) \rightleftharpoons CO_2(g) + 4H_2(g)$$

At equilibrium we find 2.20 mol of CO_2, 4.00 mol of H_2, 6.20 mol of CH_4, and 3.00 mol of H_2O in a 30.0-liter container. What is K_{eq} for the reaction?

14-7 Given the following system:

$$2HI(g) \rightleftharpoons H_2(g) + I_2(g)$$

(a) If we start with [HI] = 0.60, what are the $[H_2]$ and $[I_2]$ that would be present if all of the HI reacts?

(b) If we start with [HI] = 0.60 and $[I_2]$ = 0.20, what are the $[H_2]$ and $[I_2]$ that would be present if all of the HI reacts?

(c) If we start with only [HI] = 0.60 and 0.20 mol/liter of HI reacts, what are [HI], $[I_2]$, and $[H_2]$ at equilibrium?

*14-8 Complete the table at the bottom of this page for the reaction $N_2(g) + 3H_2(g) \rightleftharpoons 2NH_3(g)$.

14-9 Given the following system:

$$4NH_3(g) + 5O_2(g) \rightleftharpoons 4NO(g) + 6H_2O(g)$$

At the start of the reaction, $[NH_3]$ = $[O_2]$ = 1.00. At equilibrium it is found that 0.25 mol/liter of NH_3 has reacted.

(a) What is the concentration of O_2 that reacts?

(b) What is the concentration of all species at equilibrium?

(c) Write the total equilibrium constant expression and substitute the proper values for the concentrations of reactants and products.

■ **14-10** Given the following equilibrium:

$$CO(g) + Cl_2(g) \rightleftharpoons COCl_2(g)$$

At the start of the reaction, [CO] = 0.650 and $[Cl_2]$ = 0.435. At equilibrium it is found that 10.0% of the CO has reacted. What is [CO], $[Cl_2]$, and $[COCl_2]$ at equilibrium? What is the value of K_{eq}?

14-11 For the reaction

$$PCl_3(g) + Cl_2(g) \rightleftharpoons PCl_5(g)$$

K_{eq} = 0.95 at a certain temperature. If $[PCl_3]$ = 0.75 and $[Cl_2]$ = 0.40 at equilibrium, what is the concentration of PCl_5 at equilibrium?

■ **14-12** At a certain temperature, K_{eq} = 46.0 for the reaction

$$4HCl(g) + O_2(g) \rightleftharpoons 2Cl_2(g) + 2H_2O(g)$$

At equilibrium, [HCl] = 0.100, $[O_2]$ = 0.455, and $[H_2O]$ = 0.675. What is $[Cl_2]$ at equilibrium?

*14-13 Using the value for K_{eq} calculated in Problem 14-6, what is the concentration of H_2O at equilibrium if the concentrations of the other species present at equilibrium are $[CH_4]$ = 0.50, $[CO_2]$ = 0.24, and $[H_2]$ = 0.20?

	Initial Concentrations			Equilibrium Concentrations		
	$[N_2]$	$[H_2]$	$[NH_3]$	$[N_2]$	$[H_2]$	$[NH_3]$
(a)	1.00	1.00	0	_____	_____	0.20
(b)	0	0	1.00	_____	_____	0.70
■ (c)	1.00	0	1.00	_____	_____	0.80
■ (d)	0.50	0.50	0.50	0.40	_____	_____
■ (e)	0	0	0.50	_____	0.15	_____

Le Châtelier's Principle

14-14 The following equilibrium is an important industrial process used to convert N_2 to NH_3. The ammonia is used mainly for fertilizer. This method for the production of NH_3 is called the Haber process.

$$N_2(g) + 3H_2(g) \xrightarrow[500°C]{\text{Catalyst}}$$
$$2NH_3(g) + \text{heat energy}$$

Determine the direction that the equilibrium will be shifted by the following changes:

(a) Increasing the concentration of N_2
(b) Increasing the concentration of NH_3
(c) Decreasing the concentration of H_2
(d) Decreasing the concentration of NH_3
(e) Increasing the pressure in the reaction container
(f) Removing the catalyst
(g) How will the yield of NH_3 be affected by raising the temperature to 750°C?
(h) How will the yield of NH_3 be affected by lowering the temperature to 0°C? How will this affect the rate of formation of NH_3?

■ **14-15** Given the following equilibrium:

$$4NH_3(g) + 5O_2(g) \rightleftharpoons$$
$$4NO(g) + 6H_2O(g) + \text{heat energy}$$

How will this system at equilibrium be affected by

(a) Increasing the concentration of O_2
(b) Removing all of the H_2O as it is formed
(c) Increasing the concentration of NO
(d) Increasing the pressure in the reaction container

(e) Increasing the volume of the reaction container
(f) Increasing the temperature
(g) Decreasing the concentration of O_2
(h) Addition of a catalyst

14-16 The following equilibrium takes place at high temperatures:

$$N_2(g) + 2H_2O(g) + \text{heat energy} \rightleftharpoons$$
$$2NO(g) + 2H_2(g)$$

How will the yield of NO be affected by
(a) Increasing $[N_2]$
(b) Decreasing $[H_2]$
(c) Decreasing the pressure in the reaction container
(d) Decreasing the volume of the reaction container
(e) Decreasing the temperature
(f) Addition of a catalyst

14-17 Fortunately for us, the major components of air N_2 and O_2 do not react under ordinary conditions. At very high temperatures, however, like those found in an automobile engine, the equilibrium constant for the following reaction becomes larger.

$$N_2(g) + O_2(g) \rightleftharpoons 2NO(g)$$

(a) Does the information given indicate that the formation of NO is an endothermic or an exothermic process?
(b) Since NO is a serious pollutant, it is desirable to minimize its formation in automobile engines. Would a cooler- or hotter-running engine increase the yield of NO?
(c) Would a lower pressure in the engine affect the formation of NO?

Aqueous Acid–Base Equilibria

These products are concerned with pH and buffers. Both of these topics relate to the equilibrium concentration of H_3O^+ in aqueous solution.

Those who make television commercials apparently assume that the entire population has a fairly sophisticated understanding of acid–base chemistry. For example, we hear statements like "the hair rinse has a high pH" or "the medicine contains a buffer." As you will certainly notice in this chapter, discussion of pH and buffers assumes a significant background in chemistry. You now have that background.

In this chapter, we will improve and expand on our understanding of acids and bases in aqueous solution. By use of equilibrium constants, introduced in Chapter 14, the discussion will be somewhat more quantitative than the previous discussion of acids and bases in Chapter 12.

Since this chapter expands on Chapter 12 with information provided in Chapter 14, those two chapters should be singled out for a thorough review before beginning this chapter. To understand the concept of pH it is necessary to have a basic understanding of common logarithms. Refer to Appendix E for this purpose if necessary.

15-1 Equilibrium in Water

In Chapter 12, the neutralization reaction was shown to involve the net ionic equation

$$H_3O^+ + OH^- \longrightarrow 2H_2O$$

This reaction implied that the two ions were incompatible in each other's presence and reacted to form water. With a closer look, however, we find that this is not completely true, since this is a reversible reaction. Thus the following equilibrium is present in pure water:

$$2H_2O \rightleftharpoons H_3O^+ + OH^-$$

The process whereby a pure covalent compound dissociates into positive and negative ions is known as **autoionization.** Although this equilibrium lies very far to the left, there is a small but important concentration of H_3O^+ and OH^- ions that do coexist in pure water. Like other equilibria we have discussed, this is a dynamic situation, with the forward and reverse reactions occurring simultaneously. The equilibrium constant expression for this reaction is

$$K_{eq} = \frac{[H_3O^+][OH^-]}{[H_2O]^2} \qquad [H_3O^+] = \text{concentration of } H_3O^+ \text{ in } \textit{molarity}$$

In this situation the concentrations of the two ions are variable, since they can be changed relative to each other as will be shown shortly. The H_2O is a pure substance, however, and the concentration of any pure substance is a constant physical property just like density or melting point.* For water the concentration of H_2O is 55.6 moles per liter; that is,

$$\frac{10^3 \; \text{ml}}{\text{liter}} \times \frac{1.00 \; \text{g}}{\text{ml}} \times \frac{1.00 \; \text{mol}}{18.0 \; \text{g}} = 55.6 \; \text{mol/liter}$$

Since $[H_2O]^2$ is also a constant, it can be combined with the other constant, K_{eq}, to form a new constant designated K_w.

$$K_{eq}[H_2O]^2 = [H_3O^+][OH^-]$$

$$K_{eq}[H_2O]^2 = K_w$$

$$K_w = [H_3O^+][OH^-]$$

K_w *is known as the ion product of water.* In pure water, less than one out of every billion molecules is ionized. As we will see, however, even this small amount is enough to affect the concept of acids and bases in water. The actual concentration of each ion in pure H_2O at 25°C is

$$[H_3O^+] = [OH^-] = 1.00 \times 10^{-7}$$

* Even in cases where a solute is present in solution, the concentration of H_2O is essentially a constant, since it is the solvent and usually present in a large excess compared to the solute.

The value for K_w *at 25°C* is therefore

$$K_w = [H_3O^+][OH^-] = [1.00 \times 10^{-7}][1.00 \times 10^{-7}]$$
$$= 1.00 \times 10^{-14}$$

The expression for K_w gives us the relationship of $[H_3O^+]$ and $[OH^-]$ for any aqueous solution—acidic, basic, as well as neutral. It tells us that even in an acidic solution there are some OH^- ions present. Conversely, some H_3O^+ ions are present in a basic solution. The following example illustrates this relationship.

Example 15-1
In a certain solution, $[H_3O^+] = 1.50 \times 10^{-2}$. What is $[OH^-]$ in this solution?

Procedure:
Use the relationship for K_w, $[H_3O^+][OH^-] = 1.00 \times 10^{-14}$, and solve for $[OH^-]$.

Solution:

$$[H_3O^+][OH^-] = 1.00 \times 10^{-14}$$

$$[OH^-] = \frac{1.00 \times 10^{-14}}{[H_3O^+]} = \frac{1.00 \times 10^{-14}}{1.50 \times 10^{-2}}$$

$$= \underline{6.67 \times 10^{-13}}$$

See Problems 15-1 through 15-5

As you can see, there is an inverse relationship between $[H_3O^+]$ and $[OH^-]$. Although inverse relationships have been discussed and illustrated several times previously in this text, the relationship of $[H_3O^+]$ and $[OH^-]$ is illustrated in Figures 15-1 and 15-2. In the figures concentrations of $[H_3O^+]$ and $[OH^-]$ are listed between 10^{-4} and 10^{-10} M (higher and lower concentra-

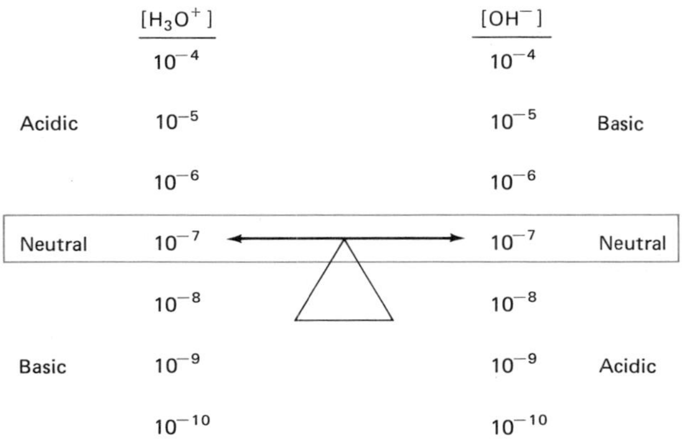

Figure 15-1 A NEUTRAL SOLUTION. In a neutral solution the $[H_3O^+]$ and the $[OH^-]$ are both equal to 10^{-7} M.

tions exist, however). Both concentrations ($[H_3O^+]$ on the left and $[OH^-]$ on the right) are shown to decrease from top to bottom. In between, there is a rigid pointer with an arrow at each end. In Figure 15-1 the situation in pure water, which is a neutral solution, is illustrated. In a **neutral solution,** the pointer is exactly balanced, indicating that *the concentrations of both ions are equal to 10^{-7} M.*

We will now add an *acid* to this neutral solution. By definition, H_3O^+ is added to the solution, and the pointer should go up on the left. At the same time, however, the rigid pointer goes *down on the right,* indicating a correspondingly smaller $[OH^-]$. If, instead, a base had been added, the situation would have been just the opposite. Figure 15-2 illustrates the relationship in an acid solution. A calculation such as we made in Example 15-1 would confirm that when $[H_3O^+]$ is *increased* from 10^{-7} to 10^{-5} *M,* the $[OH^-]$ is *decreased* from 10^{-7} to 10^{-9} *M.*

In summary, an acidic, basic, or neutral solution can now be defined in terms of concentrations of ions:

See Problems 15-6, 15-7

Neutral	$[H_3O^+] = [OH^-] = 1.00 \times 10^{-7}$
Acidic	$[H_3O^+] > 1.00 \times 10^{-7}$ and $[OH^-] < 1.00 \times 10^{-7}$
Basic	$[H_3O^+] < 1.00 \times 10^{-7}$ and $[OH^-] > 1.00 \times 10^{-7}$

In Chapter 12, the Arrhenius definition of an acid was a substance that produced H_3O^+. Since H_3O^+ is always present anyway, even in basic solution, a slight modification of the definition is helpful. *An **acid** is any substance that increases $[H_3O^+]$ in water, and a **base** is any substance that increases $[OH^-]$ in water.* With our new understanding of the equilibrium, notice that a substance can be an acid by directly donating $H^+(aq)$ to the solution (e.g., HCl, H_2S), or a substance can be an acid by reacting with OH^- ions, thus removing them from the solution. As illustrated in Figure 15-2,

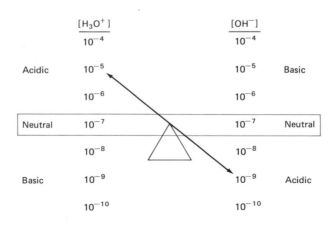

Figure 15-2 AN ACID SOLUTION. There is an inverse relationship between $[H_3O^+]$ and $[OH^-]$.

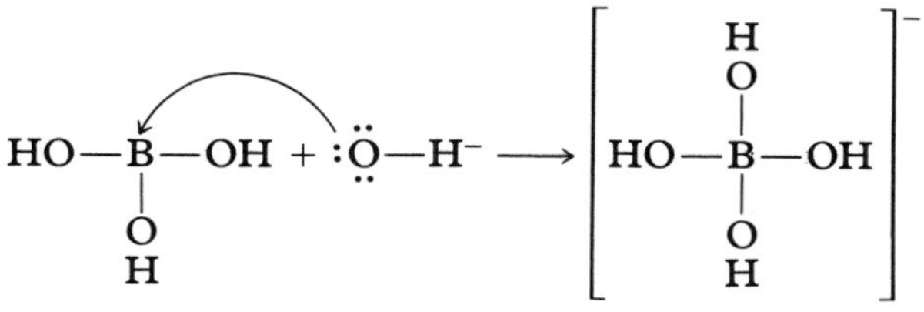

Figure 15-3 BORIC ACID. This is an example of a compound that acts as an acid by reacting with OH^- ions.

lowering $[OH^-]$ on the right has the effect of raising $[H_3O^+]$ on the left. Boric acid (H_3BO_3) is an example of an acid that reacts with OH^- rather than donating $H^+(aq)$. This is illustrated in Figure 15-3 by means of the Lewis structures of the compounds.

15-2 pH and pOH

The concentrations of H_3O^+ and OH^- in aqueous solution are usually awkward numbers, even with the use of scientific notation. To simplify the numbers, they can be expressed in terms of pH. **pH** *is a mathematical definition of* $[H_3O^+]$ *as shown below:*

Refer to Appendix E for discussion of logarithms

$$pH = \frac{1}{\log[H_3O^+]} = -\log[H_3O^+]$$

Much less popular but also valid is a similar expression called **pOH**:

$$pOH = \frac{1}{\log[OH^-]} = -\log[OH^-]$$

A simple relationship between pH and pOH can be derived from the ion product of water:

$$[H_3O^+][OH^-] = 1.00 \times 10^{-14}$$

If we now take $-\log$ of both sides of the equation, we have

$$-\log[H_3O^+][OH^-] = -\log(1.00 \times 10^{-14})$$

Since $\log(A \times B) = \log A + \log B$, the equation can be written as

$$-\log[H_3O^+] - \log[OH^-] = -\log 1.00 - \log 10^{-14}$$

Since $\log 1.00 = 0$ and $\log 10^{-14} = -14$, the equation is

$$pH + pOH = 14$$

Example 15-2
What is the pH of a neutral solution? What is the pOH?

Solution:

In a neutral solution, $[H_3O^+] = 1.00 \times 10^{-7}$.

$$pH = -\log[H_3O^+] = -\log(1.00 \times 10^{-7})$$
$$= -\log 1.00 - \log 10^{-7}$$
$$= 0.00 - (-7.00) = \underline{\underline{7.00}}^{\star}$$

Also in a neutral solution, $[OH^-] = 1.00 \times 10^{-7}$.

$$pOH = \underline{\underline{7.00}}$$

Example 15-3

What is the pH of a 0.015 M solution of $HClO_4$? What is the pOH of this solution?

Solution:

$$HClO_4 + H_2O \longrightarrow H_3O^+ + ClO_4^-$$

Since this is a strong acid, all of the $HClO_4$ is ionized to produce H_3O^+. Therefore, $[H_3O^+] = 0.015$.

$$pH = -\log[H_3O^+] = -\log(1.5 \times 10^{-2})$$
$$= -\log 1.5 - \log 10^{-2}$$
$$= -0.18 + 2.00 = \underline{\underline{1.82}}$$

$$pH + pOH = 14.00$$

$$pOH = 14.00 - pH$$
$$= 14.00 - 1.82 = \underline{\underline{12.18}}$$

Acidic, basic, and neutral solutions were previously defined in terms of concentrations. We can now do the same thing in terms of pH and pOH.

Neutral	$pH = pOH = 7.00$
Acidic	$pH < 7.00$ and $pOH > 7.00$
Basic	$pH > 7.00$ and $pOH < 7.00$

Remember, then, that a high pH (greater than 7) means basic and a low pH (less than 7) means acidic.

You've probably heard of the Richter scale for measuring earthquake intensity, especially if you're from California. If so, you are aware that an earthquake with a reading of 8.0 on the Richter scale is 10 times more powerful than one with a reading of 7.0. This is because the Richter scale, like pH, is a logarithmic scale. In such a scale, a difference of one integer actually represents a 10-fold change. For example, a solution of pH = 4 is 10 times more

---

$\star$ It is legitimate to express the pH of a three-significant-figure number (e.g., 1.00×10^{-7}) to three places to the right of the decimal point. However, since pH is rarely measured to that degree of precision, the pH is usually expressed to no more than two decimal places (e.g., 7.00).

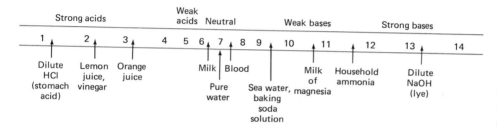

Figure 15-4 THE pH OF SOME COMMON SUBSTANCES.

acidic than a solution of pH = 5. A pH = 1 solution is actually 100,000 times more acidic than a pH = 6 solution, although both solutions are labled acidic. The pH for some common substances is given in Figure 15-4.

See Problems 15-8 through 15-14

15-3 Equilibria of Weak Acids and Bases in Water

The following equilibrium illustrates the dissociation of a weak acid in H_2O:

$$HOCl(aq) + H_2O \rightleftharpoons H_3O^+(aq) + OCl^-(aq)$$

The equilibrium constant expression is

$$K_a = \frac{[H_3O^+][OCl^-]}{[HOCl]}$$

K_a *is called the acid ionization constant.* (These reactions involve dilute solutions, which means that the $[H_2O]$ is essentially a constant. As mentioned before, constants are not included in an equilibrium constant expression.)

The values for the equilibrium constants for weak acids and bases can be calculated, as illustrated in the following examples.

Example 15-4

In a 0.20 M solution of HNO_2 it is found that 0.009 mol/liter of the HNO_2 dissociates. What is the concentration of H_3O^+, NO_2^-, and HNO_2 at equilibrium, and what is the value of K_a?

Procedure:

1 Write the equilibrium equation.

2 Calculate the $[HNO_2]$ at equilibrium from the initial concentration minus the concentration that dissociates.

$$[HNO_2]_{eq} = [HNO_2]_{initial} - [HNO_2]_{dissociated}$$

3 Notice that one H_3O^+ and NO_2^- are formed for each HNO_2 that dissociates (from the equation stoichiometry).

$$[H_3O^+]_{eq} = [NO_2^-]_{eq} = [HNO_2]_{dissociated}$$

4 Calculate K_a.

Similar problems are illustrated in Appendix B (Examples B-13, B-14)

Solution:

1
$$HNO_2 + H_2O \rightleftharpoons H_3O^+ + NO_2^-$$

2
$$[HNO_2]_{initial} = 0.20 \qquad [HNO_2]_{dissociated} = 0.009$$
$$[HNO_2]_{eq} = 0.20 - 0.009 = \underline{0.19}$$

3
$$[H_3O^+] = [NO_2^-] = 0.009$$

4
$$K_a = \frac{[H_3O^+][NO_2^-]}{[HNO_2]} = \frac{[0.009][0.009]}{[0.19]}$$
$$= \underline{\underline{4 \times 10^{-4}}}$$

Example 15-5
A 0.25 M solution of HCN has pH = 5.00. What is K_a?

Procedure:

1 Write the equilibrium reaction.

2 Convert pH to $[H_3O^+]$.

3 Notice that $[CN^-] = [H_3O^+]$ at equilibrium.

4 Calculate [HCN] at equilibrium, which is

$$[HCN]_{eq} = [HCN]_{initial} - [HCN]_{dissociated}$$

$[HCN]_{dissociated} = [H_3O^+]_{eq}$, since one H_3O^+ is produced at equilibrium for every HCN that dissociates.

5 Use these values to calculate K_a.

Solution:

$$HCN + H_2O \rightleftharpoons H_3O^+ + CN^-$$

$$[H_3O^+] = antilog\ 5.00$$

$$[H_3O^+] = 1.0 \times 10^{-5} = [CN^-]$$

$$[HCN]_{eq} = 0.25 - (1.0 \times 10^{-5}) \approx 0.25$$

($\approx$ means approximately equal)

Notice that the amount of HCN that dissociates ($10^{-5}\ M$) is negligible compared to the initial concentration of HCN (0.25 M).* therefore, $[HCN]_{eq} = [HCN]_{initial}$.

$$K_a = \frac{[H_3O^+][CN^-]}{[HCN]}$$
$$= \frac{(1.0 \times 10^{-5})(1.0 \times 10^{-5})}{(0.25)}$$
$$= \underline{\underline{4.0 \times 10^{-10}}}$$

* For the purposes of these calculations (two significant figures), a number is negligible compared to another if it is less than 10% of the larger number.

Example 15-6

A 0.10 M solution of NH_3 is 1.34% ionized. What is the value for K_b (*the base ionization consant*)?

Procedure:

Find the concentration of all species in the equilibrium constant expression at equilibrium. Substitute these values into the expression and solve for K_b.

Solution:

$$NH_3 + H_2O \rightleftharpoons NH_4^+ + OH^-$$

$$K_b = \frac{[NH_4^+][OH^-]}{[NH_3]}$$

At equilibrium, 1.34% of the NH_3 is ionized, or

$$0.0134 \times 0.10 = 1.34 \times 10^{-3} \text{ mol/liter}$$

According to the equation, for every NH_3 ionized, one NH_4^+ and one OH^- will form. Therefore, if 1.34×10^{-3} mol/liter ionize, at equilibrium

$$[NH_4^+] = [OH^-] = 1.34 \times 10^{-3}$$

The concentration of NH_3 at equilibrium is the initial concentration (0.10) minus the concentration that ionized.

$$[NH_3] = 0.10 - (1.34 \times 10^{-3}) = 0.10$$

Substituting these values into the expression,

$$K_b = \frac{(1.34 \times 10^{-3})(1.34 \times 10^{-3})}{(0.10)}$$

$$= 1.8 \times 10^{-5}$$

The values of K_a for some common weak acids are shown in Table 15-1 and K_b for some weak bases in Table 15-2. Notice that the smaller the value of K_a, the weaker the acid. For bases, the smaller the value of K_b, the weaker the base.

Table 15-1 K_a for Some Weak Acids

$$K_a = \frac{[H_3O^+][X^-]}{[HX]}$$ (HX symbolizes a weak acid, X^- its conjugate base)

Acid	Formula	K_a	
Hydrofluoric	HF	6.7×10^{-4}	
Nitrous	HNO_2	4.5×10^{-4}	
Formic	$HCHO_2$	1.8×10^{-4}	Decreasing
Acetic	$HC_2H_3O_2$	1.8×10^{-5}	acid
Hypoclorous	HOCl	3.2×10^{-8}	strength
Hypobromous	HOBr	2.1×10^{-9}	
Hydrocyanic	HCN	4.0×10^{-10}	

Table 15-2 K_b for Some Weak Bases

$$K_b = \frac{[HB^+][OH^-]}{[B]}$$ (B symbolizes a weak base, HB^+ its conjugate acid)

Base	Formula	K_b	
Dimethylamine	$NH(CH_3)_2$	7.4×10^{-4}	Decreasing
Ammonia	NH_3	1.8×10^{-5}	base
Hydrazine	N_2H_4	9.8×10^{-7}	↓ strength

See Problems 15-15 through 15-21

The values for K_a or K_b can now be used to calculate the pH from a known concentration of weak acid or base, as illustrated in the following examples.

Example 15-7

What is the pH of a 0.155 M solution of HOCl?

Procedure:

1 Write the equilibrium involved.

2 Write the appropriate equilibrium constant expression.

3 Let $x = [H_3O^+]$ at equilibrium; since $[H_3O^+] = [OCl^-]$, $x = [OCl^-]$.

4 At equilibrium, $[HOCl]_{eq} = [HOCl]_{initial} - [HOCl]_{dissociated}$

5 Using the value for K_a, solve for x.

6 Convert x to pH.

In summary:

	[HOCl]	[H₃O⁺]	[OCl⁻]
Initial	0.155	0	0
Equilibrium	$0.155 - x$	x	x

Solution:

1 $$HOCl + H_2O \rightleftharpoons H_3O^+ + OCl^-$$

2 $$K_a = \frac{[H_3O^+][OCl^-]}{[HOCl]} = 3.2 \times 10^{-8}$$

3 At equilibrium, $[H_3O^+] = [OCl^-] = x$.

4 At equilibrium, $[HOCl] = 0.155 - x$.

5
$$\frac{[x][x]}{[0.155 - x]} = 3.2 \times 10^{-8}$$

The solution of this equation appears to require the quadratic equation. However, a simplification of this calculation is possible. Notice that K_a is a small number, indicating that the degree of dissociation is small (the equilibrium lies far to the left). This means that x is a very small number. Since very small numbers make little or no difference when added to or subtracted from large numbers, they can be ignored with little or no error. (Refer to the example of the large crowd in Section 2-1.)

$$0.155 - x \approx 0.155$$

Therefore, the expression can now be simplified as follows:

$$\frac{[x][x]}{[0.155 - x]} = \frac{x^2}{0.155} = 3.2 \times 10^{-8}$$

$$x^2 = 5.0 \times 10^{-9}$$

To solve for x, take the square root of both sides of the equation:

$$\sqrt{x^2} = \sqrt{5.0 \times 10^{-9}} = \sqrt{50 \times 10^{-10}}$$

$$x = [H_3O^+] = 7.1 \times 10^{-5}$$

(Notice that x is indeed much smaller than 0.155.)

6
$$pH = -\log[H_3O^+] = \underline{4.15}.$$

Example 15-8
What is the pH of a 0.245 M solution of N_2H_4?

Solution:

1
$$N_2H_4 + H_2O \rightleftharpoons HN_2H_4^+ + OH^-$$

2
$$K_b = \frac{[HN_2H_4^+][OH^-]}{[N_2H_4]} = 9.8 \times 10^{-7}$$

3 Let $x = [OH^-] = [HN_2H_4^+]$ (at equilibrium).
4 $[N_2H_4] = 0.245 - x$ (at equilibrium)
Since K_b is very small, x is very small. Therefore,

$$[0.245 - x] \approx [0.245]$$

5
$$\frac{[x][x]}{[0.245]} = 9.8 \times 10^{-7}$$

$$x^2 = 2.4 \times 10^{-7}$$

$$x = 4.9 \times 10^{-4} = [OH^-]$$

6 $pOH = -\log[OH^-] = -\log(4.9 \times 10^{-4}) = 3.31$

$$pH = 14 - pOH = 14 - 3.31 = \underline{10.69}$$

See Problems 15-22 through 15-26

15-4 Ions as Acids and Bases: Hydrolysis

Aqueous solutions of many salts, such as $NaCl$, form neutral solutions. Other salts, such as $NaC_2H_3O_2$ and NH_4Cl, interact with water to produce a slightly basic or a slightly acidic solution, respectively. *The reaction of an anion with water to produce a basic solution and the reaction of a cation with water to produce an acidic solution is known as* **hydrolysis.**

The phenomenon of hydrolysis is readily explained by the Brønsted-Lowry definition of acids and bases introduced in Chapter 12. First let's review the role of the ions of strong acids and bases in water. In the reaction of the acid HCl with water, the HCl is completely dissociated or ionized:

$$HCl + H_2O \xrightarrow{\text{100\%}} H_3O^+ + Cl^-$$

Since this is essentially an irreversible reaction, we can conclude that the Cl^- ion has negligible basic properties in water (although it is formally identified as a "conjugate" base in the Brønsted-Lowry definition.) In summary, the following reaction does *not* occur:

$$Cl^- + H_2O \xrightarrow{\;\;/\!\!/\;\;} HCl + OH^-$$

Rule 1 *The conjugate base of a strong acid does not hydrolyze and does not affect the pH.* (An exception is the acid salt anion HSO_4^-, which is itself acidic.)

The same general rule applies to strong bases. KOH is completely dissociated in H_2O:

$$KOH(s) + H_2O \xrightarrow{\text{100\%}} K^+(aq) + OH^-(aq)$$

The cation of the base (K^+) does not interact with H_2O and does not affect the pH.

Rule 2 *The cation of a strong base does not hydrolyze and does not affect the pH* [e.g., K^+ from KOH, Ca^{2+} from $Ca(OH)_2$].

Now let's review the equilibrium involving the weak acid HCN:

$$HCN + H_2O \rightleftharpoons H_3O^+ + CN^-$$

Since the reverse reaction *does* take place in this case, the CN^- ion (the conjugate base of HCN) behaves as a base by accepting an H^+ from H_3O^+. It is also a base by the Arrhenius definition, since it raises $[OH^-]$ by removing some H_3O^+ from solution. The CN^- ion demonstrates its base behavior to a small extent even in pure water, as shown by the following hydrolysis equilibrium:

$$CN^- + H_2O \rightleftharpoons HCN + OH^-$$

Rule 3 *The conjugate base of a weak acid hydrolyzes to produce a weakly basic solution.* The general reaction is

$$X^- + H_2O \rightleftharpoons HX + OH^-$$

Acid salt anions are special cases and are discussed in the sample problems.

When NH_3 reacts with H_2O, the equilibrium produces its conjugate acid, NH_4^+:

$$NH_3 + H_2O \rightleftharpoons NH_4^+ + OH^-$$

Since the reverse reaction takes place, the NH_4^+ ion acts as an acid. The NH_4^+ demonstrates its acid behavior to a small extent even in pure water, as shown by the following hydrolysis equilibrium:

$$NH_4^+ + H_2O \rightleftharpoons NH_3 + H_3O^+$$

Rule 4 *The conjugate acid of a weak base hydrolyzes to produce a weakly acidic solution.* The general reaction is

$$MH^+ + H_2O \rightleftharpoons M + H_3O^+$$

In the following examples, indicate whether a solution of the salt is acidic, basic, or neutral, and write the reaction illustrating this behavior. To do this, both the cation and anion must be examined for hydrolysis behavior.

Example 15-9: KNO_2

Solution:

K^+ is the cation of the strong base KOH and does not hydrolyze. The NO_2^- ion, however, is the conjugate base of the weak acid HNO_2 and hydrolyzes as follows:

$$NO_2^- + H_2O \rightleftharpoons HNO_2 + OH^-$$

Since OH^- is formed in this solution, the solution is <u>basic</u>.

Example 15-10: $Ca(NO_3)_2$

Solution:

Ca^{2+} is the cation of the strong base, $Ca(OH)_2$, and does not hydrolyze. NO_3^- is the conjugate base of the strong acid HNO_3 and does not hydrolyze either. Since neither ion hydrolyzes, the solution is <u>neutral</u>.

Example 15-11: $H_2N(CH_3)_2^+Br^-$

Solution:

$H_2N(CH_3)_2^+$ is the conjugate acid of the weak base $HN(CH_3)_2$. It undergoes hydrolysis according to the equation

$$H_2N(CH_3)_2^+ + H_2O \rightleftharpoons HN(CH_3)_2 + H_3O^+$$

The Br^- ion is the conjugate base of the strong acid HBr and does not hyrolyze. Since only the cation undergoes hydrolysis, the solution is <u>acidic</u>.

Example 15-12: NH_4F

Solution:

NH_4^+ is the conjugate acid of the weak base NH_3 and the F^- ion is the conjugate base of the weak acid HF. Both ions hydrolyze as follows:

$$NH_4^+ + H_2O \rightleftharpoons NH_3 + H_3O^+$$

$$F^- + H_2O \rightleftharpoons HF + OH^-$$

Obviously, a solution cannot be acidic and basic at the same time, since we previously learned that only a very small concentration of H_3O^+ and OH^- can coexist in the same solution. However, notice that both reactions reach a point of equilibrium. The reaction whose equilibrium lies farther to the right predominates in determining whether the solution is acidic or basic. Recall that the value of the equilibrium constant (which in this case is called the hydrolysis constant) tells us about the point of equilibrium. Although hydrolysis has not been treated quantitatively in this text, it is a fact that the equilibrium constant for the first hydrolysis is larger than that for the second hydrolysis. Thus the hydrolysis reaction involving NH_4^+ lies farther to the right, which means that the solution is <u>acidic</u>. As you can see, however, the respective values of equilibrium constants for the hydrolysis reactions are needed to determine the acidity of a solution when both cation and anion hydrolyze.

Example 15-13: NaHCO₃

Solution:
Na^+ is the cation of the strong base NaOH and does not hydrolyze. HCO_3^- is the conjugate base of the weak acid H_2CO_3, so the following equilibrium is possible:

$$HCO_3^- + H_2O \rightleftharpoons H_2CO_3 + OH^-$$

Since HCO_3^- is the anion of an acid salt, a second equilibrium is possible involving the acid nature of the anion:

$$HCO_3^- + H_2O \rightleftharpoons CO_3^{2-} + H_3O^+$$

As in the previous example, both equilibria do not occur to the same extent. The equilibrium that lies farther to the right determines whether the solution is acidic or basic. In this case the constant for the hydrolysis in the first equilibrium is larger than that for the ionization of the acid in the second equilibrium. This means that the hydrolysis equilibrium lies farther to the right and the solution is <u>basic</u>. Again the values of the equilibrium constants for the two reactions are needed to determine whether a solution of an acid salt anion is acidic or basic.

See Problems 15-27 through 15-30

15-5 Buffer Solutions

The addition of small amounts of a strong acid or base to pure water causes drastic changes in the pH. For example, if 0.10 mol of HCl is added to 1 liter of pure water, the pH changes from 7.0 to 1.0. On the other hand, if 0.10 mol of KOH is added to 1 liter of pure water, the pH changes from 7.0 to 13.0. In contrast, our bloodstream is a water solution with almost the same pH as pure water (the pH of blood = 7.4), but unlike pure water, the pH of blood changes very little despite all of the acids and bases we ingest in our foods. The blood resists changes in pH because it is a buffer solution. *A* **buffer solution** *is a solution that can resist changes in pH from the addition of limited amounts of a strong acid or base.*

To understand how a buffer solution works, let's again consider the dissociation of a weak acid:

$$HCN + H_2O \rightleftharpoons H_3O^+ + CN^- \qquad K_a = 4.0 \times 10^{-10}$$

In a solution of HCN, the constant tells us that the equilibrium lies far to the left. Most of the acid is present as undissociated HCN molecules. What happens if we now add a small amount of KOH to this solution? The OH^- reacts with the H_3O^+, quickly removing it from solution:

$$H_3O^+ + OH^- \longrightarrow 2H_2O$$

However, the dissociation of the acid is a dynamic equilibrium that responds to stress according to the principle of Le Châtelier. The lost H_3O^+ is immediately replenished by further dissociation from the reservoir of undissociated HCN. Thus the equilibrium produces not only enough H_3O^+ to react with the added OH^- but restores essentially all of the original equilibrium concentration of H_3O^+. As a result, the addition of small amounts of OH^- to weak acid solutions changes the pH very little. This assumes that there is more weak acid present than added OH^-.

Could this solution also resist a change in pH if additional H_3O^+ is added to the solution? No. A solution of a weak acid *alone* could not resist a change in pH in this direction. However, notice that the added H_3O^+ could be removed if a reservoir of the conjugate base of the acid (i.e., CN^-) is also present in the solution. The CN^- reacts with the added H_3O^+ in the reverse reaction of the dissociation:

$$H_3O^+ + CN^- \longrightarrow HCN + H_2O$$

Thus a buffer solution resists changes in pH because it contains *both* a concentration of a weak acid (e.g., HCN) and a salt of its conjugate base (e.g., NaCN) *in the same solution*. A buffer solution may also be composed of a solution of a weak base (e.g., NH_3) and a salt of its conjugate acid (e.g., NH_4Cl).

As an analogy to the buffer solution, let's consider the case of a rather eccentric student who likes to maintain an equilibrium concentration of $5 in her purse. She also keeps in reserve in her room extra money, with each dollar bill in an individual envelope. If she spends some money from her purse, it is replaced by removing the appropriate number of dollars from envelopes and placing the money in her purse (see Figure 15-5a). The $5 in the student's purse is analogous to the equilibrium concentration of H_3O^+ in the buffer solution. The money in the envelopes that is available to replace spent dollars in the purse is analogous to the HCN that can ionize to replace H_3O^+ removed from the solution (by reaction with OH^-). Now, what if the student earns some dollars? Obviously, the extra money must be removed from the purse. If she also keeps a supply of empty envelopes in her room, she can place each extra dollar in an envelope, thereby increasing her dollar reserve (see Figure 15-5b). The empty envelopes are analogous to the CN^- ions in the buffer solution. Just as the empty envelopes are available to remove added

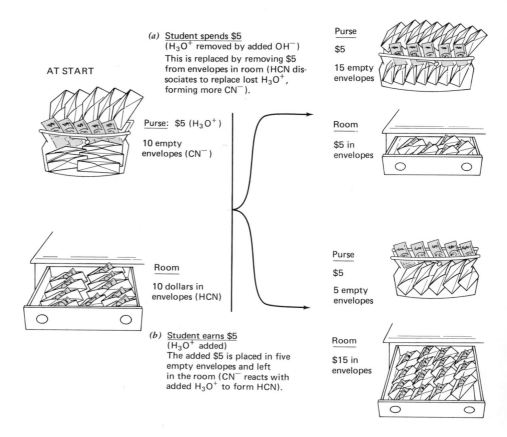

(a) <u>Student spends $5</u>
(H_3O^+ removed by added OH^-)
This is replaced by removing $5
from envelopes in room (HCN dissociates to replace lost H_3O^+,
forming more CN^-).

AT START

<u>Purse:</u> $5 ($H_3O^+$)

10 empty
envelopes (CN^-)

Purse
$5

15 empty
envelopes

Room
$5 in
envelopes

Room

10 dollars in
envelopes (HCN)

Purse
$5

5 empty
envelopes

(b) <u>Student earns $5</u>
(H_3O^+ added)
The added $5 is placed in five
empty envelopes and left
in the room (CN^- reacts with
added H_3O^+ to form HCN).

Room
$15 in
envelopes

Figure 15-5 AN
ANALOGY TO A
BUFFER
SOLUTION.

dollars, the CN^- ions are available to react with added H_3O^+ ions to form additional HCN. Thus, if the student has a supply of envelopes with dollars as well as a supply of empty envelopes, she is ready for either situation: spending or earning dollars. Likewise, if a solution contains a concentration of a weak acid and a concentration of its conjugate base it can resist change in either direction: addition of H_3O^+ or removal of H_3O^+.

As mentioned, a common buffered solution is blood, which must maintain a nearly constant pH for metabolism. This is accomplished by a combination of the following two buffer systems:

$$H_2CO_3 + H_2O \rightleftharpoons H_3O^+ + HCO_3^-$$

$$H_2PO_4^- + H_2O \rightleftharpoons H_3O^+ + HPO_4^{2-}$$

If for some reason the buffer breaks down and the pH of the blood drops lower than 7.2 (acidosis) or rises higher than 7.8 (alkalosis), serious difficulties are encountered—such as dizziness, coma, and muscle spasms. Since we ingest many acidic and basic substances, a healthy buffering system is necessary to maintain a controlled pH.

**See Problems
15-31, 15-32**

Review Questions

1 What is meant by the autoionization of water?
2 Give the value of K_w, the ion product of water.
3 Describe the relationship between $[H_3O^+]$ and $[OH^-]$ in water.
4 Define acidic, basic, and neutral solutions in terms of $[H_3O^+]$.
5 Describe how a substance can act as an acid by reaction with OH^-.
6 Define pH and pOH. How are they related?

7 Define acidic, basic, and neutral solutions in terms of pH.
8 What does the value of K_a tell us about the strength of the acid?
9 What is hydrolysis?
10 Give the rules used to determine which ions hydrolyze in water.
11 How is a buffer solution prepared?
12 How does a buffer solution resist changes in pH?

Problems

$[H_3O^+]$, $[OH^-]$, and K_w

15-1 If some ions are present in pure H_2O, why isn't pure water considered to be an electrolyte? (Consider the significance of K_w.)

15-2 (a) What is $[OH^-]$ when $[H_3O^+] = 1.50 \times 10^{-3}$?
  (b) What is $[H_3O^+]$ when $[OH^-] = 2.58 \times 10^{-7}$?
  ■ (c) What is $[H_3O^+]$ when $[OH^-] = 56.9 \times 10^{-9}$?

15-3 When 0.250 mol of the strong acid $HClO_4$ is dissolved in 10.0 liters of water, what is $[H_3O^+]$? What is $[OH^-]$?

■ 15-4 Lye is a very strong base. What is $[H_3O^+]$ in a 2.55 M solution of NaOH? In the weakly basic household ammonia, $[OH^-] = 4.0 \times 10^{-3}$; What is $[H_3O^+]$?

15-5 Why can't $[H_3O^+] = [OH^-] = 1.00 \times 10^{-2}$ in water? What would happen if we tried to make such a solution by mixing 10^{-2} mol/liter of KOH with 10^{-2} mol/liter of HCl?

Acidic and Basic Solutions

15-6 Identify the three solutions in Problem 15-2 as acidic, basic, or neutral.

15-7 Identify each of the following as an acidic, basic, or neutral solution:
  (a) $[H_3O^+] = 6.5 \times 10^{-3}$
  (b) $[H_3O^+] = 5.5 \times 10^{-10}$
  ■ (c) $[OH^-] = 10.0 \times 10^{-8}$
  ■ (d) $[OH^-] = 8.1 \times 10^{-4}$
  ■ (e) $[H_3O^+] = 518 \times 10^{-8}$

pH and pOH

15-8 What is the pH of each of the following solutions?
  (a) $[H_3O^+] = 1.0 \times 10^{-4}$
  ■ (b) $[H_3O^+] = 6.8 \times 10^{-9}$
  (c) $[H_3O^+] = 1.5$
  ■ (d) $[OH^-] = 0.058 \times 10^{-3}$
  (e) $[OH^-] = 460 \times 10^{-14}$
  Identify each of the five solutions above as acidic or basic.

15-9 What is $[H_3O^+]$ when:
  (a) pH = 3.00 (b) pH = 3.54 (c) pOH = 8.00 ■ (d) pOH = 6.38 ■ (e) pH = 12.70

15-10 What is the pH of a 0.075 M solution of the strong acid HNO_3?

■ 15-11 What is the pH of a 0.86 M solution of the strong base KOH?

15-12 What is the pH of a 0.18 M solution of $Ca(OH)_2$?

15-13 Identify each of the following solutions as strongly basic, weakly basic, neutral, weakly acidic, or strongly acidic:
  (a) pH = 1.5   (e) pOH = 7.0
  (b) pOH = 13.0 ■ (f) pH = 8.5
  (c) pH = 5.8   ■ (g) pOH = 7.5
  (d) pH = 13.0   ■ (h) pH = -1.00

15-14 Order the following substances in order of *increasing* acidity:
  (a) Household ammonia, pH = 11.4
  (b) Vinegar, pH = 2.6
  (c) Grape juice, pH = 4.0
  (d) sulfuric acid, pH = 0.40
  (e) Eggs, pH = 7.8
  (f) Rain water, pH = 6.8

K_a, K_b, and Their Values

15-15 Write the expression for K_a or K_b for each of the following equilibria. Where necessary, complete the equilibrium.
  (a) $HOBr + H_2O \rightleftharpoons H_3O^+ + OBr^-$
  (b) $NH_3 + H_2O \rightleftharpoons NH_4^+ + OH^-$
  ■ (c) $H_2SO_3 + H_2O \rightleftharpoons H_3O^+ + HSO_3^-$
  ■ (d) $HSO_3^- + H_2O \rightleftharpoons$

$$\underline{\hspace{3cm}} + SO_3^{2-}$$

  (e) $H_3PO_4 + H_2O \rightleftharpoons$
$$H_3O^+ + \underline{\hspace{2cm}}$$
  (f) $HN(CH_3)_2 + H_2O \rightleftharpoons$
$$H_2N(CH_3)_2^+ + \underline{\hspace{2cm}}$$
  ■ (g) $H_2PO_4^- + H_2O \rightleftharpoons$

$$\underline{\hspace{3cm}} + HPO_4^{2-}$$

15-16 In a 0.20 M solution of cyanic acid, $HOCN$, $[H_3O^+] = [OCN^-] = 6.2 \times 10^{-3}$.
  (a) What is [HOCN] at equilibrium?
  (b) What is K_a?
  (c) What is the pH?

15-17 A 0.58 M solution of a weak acid (HX) is 10.0% dissociated.
  (a) Write the equilibrium and the equilibrium constant expression.
  (b) What are $[H_3O^+]$, $[X^-]$, and $[HX]$ at equilibrium?
  (c) What is K_a?
  (d) What is the pH?

■ **15-18** Nicotine (Nc) is a nitrogen base in water in the same manner as ammonia. Write the equilibrium illustrating this base behavior. In a 0.44 M solution of nicotine, $[OH^-] = [NcH^+] = 5.5 \times 10^{-4}$. What is K_b for nicotine? What is the pH of the solution?

■ **15-19** In a 0.085 M solution of carbolic acid, HC_6H_5O, the pH = 5.48. What is K_a?

15-20 Novacaine (Nv) is a nitrogen base in water, like ammonia. Write the equilib-

rium. In a 1.25 M solution of novacaine, pH = 11.46. What is K_b?

15-21 In a 0.300 M solution of chloroacetic acid, HOC_2Cl_3, $[HOC_2Cl_3] = 0.277$ M at equilibrium. What is the pH? What is K_n?

$[H_3O^+]$ and pH from K_a or K_b

15-22 What is the $[H_3O^+]$ and pH of a 0.50 M solution of HOBr? For HOBr, $K_a = 2.1 \times 10^{-9}$.

15-23 (a) What is the pH of a 0.050 M solution of HOBr?
  (b) Although the solution in this problem is a 10-fold dilution of the solution from Problem 15-22, did the $[H_3O^+]$ change by the same factor?

■ **15-24** What is the $[OH^-]$ and pH of a 1.40 M solution of $HN(CH_3)_2$? $K_b = 9.8 \times 10^{-7}$.

15-25 What is the pH of a 0.660 M solution of novacaine? Use K_b from Problem 15-20.

■ **15-26** What is the pH of a 0.15 M solution of boric acid, H_3BO_3 (eye wash)? $K_a = 6.0 \times 10^{-10}$. What is the pH of a 0.15 M solution of HBr? $H_3BO_3 + H_2O \rightleftharpoons B(OH)_4^- + H_3O^+$.

Hydrolysis

15-27 Complete the following hydrolysis equilibria:
  (a) $S^{2-} + H_2O \rightleftharpoons \underline{\hspace{2cm}} + OH^-$
  (b) $N_2H_5^+ + H_2O \rightleftharpoons N_2H_4 + \underline{\hspace{1.5cm}}$
  (c) $HPO_4^{2-} + H_2O \rightleftharpoons$
$$H_2PO_4^- + \underline{\hspace{2cm}}$$
  ■ (d) $B(OH)_4^- + H_2O \rightleftharpoons$
$$H_3BO_3 + \underline{\hspace{2cm}}$$
  ■ (e) $Al(H_2O)_6^{3+} + H_2O \rightleftharpoons$
$$Al(H_2O)_5(OH)^{2+} + \underline{\hspace{2cm}}$$

15-28 Write the hydrolysis equilibrium (if any) for each of the following ions:
  (a) F^- ■ (e) HPO_4^{2-}
  (b) SO_3^{2-} ■ (f) CN^-
  (c) $H_2N(CH_3)_2^+$ ■ (g) Li^+
  (d) Ca^{2+} ■ (h) ClO_4^-

15-29 Tell whether each of the following compounds forms acidic, basic, or neutral solutions when added to pure water. Write the reaction illustrating the acidic or basic behavior where appropriate. (Review oxides as acids and bases in Chapter 12 if necessary.)

(a) H_2S

(b) $KClO$

(c) NaI

(d) NH_3

(e) $HN_2H_4^+Br^-$

(f) $HOBr$

(g) CaO

(h) $Ba(OH)_2$

■ (i) $Sr(NO_3)_2$

■ (j) $LiNO_2$

■ (k) H_2SO_3

■ (l) Na_2SO_3

■ (m) SO_3

■ (n) NH_4ClO_4

15-30 In a lab there are five different solutions with pH's of 1.0, 5.2, 7.0, 10.2, and 13.0. The solutions are LiOH, $SrBr_2$, KOCl, NH_4Cl, and HI, all at the same concentration. Which pH corresponds to which compound? What is the concentration of these compounds?

Buffers

15-31 In a certain solution, $[HCl] = 0.20\ M$ and $[NaCl] = 0.20\ M$. Why can't this solution act as a buffer solution?

15-32 Identify which of the following form buffer solutions when 0.50 mol of both compounds are dissolved in 1 liter of water.

(a) HNO_2 and KNO_2

(b) NH_4Cl and NH_3

(c) HNO_3 and KNO_2

(d) HNO_3 and KNO_3

(e) $HOBr$ and $Ca(OBr)_2$

■ (f) HCN and $KClO$

■ (g) NH_3 and $BaBr_2$

■ (h) H_2S and $LiHS$

(i) KH_2PO_4 and K_2HPO_4

16 Organic Chemistry

Petroleum from a well like this not only supplies us with energy but is the ultimate source of most of the organic chemicals that we now take for granted.

Organic compounds are so named because it was once thought that they originated only from living matter (organisms). Although this is no longer considered to be true, the name remains. The principal element in all organic compounds is carbon (although some carbon compounds, such as $CaCO_3$ and $LiCN$, are usually not included). Organic compounds are certainly central to our lives. Hydrocarbons (compounds composed only of carbon and hydrogen) are used as fuel to power our cars and to heat our homes. Our bodies are fueled with organic compounds obtained from the food we eat in the form of sugars, carbohydrates, fats, and proteins. This food is made more palatable by organic flavorings, wrapped in organic plastic, and kept from spoiling with organic preservatives. Our clothes are made of organic compounds, whether these compounds come from plant and animal sources (cotton and wool) or are synthetic (nylon and Dacron®). These fabrics are made colorful with organic dyes. When we are ill, we take drugs which may also be organic: Aspirin relieves headaches, codeine suppresses coughs, and diazepam (Valium®) calms nerves. These are just a few examples of how we use organic chemicals daily. Although it is impossible to cover all of organic chemistry in one chapter (or even one course) we will discuss some of the various kinds of organic compounds, how they are made, and their uses.

As background to this chapter, you should review:

1 Lewis structures (Section 6-9)

2 Polarity of covalent bonds and molecules (Sections 6-11 and 6-12)

3 The solubility of polar and nonpolar compounds in solvents (Section 11-2)

16-1 Carbon and Its Chemical Bonds

Carbon has a unique ability among the elements to share its electrons with other carbon atoms. Carbon may share its electrons with one, two, three, or four other carbons, and each of these may, in turn, share their electrons with additional carbons. Other elements do not do this to a large extent. Carbon may also share electrons with hydrogen, oxygen, nitrogen, sulfur, the halogens, and other elements. For the most part, we will be concerned only with those organic compounds containing carbon, hydrogen, oxygen, and nitrogen.

Notice that we said that carbon *shares* its electrons with other carbons or other elements. Carbon forms *covalent* rather than ionic bonds, and this gives rise to many of the characteristic properties of organic molecules. Although some organic compounds are somewhat polar (especially when combined with oxygen), they are usually much less polar than inorganic compounds. Because of this, most organic chemicals are insoluble or only slightly soluble in water. Although most inorganic compounds have high melting and boiling points, most organic compounds melt and boil at much lower temperatures.

Since carbon is in group IVA and has four valence electrons, it must form a total of *four bonds* to achieve an octet of outer electrons. Carbon forms single, double, and triple bonds. Let's review some simple organic compounds to see how these molecules appear when written as Lewis structures.

Single bonds

$$CH_4 \qquad H-\overset{\displaystyle H}{\underset{\displaystyle H}{\overset{\displaystyle |}{\underset{\displaystyle |}{C}}}}-H \qquad CCl_4 \qquad :\overset{..}{\underset{..}{Cl}}-\overset{\displaystyle :\overset{..}{Cl}:}{\underset{\displaystyle :\overset{..}{Cl}:}{\overset{\displaystyle |}{\underset{\displaystyle |}{C}}}}-\overset{..}{\underset{..}{Cl}}:$$

$$C_2H_6 \qquad \begin{array}{c} \text{H} \quad \text{H} \\ | \quad | \\ \text{H}-\text{C}-\text{C}-\text{H} \\ | \quad | \\ \text{H} \quad \text{H} \end{array} \qquad C_2H_6O \qquad \begin{array}{c} \text{H} \quad \text{H} \\ | \quad | \\ \text{H}-\text{C}-\text{C}-\ddot{\text{O}}-\text{H} \\ | \quad | \quad \ddot{} \\ \text{H} \quad \text{H} \end{array}$$

Double bonds

$$C_2H_4 \qquad \begin{array}{c} \text{H} \qquad \text{H} \\ \diagdown \qquad \diagup \\ \text{C}=\text{C} \\ \diagup \qquad \diagdown \\ \text{H} \qquad \text{H} \end{array} \qquad CH_2O \qquad \begin{array}{c} \text{H} \\ \diagdown \\ \text{C}=\ddot{\text{O}}\vdots \\ \diagup \\ \text{H} \end{array}$$

Triple bonds C_2H_2 $\text{H}-\text{C}\equiv\text{C}-\text{H}$

Example 16-1

Give the Lewis structures for the following compounds:

(a) C_3H_8 (b) C_3H_6 (c) H_3C_2N (all H's on one C)

Answers:

(a) $\begin{array}{c} \text{H} \quad \text{H} \quad \text{H} \\ | \quad | \quad | \\ \text{H}-\text{C}-\text{C}-\text{C}-\text{H} \\ | \quad | \quad | \\ \text{H} \quad \text{H} \quad \text{H} \end{array}$ (b) $\begin{array}{c} \text{H} \quad \text{H} \\ | \quad | \\ \text{H}-\text{C}-\text{C}=\text{C}-\text{H} \\ | \quad | \\ \text{H} \quad \text{H} \end{array}$ or $\begin{array}{c} \text{H} \qquad \text{H} \\ | \qquad | \\ \text{H}-\text{C}\text{---}\text{C}-\text{H} \\ \diagdown \text{C} \diagup \\ \diagup \quad \diagdown \\ \text{H} \qquad \text{H} \end{array}$

(c) $\begin{array}{c} \text{H} \\ | \\ \text{H}-\text{C}-\text{C}\equiv\text{N}\vdots \\ | \\ \text{H} \end{array}$

See Problem 16-1

Many times we find that there is more than one Lewis structure for a given formula. Such is the case with butane, C_4H_{10}, since two correct structures can be drawn. *Compounds with different structures but the same molecular formula are called* **isomers.**

$$\begin{array}{c} \text{H} \quad \text{H} \quad \text{H} \quad \text{H} \\ | \quad | \quad | \quad | \\ \text{H}-\text{C}-\text{C}-\text{C}-\text{C}-\text{H} \\ | \quad | \quad | \quad | \\ \text{H} \quad \text{H} \quad \text{H} \quad \text{H} \end{array} \quad \begin{array}{c} \text{Isomers} \\ \longleftarrow \quad \text{of} \quad \longrightarrow \\ \text{each other} \end{array} \quad \begin{array}{c} \text{H} \quad \text{H} \quad \text{H} \\ | \quad | \quad | \\ \text{H}-\text{C}-\text{C}-\text{C}-\text{H} \\ | \quad | \quad | \\ \text{H} \quad | \quad \text{H} \\ \quad \text{H}-\text{C}-\text{H} \\ \quad | \\ \quad \text{H} \end{array}$$

$\qquad\qquad$ *n*-Butane $\qquad\qquad\qquad\qquad\qquad\qquad\qquad\qquad$ Isobutane

We refer to isomers by different names and by drawing out the structures rather than just giving the molecular formula. When drawing the structure, it can become awkward writing out all of the carbons and hydrogens, especially when there are many of them. A **condensed formula** is used, where

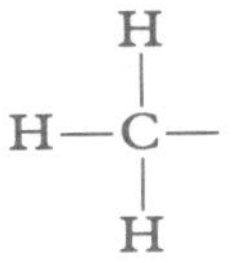

$$
\begin{array}{c}
\text{H} \\
| \\
\text{H} - \text{C} - \\
| \\
\text{H}
\end{array}
$$

is represented as

$$CH_3 -$$

and

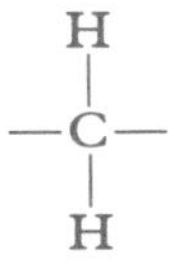

$$
\begin{array}{c}
\text{H} \\
| \\
- \text{C} - \\
| \\
\text{H}
\end{array}
$$

is represented as

$$-CH_2-$$

Depending on what we are trying to show, the structure may be partially or fully condensed.

	Partially condensed	Fully condensed
n-Butane*	$CH_3 - CH_2 - CH_2 - CH_3$	$CH_3(CH_2)_2CH_3$

Isobutane*
$$
CH_3 - \overset{\displaystyle \overset{\text{H}}{|}}{\underset{\displaystyle \underset{\text{CH}_3}{|}}{C}} - CH_3
\qquad (CH_3)_3CH
$$

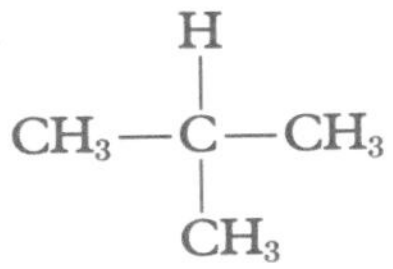

A few other compounds and their isomers are shown in Table 16-1. As you can see, the number of isomers increases as the number of carbons increase, and the addition of a **hetero atom** (*any atom other than carbon or hydrogen*) also increases the number of isomers.

16-2 Alkanes

The most basic types of organic compounds are called hydrocarbons. **Hydrocarbons** *contain only carbon and hydrogen.* **Alkanes** *are the simplest of all hy-*

* The *n* refers to the isomer in which all the C's are bound consecutively in a straight chain. The *iso* refers to an isomer in which there are branches along the chain of carbon atoms.

Table 16-1   Isomers

Formula	Isomers (Names)		
C_5H_{12}	$CH_3CH_2CH_2CH_2CH_3$ *n*-Pentane	$CH_3CH_2\overset{\displaystyle CH_3}{\overset{\displaystyle \vert}{CH}}{-}CH_3$ Isopentane	$CH_3{-}\overset{\displaystyle CH_3}{\underset{\displaystyle CH_3}{\overset{\displaystyle \vert}{\underset{\displaystyle \vert}{C}}}}{-}CH_3$ Neopentane
C_3H_6	$H_2C{=}CHCH_3$ Propene	$H_2C\underset{\displaystyle CH_2}{\overset{\displaystyle \diagdown \diagup}{\overline{\qquad}}}CH_2$ Cyclopropane	Notice that the carbons can also be arranged in a ring or *cyclic* structure.
C_2H_6O	CH_3CH_2OH Ethanol	$H_3C{-}O{-}CH_3$ Dimethyl ether	
C_3H_6O	$CH_3CH_2\overset{\displaystyle O}{\overset{\displaystyle \|}{C}}H$ Propanal	$H_3C{-}\overset{\displaystyle O}{\overset{\displaystyle \|}{C}}{-}CH_3$ Propanone	$H_2C{=}CHCH_2OH$ Allyl alcohol and others

See Problems 16-2, 16-3

drocarbons, since they contain only carbon and hydrogen joined together by single bonds. The general formula for open-chained alkanes is C_nH_{2n+2}. The simplest alkane ($n = 1$) is methane, in which the carbon shares a pair of electrons with four different hydrogen atoms. The next alkane is ethane (C_2H_6), and the third alkane is propane (C_3H_8). These alkanes are members of a homologous series. *In a* **homologous series,** *the next member differs from the previous one by a constant amount, which in this case is CH_2.*

The homologous series of alkanes up to 10 carbons is given in Table 16-2 together with the names of the compounds.

The names in Table 16-2 are the basis of the names of all organic compounds. By slightly altering them, we can name other classes of organic compounds that are discussed later. There are two systems of nomenclature used in organic chemistry. The most systematic is the one devised by the International Union of Pure and Applied Chemists (the IUPAC system). Although the rules for naming complex molecules can be extensive, we will be concerned just with the basic concept. Compounds are also known by *common* or

Table 16-2 Alkanes

Formula	Name	Formula	Name
CH_4	<u>Meth</u>ane	C_6H_{14}	<u>Hex</u>ane
C_2H_6	<u>Eth</u>ane	C_7H_{16}	<u>Hept</u>ane
C_3H_8	<u>Prop</u>ane	C_8H_{18}	<u>Oct</u>ane
C_4H_{10}	<u>But</u>ane	C_9H_{20}	<u>Non</u>ane
C_5H_{12}	<u>Pent</u>ane	$C_{10}H_{22}$	<u>Dec</u>ane

trivial names. Sometimes these names follow a pattern, sometimes they do not. They have been used for so many years that it is hard to break the habit of using them. When a chemical which is frequently known by its common name is encountered, that name is given in parentheses.

The name of a simple organic compound has two parts; the *prefix* gives the number of carbons in the longest carbon chain and the *suffix* tells what kind of a compound it is. The underlined portions of the names in Table 16-2 are the prefixes used for compounds containing one through 10 carbons in the longest chain; *meth-* stands for one carbon, *eth-* for two carbons, and so on. The ending used for alkanes is *-ane*. Therefore, the one-carbon alkane is methane, the two-carbon alkane is ethane.

Organic compounds can exist as unbranched compounds (all carbons bound to each other in a continuous chain) or as branched compounds. Previously, we indicated that *n*-butane is an unbranched alkane and that isobutane is a branched alkane. The IUPAC system bases its names on the longest carbon chain in the molecule, whereas the common names frequently include all of the carbons in the name (e.g., isobutane). The longest carbon chain in isobutane is three carbons long and is therefore considered a propane in the IUPAC system. Notice in isobutane that there is a CH_3— group attached to the propane chain. In this system of nomenclature, the branches are named separately. *Since the branches can be considered as groups of atoms substituted for a hydrogen, the branches are called* **substituents.** *Substituents that contain one less hydrogen than an alkane are called* **alkyl groups.** Alkyl groups are not compounds by themselves; they must always be attached to some other group or atom. They are named by taking the alkane name, dropping the *-ane* ending, and substituting *-yl.* The most common alkyl groups are given in Table 16-3.

Thus the alkyl sustituent in isobutane is a methyl group. The IUPAC name for isobutane is *methyl propane.*

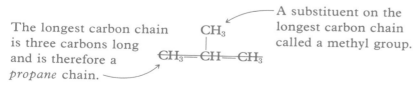

The longest carbon chain is three carbons long and is therefore a *propane* chain. ⟶

A substituent on the longest carbon chain called a methyl group.

Table 16-3 Alkyl Groups

Alkyl Group[a]	Name	Alkyl Group[a]	Name
CH_3—	Methyl	$(CH_3)_2CH$—	Isopropyl
CH_3CH_2—	Ethyl	$CH_3CH_2CH_2CH_2$—	n-Butyl
$CH_3CH_2CH_2$—	n-Propyl	$(CH_3)_3C$—	tert-Butyl
			(or t-Butyl)

[a] The dashed line shows where the alkyl group is attached to a carbon chain or to another atom such as a halogen, oxygen, or a nitrogen.

When the carbons form a ring, the name is prefixed with *cyclo-*. Therefore, the three-carbon ring compound

$$
\begin{array}{ccc}
H & & H \\
\diagdown & C & \diagup \\
& | & \\
H-C & \!\!\!\!-\!\!\!\! & C-H \\
| & & | \\
H & & H
\end{array}
$$

is called cyclopropane. Cycloalkanes have the general formula C_nH_{2n}.

There is much more that could be said about naming compounds, but we will limit ourselves to recognizing the number of carbons in the longest chain, the type of compound, and the presence of substituents when given the name of a compound or its structure.

Example 16-2
Draw a line through the longest carbon chain in the following compounds and circle the substituents. Name the longest chain.

$$
\begin{array}{c}
\quad\; CH_3 \quad CH_3 \;\; CH_3 \\
\quad\;\; | \qquad | \qquad | \\
\text{(a)}\; CH_2-CH-CH \\
\qquad\qquad\quad\; | \\
\qquad\qquad\quad CH_2-CH_3
\end{array}
$$

$$
\begin{array}{c}
\qquad\qquad\quad CH_2-CH_2-CH_3 \\
\qquad\qquad\quad | \\
\text{(c)}\; CH_3CH_2-CH-CH_2-CH_2-CH_2-CH_3
\end{array}
$$

$$
\begin{array}{c}
\qquad\quad CH-CH_3 \\
\qquad\quad\; | \\
\text{(b)}\; CH_2-CH-CH_2-CH_3 \\
\quad\; | \\
\quad CH_3
\end{array}
$$

$$
\begin{array}{c}
\qquad\; CH_3 \\
\qquad\;\; | \\
\text{(d)}\; H-C-CH_2-CH_3 \\
\qquad\;\; | \\
\qquad\; CH_3
\end{array}
$$

Answers:

$$
\begin{array}{c}
\quad CH_3 \;\; \boxed{CH_3}\;\boxed{CH_3} \\
\quad\;\; | \qquad | \qquad | \\
\text{(a)}\; CH_2-CH-CH \\
\qquad\qquad\quad | \\
\qquad\qquad\quad CH_2-CH_3 \\
\qquad\qquad \text{Hexane}
\end{array}
$$

$$
\begin{array}{c}
\qquad\qquad\qquad CH_2-CH_2-CH_3 \\
\qquad\qquad\qquad | \\
\text{(c)}\; \boxed{CH_3CH_2}-CH-CH_2-CH_2-CH_2-CH_3 \\
\qquad\qquad\qquad\quad \text{Nonane}
\end{array}
$$

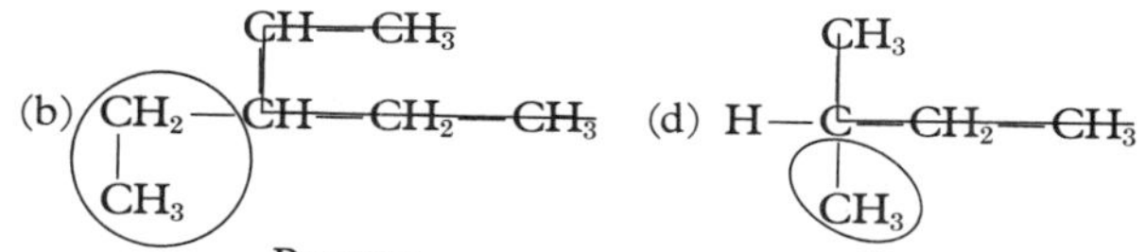

(b) CH₂—CH—CH₂—CH₃ with CH—CH₃ and CH₃ branches
Pentane

(d) H—C—CH₂—CH₃ with CH₃ above and CH₃ below
Butane

See Problems 16-4, 16-5

Alkanes are nonpolar, since they are made up of only carbon and hydrogen, which have approximately the same electronegativities. Because of this, they are insoluble in water and have low boiling points. Alkanes containing four or fewer carbons are gases, and those with more than 18 carbons are solids (resembling candle wax); the rest are liquids. Alkanes have no odor and are colorless. All are extremely flammable, and the lighter alkanes are also volatile (they evaporate easily).

There are two major sources of alkanes, natural gas and crude oil (petroleum). Natural gas is mainly methane with smaller amounts of ethane, propane, and butanes. Unlike natural gas, petroleum contains hundreds of compounds, the majority of which are open-chain and cyclic alkanes. Before we can make use of petroleum, it must be separated into groups of compounds with similar properties. Further separation may or may not be carried out depending on the final use of the hydrocarbons.

Crude oil is separated into groups of compounds according to boiling points by distillation in a refinery (see Figure 16-1). In such a distillation, the liquid is boiled and the gases move up a large column which becomes cooler and cooler toward the top. Compounds condense (become liquid) at different places in the column, depending on their boiling points. As the liquids condense they are drawn off, providing a rough separation of the crude oil. Some of the material is too high boiling to vaporize and remains in the bottom of the column. A drawing showing this process and the various fractions ob-

Figure 16-1 AN OIL REFINERY. The crude oil is separated into various fractions in large refineries.

tained is shown in Figure 16-2. Notice that the fewer the carbon atoms in the alkane, the lower the boiling point (the more volatile it is).

The composition of crude oil itself varies somewhat depending on where it is found. Certain crude oil, such as that found in Nigeria and Libya, is called "light" oil because it is especially rich in the hydrocarbons that are present in gasoline. Otherwise, one fraction can be converted into another by three processes, known as cracking, reforming, and alkylation. **Cracking** *changes large molecules into small molecules.* **Reforming** *removes hydrogens from the carbons and/or changes unbranched hydrocarbons into branched hydrocarbons.* (Branched hydrocarbons perform better in gasoline; that is, they have a higher "octane" rating.) **Alkylation** *takes small molecules and puts them together to make larger molecules.* In all of these processes, catalysts are used, but there is a different catalyst for each process.

About 96% of all oil and gas is burned as fuel, whereas only 4% is used to make other organic chemicals. As a fuel, hydrocarbons burn to give carbon dioxide, water, and a great deal of heat energy.

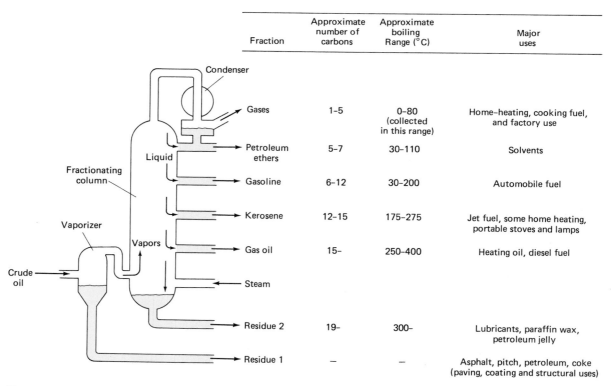

Fraction	Approximate number of carbons	Approximate boiling Range (°C)	Major uses
Gases	1–5	0–80 (collected in this range)	Home–heating, cooking fuel, and factory use
Petroleum ethers	5–7	30–110	Solvents
Gasoline	6–12	30–200	Automobile fuel
Kerosene	12–15	175–275	Jet fuel, some home heating, portable stoves and lamps
Gas oil	15–	250–400	Heating oil, diesel fuel
Steam			
Residue 2	19–	300–	Lubricants, paraffin wax, petroleum jelly
Residue 1	–	–	Asphalt, pitch, petroleum, coke (paving, coating and structural uses)

Figure 16-2 THE REFINING OF PETROLEUM. Oil is separated into fractions according to boiling points.

$$CH_4 + 2O_2 \xrightarrow{\text{Spark}} CO_2 + 2H_2O + \text{energy}$$

$$2C_8H_{18} + 25O_2 \xrightarrow{\text{Spark}} 16CO_2 + 18H_2O + \text{energy}$$

Industrially, most synthetic organic chemicals have their ultimate origin in *the alkanes obtained from crude oil.* These **petrochemicals** are put to a wide range of uses in the manufacture of fibers, plastics, coatings, adhesives, synthetic rubber, some flavorings, perfumes, and pharmaceuticals.

16-3 Alkenes

Alkenes *are hydrocarbons that contain a double bond. Organic compounds with multiple bonds are said to be* **unsaturated.** The general formula for an open-chain alkene with one double bond is C_nH_{2n}. The simplest alkene is ethene (common name ethylene), which has the structure

Alkenes are named by dropping the *-ane* ending of the corresponding alkane name and substituting *-ene.* The double bond is located by numbering the carbon chain in such a way that the double bond is assigned the lowest possible number. Thus, the name of $CH_3CH_2CH_2CH{=}CHCH_3$ is 2-hexene.

Only small amounts of alkenes are found naturally in crude oil; the majority are made from alkanes by the reforming process during the refining of crude oil. When alkanes are heated over a catalyst, hydrogen is lost from the molecule and alkenes together with hydrogen are formed.

$$C_nH_{2n+2} \xrightarrow[\text{Heat}]{\text{Catalyst}} C_nH_{2n} + H_2$$

$$C_2H_6 \xrightarrow[\text{Heat}]{\text{Catalyst}} CH_2{=}CH_2 + H_2$$

Most of the alkenes produced industrially are used to make polymers. When a catalyst is added to an alkene or a mixture of alkenes, the double bond is broken and the alkenes become joined to each other by single bonds. *This produces a high-molecular-weight molecule called a* **polymer,** *which has repeating units of the original alkene (called the* **monomer**).

$$CH_2{=}CH_2 \xrightarrow{\text{Catalyst}}$$
Monomer (ethylene)

etc. $\;-\!\!\!\Big[CH_2{-}CH_2\Big]\!\!\Big[CH_2{-}CH_2\Big]\!\!\Big[CH_2{-}CH_2\Big]\!\!\Big[CH_2{-}CH_2\Big]\!-\;$ etc.

Written as $-(CH_2{-}CH_2)_n$
Polymer (polyethylene)

Figure 16-3 POLYMERS. The plastic bottles, the bag, and the covering on the walkway are made of polymers.

Polymers are named by adding *poly-* to the name of the alkene used to form the polymer. In the example shown above, the polymer was made from ethylene (usually common names are used for polymers), and so the polymer is called polyethylene. Since it would be impossible to write out the structure of a polymer, which may contain thousands of carbons, we abbreviate the structure by giving the repeating unit in parentheses along with a subscript n to indicate that the monomer is repeated many times. Groups attached to the double bond affect the properties of the polymer, and by varying the group, we can vary the uses for which a polymer is suited (see Figure 16-3).

See Problem 16-6 Some commonly used polymers and their uses are given in Table 16-4.

Chemicals can add to the double bond to form new single bonds and therefore new compounds. A test based on such a reaction, the addition of bromine to an alkene, is used to show the presence of alkenes. It is a very simple test to perform. If the red color of bromine disappears when it is added to a liquid, this means that multiple bonds are present (the dibromide formed is colorless). We will talk more about this test when we discuss aromatic compounds.

$$CH_2{=}CH_2 + Br_2 \longrightarrow \underset{\substack{| \quad\quad |\\ Br \quad\; Br}}{CH_2{-}CH_2}$$

Red   Colorless

16-4 Alkynes

Alkynes *are hydrocarbons that contain a triple bond.* The general formula for an open-chain alkyne with one triple bond is C_nH_{2n-2}. The simplest alkyne is ethyne (acetylene), which has the structure

$$H{-}C{\equiv}C{-}H$$

Alkynes are named just like alkenes except that *-yne* is substituted for the alkane ending instead of *-ene*. Thus, $CH_3CH_2CH_2CH_2C{\equiv}CCH_2CH_3$ is named 3-octyne.

Alkynes are not found in nature, but can be prepared synthetically. Acetylene can be made from coal by first reacting the coal with calcium oxide at high temperature and then treating the calcium carbide formed with water:

$$\underset{\text{Coal}}{C} + \underset{\substack{\text{Calcium}\\\text{oxide}}}{CaO} \xrightarrow{\Delta} \underset{\substack{\text{Calcium}\\\text{carbide}}}{CaC_2} \xrightarrow{H_2O} \underset{\text{Acetylene}}{H{-}C{\equiv}C{-}H} + \underset{\substack{\text{Calcium}\\\text{hydroxide}}}{Ca(OH)_2}$$

A more common method of making acetylene is through the reforming process. Methane is heated in the presence of a catalyst, forming acetylene and hydrogen:

$$2CH_4 \xrightarrow[\Delta]{\text{Catalyst}} H{-}C{\equiv}C{-}H + 3H_2$$

Table 16-4 Polymers

Monomer Name, Structure	Polymer Name, Structure	Some Common Trade Name	Uses
Ethylene $CH_2=CH_2$	Polyethylene $+CH_2-CH_2\frac{}{}_n$	Polyfilm[a] Marlex[b]	Electrical insulation, packaging (plastic bags), floor covering, plastic bottles, pipes, tubing
Propylene $CH_2=CH$ $\quad\mid$ $\quad CH_3$	Polypropylene $\left(CH_2-CH\atop\qquad\mid\atop\qquad CH_3\right)_n$	Herculon[c]	Pipes, carpeting, artificial turf, molded auto parts, fibers
Vinyl chloride $CH_2=CH$ $\quad\mid$ $\quad Cl$	Polyvinyl chloride (PVC) $\left(CH_2-CH\atop\qquad\mid\atop\qquad Cl\right)_n$	Tygon[d]	Wire and cable coverings, pipes, rainwear, shower curtains, tennis court playing surfaces
Styrene $CH_2=CH$	Polystyrene $\left(CH_2-CH\right)_n$	Styrofoam[a] Styron[a]	Molded objects (combs, toys, brush and pot handles), refrigerator parts, insulating material, phonograph records, clock and radio cabinets
Tetrafluoroethylene $CF_2=CF_2$	Polytetrafluoroethylene $+CF_2-CF_2\frac{}{}_n$	Teflon[e] Halon[f]	Gaskets, valves, tubing, coatings for cookware
Methyl methacrylate $\quad CH_3$ $\quad\mid$ $CH_2=C$ $\qquad\backslash CO_2CH_3$	Polymethyl methacrylate $\left(CH_2-C\atop\qquad\mid\atop\qquad CO_2CH_3\right)_n$ with CH_3	Plexiglas[g] Lucite[e]	Glass substitute, lenses, aircraft glass, dental fillings, artificial eyes, braces
Acrylonitrile $CH_2=CH$ $\quad\mid$ $\quad CN$	Polyacrylonitrile $\left(CH_2-CH\atop\qquad\mid\atop\qquad CN\right)_n$	Orlon[e] Acrilan[b]	Fibers for clothing, carpeting

[a] Dow Chemical Co.
[b] Phillips Petroleum Co.
[c] Hercules, Inc.
[d] U.S. Stoneware Co.
[e] E. I. du Pont de Nemours & Co.
[f] Allied Chemical Corp.
[g] Rohm & Haas Co.
[h] Monsanto Industrial Chemicals, Inc.

Although acetylene has some use in oxyacetylene torches, its biggest use is in the manufacture of other organic compounds used as monomers for polymers. For example, acetylene is used to make vinyl chloride (which is then used to make polyvinyl chloride) and acrylonitrile (used to make Acrilan® and Orlon® (see Table 16-4)].

$$H-C\equiv C-H + HCl \longrightarrow \underset{H}{\overset{H}{\diagdown}}C=C\underset{Cl}{\overset{H}{\diagup}}$$

Vinyl chloride

16-5 Aromatic Compounds

Another class of hydrocarbons is known as aromatics. *An **aromatic compound** is a cyclic hydrocarbon containing alternating single and double bonds between adjacent carbon atoms.* This discussion will concentrate on one aromatic compound, benzene (C_6H_6), and its derivatives. (*A **derivative** of a compound is produced by the substitution of a group on the molecules of the original compound.*) At first glance, benzene looks like a cyclic alkene with three double bonds. It turns out, however, that the three double bonds do not act

[Simplified structures can be written by omitting carbons (B) or by omitting carbons and hydrogens (C)]

(A)   (B)   (C)

like alkene double bonds. For example, benzene does not "decolorize" a solution of bromine as simple alkenes do.

$$CH_2{=}CH_2 + Br_2 \longrightarrow \underset{\underset{Br}{|}}{CH_2}-\underset{\underset{Br}{|}}{CH_2}$$

Red   Colorless

$$+ Br_2 \longrightarrow \text{No reaction}$$

Red   (Still red)

Benzene is less chemically reactive than a simple alkene, and it is this property that sets it apart from alkenes.

Compounds which contain the benzene ring can usually be recognized by name because they are named as derivatives of benzene.

Chlorobenzene   Ethylbenzene   Methylbenzene
                                                     (toluene)

Hydroxybenzene   Aminobenzene   Benzoic acid
     (phenol)   (aniline)

Although some benzene and toluene is present in crude oil, additional quantities can be obtained by the reforming process. When cyclohexane and/or hexane are heated with a catalyst, benzene is formed. Coal is another source of benzene. When coal is heated to high temperatures in the absence of air, some benzene is formed.

Benzene and toluene are used mainly as solvents and as starting materials to make other aromatic compounds. Benzene must be used with care, however, because it has been found to be a potent carcinogen (it can cause cancer). Phenol and its derivatives have been used as disinfectants, in the manufacture of dyes, explosives, drugs, and plastics, and as preservatives.

16-6 Organic Functional Groups

In contrast to the hydrocarbons, many organic compounds contain alkyl or other hydrocarbon groups attached to other elements, particularly nitrogen and oxygen. This, of course, alters the chemistry of these compounds significantly from that of the basic hydrocarbons. *The part of the molecule that contains the elements other than C is called the* **functional group.** It is called a functional group because that is where most chemical reactions take place. In this respect, *the double bond in alkenes and the triple bond in alkynes are also considered functional groups.* The various functional groups that we will mention are shown in Table 16-5. Each group establishes certain basic chemical characteristics for the compounds that contain them.

Table 16-5 Functional Groups

Name of Class	Functional Group	General Formula[a]	Examples
Alcohols	$-\overset{\displaystyle \mid}{\underset{\displaystyle \mid}{C}}-\overset{\displaystyle \cdot\cdot}{\underset{\displaystyle \cdot\cdot}{O}}-H$	$R-OH$	$CH_3OH,\ CH_3CH_2OH$
Ethers	$-\overset{\displaystyle \mid}{\underset{\displaystyle \mid}{C}}-\overset{\displaystyle \cdot\cdot}{\underset{\displaystyle \cdot\cdot}{O}}-\overset{\displaystyle \mid}{\underset{\displaystyle \mid}{C}}-$	$R-O-R'$	$CH_3OCH_3,\ CH_3OCH_2CH_3$
Amines	$-\overset{\displaystyle \mid}{\underset{\displaystyle \mid}{C}}-\overset{\displaystyle \cdot\cdot}{\underset{\displaystyle \mid}{N}}-$	$R-NH_2$	$CH_3NH_2,\ CH_3CH_2NH_2$
Aldehydes	$-\overset{\displaystyle :O:}{\overset{\displaystyle \|}{C}}-H$	$\overset{O}{\overset{\|}{R-C}}-H$	$\overset{O}{\overset{\|}{H-C}}H,\quad \overset{O}{\overset{\|}{CH_3-C}}H$
Ketones	$-\overset{\displaystyle \mid}{\underset{\displaystyle \mid}{C}}-\overset{\displaystyle :O:}{\overset{\|}{C}}-\overset{\displaystyle \mid}{\underset{\displaystyle \mid}{C}}-$	$\overset{O}{\overset{\|}{R-C}}-R'$	$\overset{O}{\overset{\|}{CH_3-C}}CH_3,\quad \overset{O}{\overset{\|}{CH_3CH_2-C}}CH_3$
Carboxylic acids	$-\overset{\displaystyle :O:}{\overset{\|}{C}}-\overset{\cdot\cdot}{O}H$	$RCOH$	$HCOH,\ CH_3COH$
Esters	$-\overset{\displaystyle :O:}{\overset{\|}{C}}-\overset{\cdot\cdot}{O}-\overset{\displaystyle \mid}{\underset{\displaystyle \mid}{C}}-$	$\overset{O}{\overset{\|}{R-C}}-OR'$	$\overset{O}{\overset{\|}{CH_3-C}}OCH_3,\quad \overset{O}{\overset{\|}{CH_3CH_2-C}}OCH_3$
Amides	$-\overset{\displaystyle :O:}{\overset{\|}{C}}-\overset{\cdot\cdot}{\underset{\displaystyle \mid}{N}}-$	$\overset{O}{\overset{\|}{R-C}}-NH_2$	$\overset{O}{\overset{\|}{CH_3-C}}NH_2,\ HCNHCH_3$

See Problems 16-7, 16-8

[a] R and R' stand for hydrocarbon groups (e.g., alkyl). They may be different or the same group.

16-7 Alcohols

The simplest alcohol, methanol (methyl alcohol), has the formula CH_4O and the structure CH_3-O-H. *The functional group of alcohol is the OH group* (**hydroxyl group**).

Alcohols are named by taking the alkane name, dropping the $-e$ and substituting $-ol$. Common names are obtained by just naming the alkyl group attached to the $-OH$ followed by *alcohol*. Some very useful alcohols (see figure 16-4) have more than one hydroxyl group. Two of them are shown below.

$$CH_3CH_2OH \qquad CH_3CH_2CH_2OH$$

$$
\begin{array}{cc}
CH_2 - CH_2 \\
| \qquad | \\
OH \quad OH
\end{array}
\qquad
\begin{array}{ccc}
CH_2 - CH - CH_2 \\
| \qquad | \qquad | \\
OH \quad OH \quad OH
\end{array}
$$

Ethanol   Propanol   1,2-Ethane<u>di</u>ol   1,2,3-Propane<u>tri</u>ol
(ethyl alcohol)   (*n*-propyl alcohol)   (ethylene glycol)   (glycerol)
                                     [*di* means two   [*tri* means three
                                       hydroxyl groups]   hydroxyl groups]

Methanol and ethanol can both be obtained from natural sources. Methanol can be prepared by heating wood in the absence of oxygen to about 400°C; at such high temperatures, methanol, together with other organic compounds, is given off as a gas. Since methanol was once made exclusively by this process, it is often called wood alcohol. Currently, methanol is prepared from synthesis gas, a mixture of CO and H_2. When CO and H_2 are passed over a catalyst at the right temperature and pressure, methanol is formed:

$$CO + 2H_2 \xrightarrow[\Delta]{\text{Catalyst}} CH_3OH$$

Ethanol (often known simply as alcohol) is formed in the fermentation of various grains (therefore it is also known as grain alcohol) and is the "active

Figure 16-4 ALCOHOLS. A major ingredient in each of these products is an alcohol.

ingredient" in all alcoholic beverages. In fermentation, the sugars and carbohydrates in grains are converted to ethanol and carbon dioxide by the enzymes in yeast.

$$C_6H_{12}O_6 \xrightarrow[\text{enzymes}]{\text{Yeast}} 2CH_3CH_2OH + 2CO_2$$

Glucose (a sugar)

Industrially, ethanol is prepared by reacting ethylene with water in the presence of an acid catalyst.

$$CH_2{=}CH_2 + H_2O \xrightarrow{H^+} CH_3CH_2OH$$

Isopropanol (isopropyl alcohol or rubbing alcohol) can be prepared from propene (propylene) in the same way.

$$CH_3CH{=}CH_2 + H_2O \xrightarrow{H^+} CH_3-\overset{\displaystyle CH_3}{\underset{\displaystyle OH}{\overset{|}{\underset{|}{C}}}}-H$$

Methanol has been used as a solvent for shellac, as a denaturant for ethanol (it makes the ethanol undrinkable), and as an antifreeze for automobile radiators. It is very toxic. In small doses it causes blindness, and in large doses it can cause death.

Ethanol is present in alcoholic beverages such as beer, wine, and liquor. The "proof" of an alcoholic beverage is two times the percent by volume of alcohol. If a certain brand of bourbon is 100 proof, it contains 50% ethanol. Recently, there has been much interest in the use of ethanol as an additive to gasoline, called "gasohol." Ethanol is an excellent solvent and has been used as such in perfumes, medicines, and flavorings. It has also been used as an antiseptic and as a rubbing compound to cleanse the skin and lower a feverish person's temperature. While ethanol is not as toxic as methanol, it can cause coma or death when ingested in large quantities.

Ethylene glycol is the major component of antifreeze and coolant used in automobiles. It is also used to make polymers, the most common of which is Dacron®, a polyester.

Glycerol, which can be obtained from fats, is used in many applications where a lubricant and/or softener is needed. It has been used in pharmaceuticals, cosmetics, foodstuffs, and in some liqueurs. When glycerol reacts with nitric acid, it produces nitroglycerin. Nitroglycerin is a powerful explosive,

$$\underset{\substack{| \\ OH}}{CH_2}-\underset{\substack{| \\ OH}}{CH}-\underset{\substack{| \\ OH}}{CH_2} + 3HNO_3 \longrightarrow \underset{\substack{| \\ ONO_2}}{CH_2}-\underset{\substack{| \\ ONO_2}}{CH}-\underset{\substack{| \\ ONO_2}}{CH_2} + 3H_2O$$

Glycerol   Nitroglycerin

but it is also a strong smooth-muscle relaxant and vasodilator and has been used to lower the blood pressure and to treat angina pectoris.

16-8 Ethers

*An **ether** contains an oxygen bonded to two hydrocarbon groups* (rather than one hydrocarbon group and one oxygen as in alcohols). The simplest ether, dimethyl ether, is an isomer of ethanol but has very different properties.

Ethers are named by giving the alkyl groups on either side of the oxygen and adding *ether*.

$H_3C-O-CH_3$
Dimethyl ether

$CH_3OCH_2CH_3$
Methyl ethyl ether

$CH_3CH_2OCH_2CH_3$
Diethyl ether

$CH_3CH_2OCH_2CH_2CH_3$
Ethyl propyl ether

Diethyl ether is made industrially by reacting ethanol with sulfuric acid. In this reaction, two ethanol molecules are joined together with the loss of a H_2O molecule.

$$2CH_3CH_2OH \xrightarrow{H_2SO_4} CH_3CH_2-O-CH_2CH_3 + H_2O$$

The most commonly known ether, diethyl ether (or ethyl ether or just ether) has, in the past, been used extensively as an anesthetic. It has the advantages of being an excellent muscle relaxant that doesn't affect the blood pressure, pulse rate, or the rate of respiration greatly. On the other hand, ether has an irritating effect on the respiratory passages and often causes nausea. Its flammability also is a drawback due to dangers of fire and explosions. Diethyl ether is rarely used now as an anesthetic. Other anesthetics that do not have its disadvantages have taken its place.

16-9 Amines

*An **amine** contains a nitrogen with single bonds to a hydrocarbon group and a total of two other hydrocarbon groups or hydrogens.* The nitrogen in amines has one pair of unshared electrons similar to ammonia, NH_3. As we learned in Chapter 12, NH_3 utilizes the unshared pair of electrons to form weakly basic solutions in water. In a similar manner, amines are characterized by their ability to act as bases in water.

$$\overset{..}{N}H_3 + H_2O \rightleftharpoons NH_4^+ + OH^-$$

$$CH_3\overset{..}{N}H_2 + H_2O \rightleftharpoons CH_3NH_3^+ + OH^-$$

Basic solutions

Amines are named by listing the alkyl groups attached to the nitrogen and adding *amine*.

$$CH_3—\overset{\overset{\displaystyle\cdot\cdot}{}}{\underset{\underset{\displaystyle H}{|}}{N}}—H \qquad CH_3—\overset{\overset{\displaystyle\cdot\cdot}{}}{\underset{\underset{\displaystyle H}{|}}{N}}—CH_3 \qquad CH_3—\overset{\overset{\displaystyle\cdot\cdot}{}}{\underset{\underset{\displaystyle CH_2CH_3}{|}}{N}}—CH_2CH_2CH_3$$

Methyl amine   Dimethyl amine   Methyl   ethyl   propyl amine
                   (*di* = two methyls)

Simple amines are prepared by the reaction of ammonia with alkyl halides (e.g., CH_3Cl). In the reaction the alkyl group substitutes for the hydrogen on ammonia. The hydrogen and chlorine combine to form HCl.

$$H—\overset{\overset{\displaystyle\cdot\cdot}{}}{\underset{\underset{\displaystyle H}{|}}{N}}—\boxed{H + Cl}—CH_3 \longrightarrow H—\overset{\overset{\displaystyle\cdot\cdot}{}}{\underset{\underset{\displaystyle H}{|}}{N}}—CH_3 + HCl$$

Ammonia   Methylchloride   methylamine

Further reaction with the same or different alkyl halides leads to substitution of one or both remaining hydrogens on the methylamine.

Amines are used in the manufacture of dyes, drugs, disinfectants, and insecticides. They also occur naturally in biological systems and are important in many biological processes.

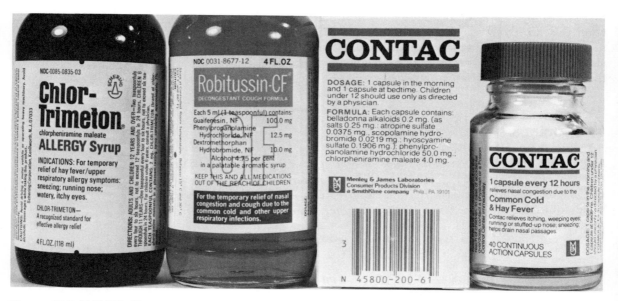

Figure 16-5 AMINES. The active ingredient in an antihistamine is an organic amine.

Amine groups are present in many synthetic and naturally occurring drugs (see Figure 16-5). They may be useful as antidepressants, antihistamines, antibiotics, antiobesity preparations, antinauseants, analgesics, antitussives, diuretics, and tranquilizers, among others. Frequently, a drug may have more than one use [codeine is both an analgesic (pain reliever) and an antitussive agent (cough depressant).] Often a drug may be obtained from plants or animals. A class of compounds which falls into this type of drug are the alkaloids, which are amines found in plants. Some common drugs are shown below.

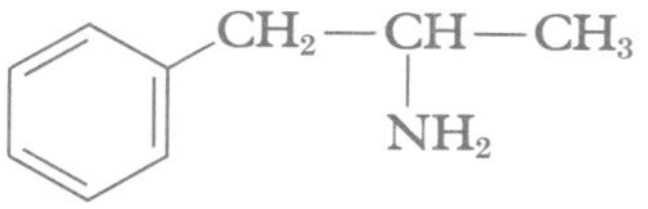

Amphetamine
(Benzedrine®;
synthetic
appetite depressant,
stimulant)

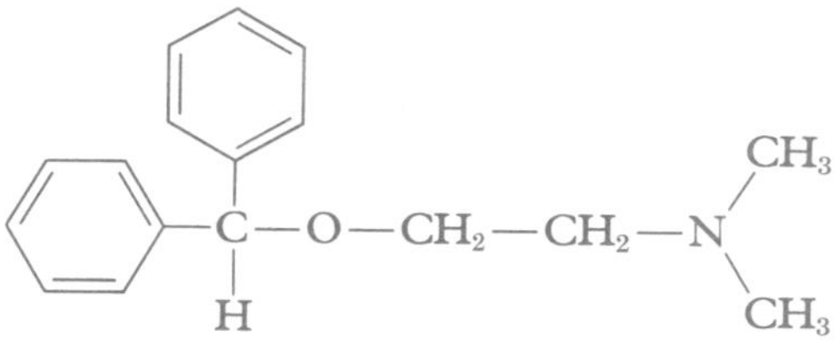

Diphenyldramine
(synthetic
antihistamine)

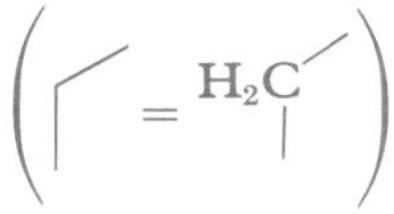

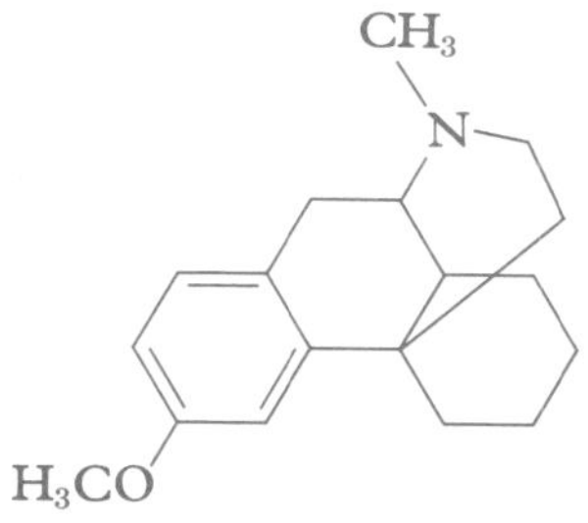

Dextromethorphan
(synthetic
analgesic, antitussive)

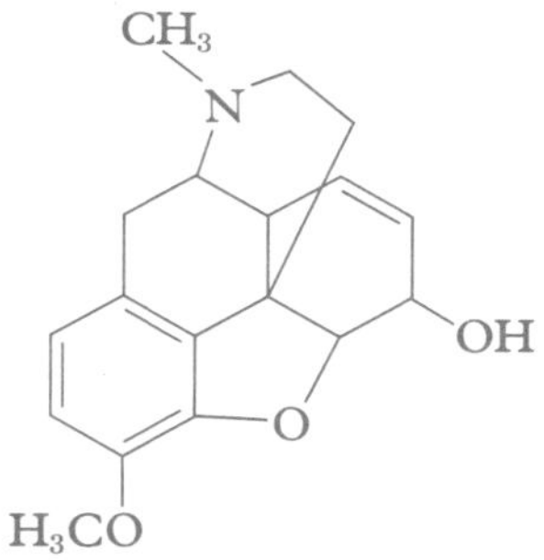

Codeine
(from opium;
analgesic, antitussive)

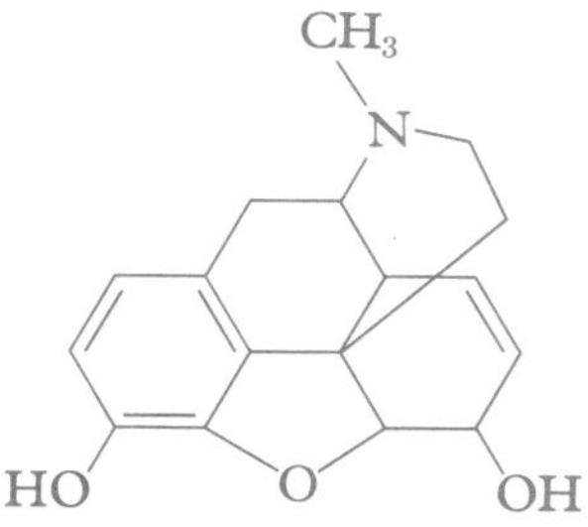

Morphine
(from opium;
analgesic)

16-10 Aldehydes and Ketones

The functional group of aldehydes and ketones is a carbonyl group. *A* **car-**

bonyl group *is a carbon with a double bond to an oxygen (i.e.,* $-\overset{O}{\overset{\|}{C}}-$ *). In* **alde-**

hydes *the carbonyl group is bound to one hydrocarbon group and one hydrogen, whereas in* **ketones** *the carbonyl group is bound to two hydrocarbon groups.*

Aldehydes are named by dropping the *-e* of the corresponding alkane name and substituting *-al*. Therefore the two-carbon aldehyde is ethanal. Ketones are named by taking the alkane name, dropping the *-e*, and substituting *-one*. The common names of ketones are obtained by naming the alkyl groups on either side of the C=O and adding *ketone*. Thus, the four-carbon ketone can be called butanone (IUPAC) or methyl ethyl ketone (common).

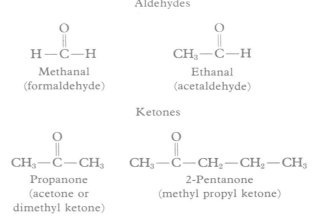

Aldehydes

O
‖
H—C—H
Methanal
(formaldehyde)

O
‖
CH₃—C—H
Ethanal
(acetaldehyde)

Ketones

O
‖
CH₃—C—CH₃
Propanone
(acetone or
dimethyl ketone)

O
‖
CH₃—C—CH₂—CH₂—CH₃
2-Pentanone
(methyl propyl ketone)

Many aldehydes and ketones are prepared by the oxidation of alcohols. Industrially, the alcohol is oxidized by heating it in the presence of oxygen and a catalyst. About half of the methanol made industrially is used to make formaldehyde by oxidation. Acetone can be prepared in the same fashion by oxidation of isopropanol.

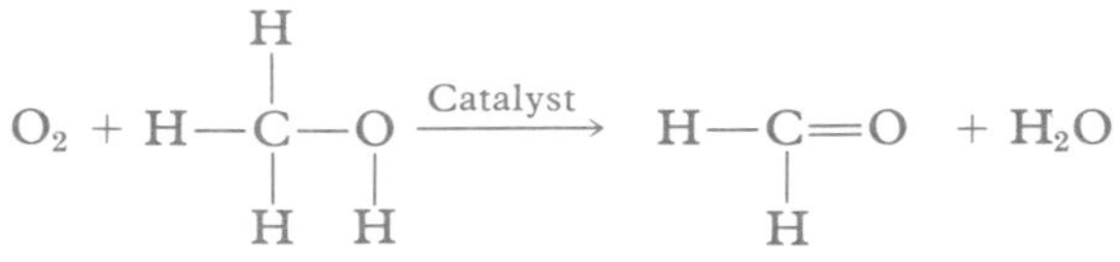

$$O_2 + H-\underset{\underset{H}{|}}{\overset{\overset{H}{|}}{C}}-O-H \xrightarrow{\text{Catalyst}} H-C{=}O + H_2O$$

Methanol — Formaldehyde

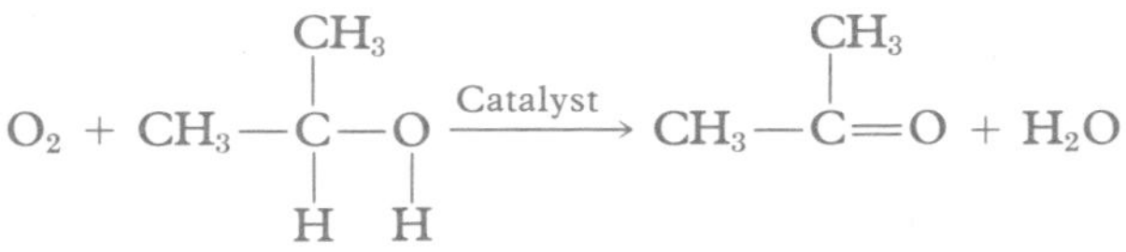

$$O_2 + CH_3-\underset{\underset{H}{|}}{\overset{\overset{CH_3}{|}}{C}}-O-H \xrightarrow{\text{Catalyst}} CH_3-C{=}O + H_2O$$

Isopropanol — Acetone

Aldehydes and ketones have uses as solvents, the preparation of polymers, as flavorings, and in perfumes. The simplest aldehyde, formaldehyde, has been used as a disinfectant, antiseptic, germicide, fungicide, and embalming fluid (as a 37%-by-weight water solution). It has also been used in the preparation of polymers such as Bakelite (the first commercial plastic) and Melmac® (used to make dishes). Formaldehyde polymers have also been used as coatings on fabrics to give "permanent press" characteristics.

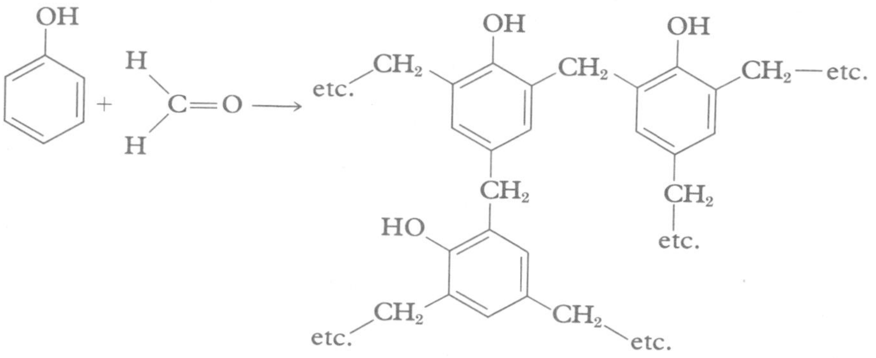

Bakelite

The simplest ketone, acetone, has been used mainly as a solvent. It is soluble in water and dissolves relatively polar and nonpolar molecules. It is an excellent solvent for paints and coatings.

More complex aldehydes and ketones, such as the following two examples, are used in flavorings and perfumes.

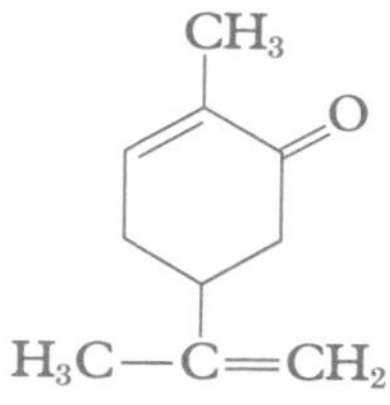

Carvone
(in oil of spearmint,
a flavoring)

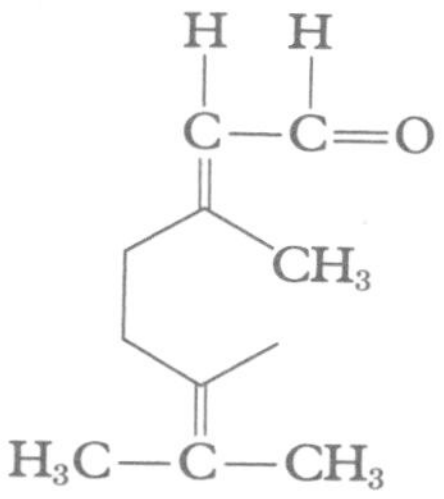

Citral
(in oil of lemon grass;
a fragrance)

16-11 Carboxylic Acids, Esters, and Amides

Carboxylic acids *contain the functional group* $-\overset{\overset{\displaystyle O}{\|}}{C}-O-H$ *(also written* $-COOH$), *which is known as a* **carboxyl group.** They are the most acidic of organic compounds but not as acidic as inorganic acids such as nitric or sulfuric. Carboxylic acids are easily changed into two derivatives: amides and esters. **Amides** *have an amine group substituted for the hydroxyl group of the acid.* **Esters** *have a hydrocarbon group substituted for the hydrogen in the carboxyl group.*

Carboxylic acids are named by dropping the -*e* from the alkane name and substituting -*oic acid*. Esters are named by giving the name of the alkyl group attached to the oxygen, followed by the acid name minus the -*ic* acid ending and substituting -*ate*. Amides are named by dropping the -*oic acid* portion of the carboxylic acid name and substituting -*amide*.

A few examples of carboxylic acids, esters, and amides are as follows.

Acids

$$\underset{\substack{\text{Methanoic acid}\\(\text{formic acid})}}{H-\overset{\overset{\textstyle O}{\|}}{C}-OH} \qquad \underset{\substack{\text{ethanoic acid}\\(\text{acetic acid})}}{CH_3-\overset{\overset{\textstyle O}{\|}}{C}-OH} \qquad \underset{\text{Benzoic acid}}{\bigcirc-\overset{\overset{\textstyle O}{\|}}{C}-OH} \qquad \underset{\text{Oxalic acid}}{HO-\overset{\overset{\textstyle O}{\|}}{C}-\overset{\overset{\textstyle O}{\|}}{C}-OH}$$

Esters

$$\underset{\substack{\text{Ethyl methanoate}\\(\text{ethyl formate})}}{H-\overset{\overset{\textstyle O}{\|}}{C}-O-CH_2CH_3} \qquad \underset{\substack{\text{Ethyl ethanoate}\\(\text{ethyl acetate})}}{CH_3-\overset{\overset{\textstyle O}{\|}}{C}-O-CH_2CH_3} \qquad \underset{\text{Methyl benzoate}}{\bigcirc-\overset{\overset{\textstyle O}{\|}}{C}-O-CH_3}$$

Amides

$$\underset{\substack{\text{Methanamide}\\(\text{formamide})}}{H-\overset{\overset{\textstyle O}{\|}}{C}-NH_2} \qquad \underset{\substack{\text{Ethanamide}\\(\text{acetamide})}}{CH_3-\overset{\overset{\textstyle O}{\|}}{C}-NH_2} \qquad \underset{\text{Benzamide}}{\bigcirc-\overset{\overset{\textstyle O}{\|}}{C}-NH_2}$$

Carboxylic acids are made by oxidizing either alcohols or aldehydes. Formic acid, which was first isolated by distilling red ants, can be made by oxidizing either methanol or formaldehyde.

$$CH_3-OH \xrightarrow[\text{agent}]{\text{Oxidizing}} H-\overset{\overset{\textstyle O}{\|}}{C}-OH \xleftarrow[\text{agent}]{\text{Oxidizing}} \overset{\overset{\textstyle O}{\|}}{\underset{H \qquad H}{C}}$$

Acetic acid can be made by oxidizing ethanol; this in fact happens when wine becomes "sour"—the ethanol is oxidized by air and we end up with wine vinegar. The pungent flavor of vinegar is due to the acetic acid, which is present in a concentration of about 4 to 5%.

Esters are made from the reaction of alcohols and carboxylic acids. In the reaction, H_2O is split off from the two molecules (OH from the acid and H from the alcohol), leading to the union of the remnants of the two molecules.

$$H-\overset{O}{\underset{||}{C}}-\boxed{OH} + \boxed{H}-O-CH_3 \xrightarrow{H^+} H-\overset{O}{\underset{||}{C}}-O-CH_3 + H_2O$$

Amides are made in a similar fashion, except that the acid is mixed with ammonia or an amine instead of an alcohol.

$$H-\overset{O}{\underset{||}{C}}-\boxed{OH} + \boxed{H}-NH_2 \xrightarrow{\Delta} H-\overset{O}{\underset{||}{C}}-NH_2 + H_2O$$

$$H-\overset{O}{\underset{||}{C}}-\boxed{OH} + \boxed{H}-NHCH_3 \xrightarrow{\Delta} H-\overset{O}{\underset{||}{C}}-NHCH_3 + H_2O$$

Carboxylic acids, esters, and amides are frequently present in compounds which have medicinal uses. Salicylic acid has both antipyretic (fever-reducing) and analgesic (pain-relieving) properties. It has the disadvantage, however, of causing severe irritation of the stomach lining. Acetylsalicylic acid (aspirin), which is both an acid and an ester, doesn't irritate the stomach as much. Aspirin is broken up in the small intestine, to form salicylic acid, where it is absorbed. Some people are allergic to aspirin, and must take "aspirin substitutes." The common aspirin substitutes are amides, phenacetin and acetaminophen. Acetaminophen is the active ingredient in Tylenol® and Datril®.

Salicylic acid   Acetylsalicylic acid

Phenacetin   Acetaminophen

Just as aldehydes and ketones are used as fragrances and flavorings, so are esters. Some of these are

$$\underset{\substack{\displaystyle \text{Ethyl formate} \\ \text{(rum)}}}{H-\overset{\displaystyle O}{\overset{\|}{C}}OCH_2CH_3}$$

$$\underset{\substack{\displaystyle \text{Pentyl acetate} \\ \text{(bananas)}}}{CH_3\overset{\displaystyle O}{\overset{\|}{C}}O(CH_2)_4CH_3}$$

$$\underset{\substack{\displaystyle \text{Ethyl butanoate} \\ \text{(pineapple)}}}{CH_3CH_2CH_2\overset{\displaystyle O}{\overset{\|}{C}}OCH_2CH_3}$$

Methyl salicylate
(oil of wintergreen)

One of the biggest uses for carboxylic acids, esters, and amides is in the formation of polymers. These are different from the polymers we discussed earlier, in that a small molecule (usually H_2O) is given off during the forma-

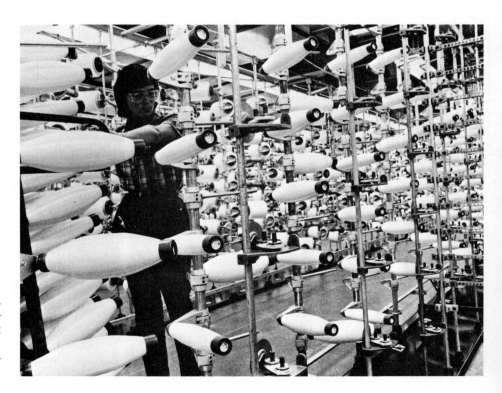

Figure 16-6 NYLON. Nylon was first synthesized by DuPont chemists. It was one of the first synthetic fabrics.

tion of the polymer. Two of the most widely known polymers used today are nylon 66 (see Figure 16-6) and Dacron®. The first is a poly<u>amide</u>, made from a <u>di</u>acid and a <u>di</u>amine; the second is a poly<u>ester</u>, made from a <u>di</u>acid and a <u>di</u>alcohol (diol). Both of these polymers are used to make fibers.

Nylon 66

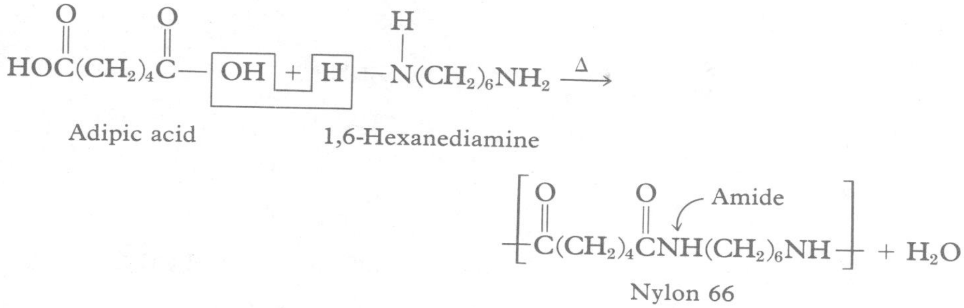

Adipic acid   1,6-Hexanediamine

Nylon 66

Dacron®

Terephthalic acid   Ethylene glycol

Dacron®

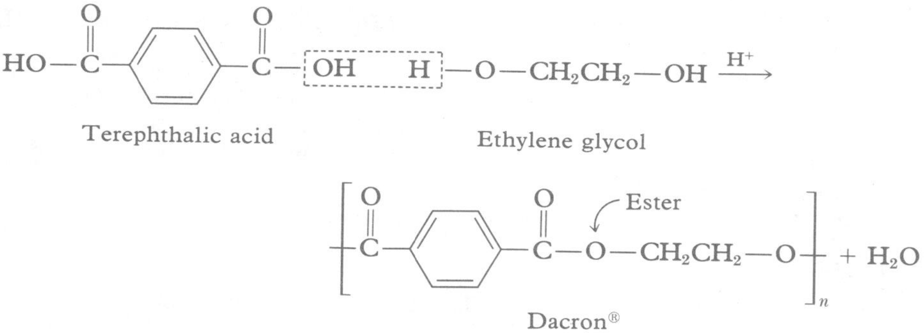

See Problems 16-9 through 16-12

Review Questions

1 How many bonds does carbon form in its compounds?
2 Define isomers.
3 What is a hydrocarbon? What is the general formula for an alkane?
4 Give the meaning of a homologous series.
5 What is meant by a substituent group?
6 How does an alkene differ from an alkane?
7 What is a polymer? From what types of hydrocarbons are they made?
8 How does an alkyne differ from an alkane?
9 Describe an aromatic hydrocarbon.

10 What is meant by a functional group?
11 Give the functional group of an alcohol, ether, amine, aldehyde, ketone, carboxylic acid, ester, and amide.
12 How is ethanol prepared (two ways)?
13 What is a carbonyl group?
14 Differentiate between an aldehyde and a ketone.
15 What is meant by a derivative of a carboxylic acid?
16 What are the two derivatives of carboxylic acid discussed?

Problems

16-1 Draw Lewis structures for each of the following compounds (more than one structure may be possible):

(a) CH_3Br ■ (e) C_2H_7N
(b) C_3H_4 ■ (f) C_3H_8O
(c) C_4H_8 ■ (g) C_2H_4O
(d) CH_5N

16-2 Write all of the isomers for C_6H_{14}.

16-3 Which of the following pairs of compounds are isomers of each other? Why or why not?

(a) $CH_3CH_2OCH_2CH_3$ and $CH_3CH_2CH_2CH_3$
(b) $CH_3CH_2CH_2OH$ and $CH_3CH_2OCH_2CH_3$
■ (c) $CH_3CH_2NH_2$ and CH_3NHCH_3
■ (d) $CH_3CH_2CH_2NHCH_3$ and $CH_3CH_2CH_2NHCH_2CH_3$

16-4 How many carbons are in the longest chain of each of the following compounds? What is the prefix used for the number of carbons in the longest carbon chain in each of these compounds?

(a) $CH_3CH-CH_2-\overset{\displaystyle CH_3-CH-CH_3}{CH}-CH_3$
  $\underset{CH_3}{\vert}$

(b) $CH_2-CH_2-\overset{\displaystyle CH_3}{\underset{\displaystyle CH_2CH_3}{CH}}\quad \overset{\displaystyle CH_2CH_2CHCH_3}{CH_3}$

(c) $CH_2-\overset{\displaystyle CH_2CH_2CH_2}{\underset{\displaystyle CH_3}{C}}-CH_3 \quad CH_2CH_2CH_3$
  $\underset{CH_3}{\vert}$

■ (d) $CH_2-\overset{\displaystyle CH_3}{CH}-CH_2$
  $\underset{CH_2-CH_3}{\vert}\quad \underset{CH_2-CH_3}{\vert}$

16-5 Circle the substituents attached to the longest chain in each of the following compounds. Name each alkyl substituent.

(a) $CH_3CH_2CH-\overset{\displaystyle CH_3}{\underset{\displaystyle CH_2CH_3}{CH}}CHCH_2CH_2CH_3$

(b) $CH_3CH_2CH_2CH-(CH_2)_2CH_3$
  $\underset{CH_3-CH-CH_3}{\vert}$

(c) $CH_3CH_2CH-\overset{\displaystyle CH_3CCH_3}{\underset{\displaystyle CH_2CH_2CH_3}{CH}}-CH_2CH_2CH_3$

■ (d) $CH_3CH_2CH_2CH_2-\overset{\displaystyle CH_2CH_2CH_3}{\underset{\displaystyle CH_2CH_2CH_2CH_3}{C}}-CH_2CH_2CH_3$

16-6 Give the abbreviated structure of the polymer which would be formed from each of the following alkenes. Name the polymer.

(a) $CH_2=CH$ (propylene)
  $\underset{CH_3}{\vert}$

(b) $CH_2=CH$ (vinyl fluoride)
  $\underset{F}{\vert}$

■ (c) $CH_2=CF_2$ (difluoroethylene)

■ (d) $CH_2=CH$ (methylacrylate)
  $\underset{CO_2CH_3}{\vert}$

16-7 Circle and name the functional groups in each of the following compounds:

(a) $CH_3CH=CHCH_2\overset{\displaystyle O}{\overset{\Vert}{C}}H$

(b)

(c) 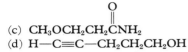 CH₃OCH₂CH₂$\overset{\overset{\displaystyle O}{\|}}{C}$NH₂

(d) H—C≡C—CH₂CH₂CH₂OH

■ (e)

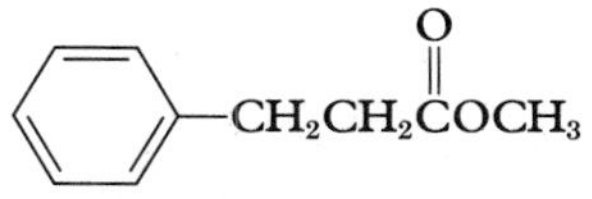

$\overset{\overset{\displaystyle O}{\|}}{}$

—CH₂CH₂COCH₃

■ (f) CH₃CH₂CH₂COOH

16-8 Tell how the following differ in structure:
(a) Alcohols and ethers
(b) Aldehydes and ketones
(c) Amines and amides
■ (d) Carboxylic acids and esters

16-9 To what class does each of the following compounds belong?
(a) 3-Heptanone
(b) 3-Nonene
(c) 2-Methylpentane
(d) 2-Ethylhexanal
(e) Bromobenzene

■ (f) 2-Octanol
■ (g) 2-Butyne
■ (h) Ethyl pentanoate

■ **16-10** Give a general method for making each of the following compounds.
(a) Alcohols   (e) Amides
(b) Ketones   (f) Aldehydes
(c) Carboxylic acids (g) Amines
(d) Esters

■ **16-11** Tell how the following can be obtained from natural sources.
(a) Ethanol   (d) Methane
(b) Acetic acid   (e) Gasoline
(c) Methanol

■ **16-12** Give one possible use for each of the following compounds or class of compounds.
(a) Formaldehyde   (e) Amines
(b) Acetic acid   (f) Esters
(c) Ethylene glycol   (g) Alkenes
(d) Acetylsalicylic acid (h) Alkanes

Forward to the Appendixes

The successful athlete must stay "in shape" physically. The successful chemistry student must stay "in shape" mathematically.

It is amazing how students who have *confidence* enjoy and do well in chemistry. Confidence, of course, comes from being prepared with the proper background. Just as a professional football player has the proper background of a successful college career, the serious chemistry student needs a proper mathematical background. Many, if not most, students who take this course need to give some early attention to reviewing their math. The reason is not important, regardless of whether it's because of a deficient secondary school background, a change in career intentions, or a lengthy interruption of studies. If the deficiencies are serious, the student should consider taking math *before* this course. More often, a review like that provided by the following appendixes can patch the weak spots so that math is not a hindrance to understanding chemistry.

Appendixes A, B, and C are intended to aid your review of the math and algebra required in the study of college or university chemistry. The first three appendixes begin with a short test, so that you can quickly gauge if the review and exercises that follow are needed.

Appendix D complements and expands on the introduction to problem solving by the unit-factor method in Chapter 2. Students who can become comfortable with the types of problems in this appendix will find themselves in excellent condition for the quantitative aspects of chemistry. Appendix E is an introduction to the use of common logarithms, which are needed to understand pH in Chapter 15. Appendix F is a discussion of the interpretation and construction of graphs, a topic of prime importance in both the social and the natural sciences. Appendix G is a glossary of terms used throughout the text, and finally, Appendix H contains answers to about two-thirds of the problems at the ends of the chapters.

A

Basic Mathematics

A Test of Basic Mathematics

Following is a brief test on some basic mathematical skills necessary in chemistry. Work the test as carefully as you can *without a calculator*. Compare your answers with the answers given at the end of the test. If you have 100% of the answers correct, proceed to Appendix B on algebra. If you don't get 100, recheck your calculations. If a mistake or two was due to minor computational errors, then you should move on to Appendix B. If you make more than two mistakes or can't identify the source of your error, work through Appendix A before proceeding. After you have finished, take the test again. Hopefully, you will then get 100. If not, consult with your instructor for further review.

Math Test

1 Carry out the following calculations:

    (a) $43.76 + 6.90 + 134.88$ **1(a)** _____

    (b) $0.668 - 9.232 + 3.445$ **1(b)** _____

    (c) $5.65 - (-2.34) + (-0.89)$ **1(c)** _____

2 Carry out the following calculations. Include units and the proper number of significant figures in (c) and (d).

    (a) 10.7×9.755 **2(a)** _____

    (b) $(-0.56) \times 234$ **2(b)** _____

    (c) 16 in. $\times$ 34 in.   **2(c)** _____

    (d) 13.6 g/ml $\times$ 187 ml   **2(d)** _____

3 Express the following answers in decimal fraction form (e.g. $\frac{3}{5} = 0.6$). Include units and the proper number of significant figures in the answer.

    (a) 465 miles $\div$ 5.7 hr   **3(a)** _____

    (b) $\dfrac{115 \text{ g}}{7.82 \text{ ml}}$ **3(b)** _____

    (c) $\dfrac{186 \text{ ft}^3}{4.5 \text{ ft}}$ **3(c)** _____

(d) $\dfrac{1580 \text{ ft}}{432 \text{ sec}} \times 45.0 \text{ sec}$ **3(d)** _____

4 Express the following answers in decimal fraction form. When problems have units, include the units and the proper number of significant figures in the answer.

(a) $\left(-\frac{3}{4}\right) \times \frac{6}{7} \times \left(-\frac{10}{18}\right)$ **4(a)** _____

(b) $\frac{3}{4} \div \left(\frac{6}{7} \times \frac{1}{2}\right)$ **4(b)** _____

(c) $\dfrac{187 \text{ torr}/7.45}{760 \text{ torr}/\text{atm}}$ **4(c)** _____

(d) $\sqrt{81 \text{ cm}^2}$ **4(d)** _____

(e) $\left(\frac{2}{3} \text{ cm}\right)^3$ **4(e)** _____

5 There are 156 apples in a crate. Seventy apples are Johnathans, 45 are Golden delicious, and the others are rotten. What is the decimal fraction of rotten apples?

         5 _____

6 Express the following fractions as percentages:

(a) $\frac{3}{7}$ **6(a)** _____

(b) 0.0875   **6(b)** _____

7 Express the following percentages as decimal fractions:

(a) 98.9%   **7(a)** _____

(b) 0.74%   **7(b)** _____

Answers to Test

1(a) 185.54   **1(b)** −5.119   **1(c)** 7.10   **2(a)** 104.4   **2(b)** −131.04
2(c) 540 in.² **2(d)** 2540 g **3(a)** 82 miles/hr, **3(b)** 14.7 g/ml **3(c)** 41 ft²
3(d) 165 ft   **4(a)** 0.357   **4(b)** 1.75   **4(c)** 0.0330 atm   **4(d)** 9.0 cm
4(e) 0.3 cm³   **5** 0.263   **6(a)** 42.9%   **6(b)** 8.75%   **7(a)** 0.989
7(b) 0.0074

B A Review of Basic Mathematics

The following is a quick (very quick) refresher of fundamentals of math. This may be sufficient to aid you if you are just a little rusty on some of the basic concepts. For more thorough explanations and practice, however, you are urged to use a more comprehensive math review workbook or consult with your instructor.

1 Addition and Subtraction

Since most of the calculations in this text use numbers expressed in decimal form, we will emphasize the manipulation of these types of numbers. In addition and subtraction, it is important to line up the decimal point carefully before doing the math.

Example A-1
Add the following numbers: 16.75, 13.31, and 175.67.

$$\begin{array}{r} \downarrow \\ 16.75 \\ 13.31 \\ \underline{175.67} \\ \underline{205.73} \end{array}$$

Subtraction is simply the addition of a negative number. Remember that subtraction of a negative number changes the sign to a plus (two negatives make a positive). For example, $4 - 7 = -3$, but $4 - (-7) = 4 + 7 = 11$.

Example A-2
Carry out the following calculations:
(a) $11.8 + 13.1 - 6.1$

$$\begin{array}{r} 11.8 \\ +13.1 \\ \underline{24.9} \end{array} \qquad \begin{array}{r} 24.9 \\ -6.1 \\ \underline{18.8} \end{array}$$

(b) $47.82 - 111.18 - (-12.17)$
This is the same as $47.82 - 111.18 + 12.17$.

$$\begin{array}{r} 47.82 \\ +12.17 \\ \underline{59.99} \end{array} \qquad \begin{array}{r} -111.18 \\ +59.99 \\ \underline{-51.19} \end{array}$$

Exercises

A-1 Carry out the following calculations:

(a) $47 + 1672$ (d) $-97 + 16 - 118$

(b) $11.15 + 190.25$ (e) $0.897 + 1.310 - 0.063$

(c) $114 + 26 - 37$ (f) $-0.377 - (-0.101) + 0.975$

(g) $17.489 - 318.112 - (-0.315) + (-3.330)$

Answers: (a)   1719   (b)   201.40   (c)   103   (d)   −199   (e)   2.144
(f)   0.699   (g)   −303.638

2 Multiplication

Multiplication is expressed in various ways as follows:

$$13.7 \times 115.35 = 13.7 \cdot 115.35 = (13.7)(115.35) = 13.7(115.35)$$

If it is necessary to carry out the multiplication in longhand, you must be careful to place the decimal point correctly in the answer. Count the *total* number of digits to the right of the decimal point in both multipliers (three in this example). The answer has that number of digits to the right of the decimal point in the answer.

$$13.7 \times 115.35 = 1580.295$$
$$1 + \quad 2 \quad = \quad 3$$

Numbers without units are assumed to be exact numbers in these exercises. In most numbers used in chemistry calculations, however, the numbers include specific units. In these cases, the result of any mathematical manipulation must be expressed to the proper number of significant figures. We will observe the rules for significant figures (See Section 2-1) for exercises where units are included.

A number raised to a power (which is known as an exponent) is multiplied by itself the number of times indicated by the exponent. For example,

$$4^2 = 4 \times 4 = 16 \qquad \text{(four "squared")}$$

$$2^4 = 2 \times 2 \times 2 \times 2 = 16 \qquad \text{(two to the fourth power)}$$

$$4^3 = 4 \times 4 \times 4 = 64 \qquad \text{(four "cubed")}$$

$$(14.1)^2 = 14.1 \times 14.1 = 198.81$$

In the calculations used in this book, most numbers have specific units. In multiplication, the units are multiplied as well as the numbers. For example,

$$3.7 \text{ cm} \times 4.61 \text{ cm} = 17 \text{ (cm} \times \text{cm)} = 17 \text{ cm}^2$$

$$(4.5 \text{ in.})^3 = 91 \text{ in.}^3$$

In the multiplication of a series of numbers, grouping is possible:

$$(a \times b) \times c = a \times (b \times c)$$

$$3.0 \text{ cm} \times 148 \text{ cm} \times 3.0 \text{ cm} = (3.0 \times 3.0) \times 148 \times (\text{cm} \times \text{cm} \times \text{cm})$$
$$= \underline{1300 \text{ cm}^3}$$

When multiplying signs, remember:

$$+ \times - = - \qquad + \times + = + \qquad - \times - = +$$

For example, $(-3) \times 2 = -6$; $(-9) \times (-8) = +72$.

Exercises

A-2 Carry out the following calculations:

(a) $16.2 \times (-118)$ (d) 17.7 in. $\times$ (13.2 in. $\times$ 25.0 in.)

(b) $(4 \times 2) \times 879$ (e) $(-8) \times (-2) \times (-37)$

(c) 3.0 ft $\times$ 18 lb   (f) $(-47.8) \times -9.6$

(g) 0.67 atm $\times$ (1.22 liters $\times$ 0.50 atm)

Answers: (a) -1911.6 (b) 7032   (c) 54 ft · lb   (d) 5840 in.^{3}
(e) -592 (f) 458.88 (g) 0.41 liter · atm^2

3 Division and Fractions

Division problems are expressed in fraction form as follows:

Division form	Common fraction form	Decimal fraction form

$$88.8 \div 2.44 = \frac{88.8}{2.44} = 88.8/2.44 = \qquad 36.4$$

In most cases in chemistry, answers are expected in decimal fraction form rather than common fraction form. Therefore, to obtain the answer, the numerator (the number on the top) is divided by the denominator (the number on the bottom). Before doing the actual calculation, it helps to have a feeling for how the fractional number should look in decimal form. If the numerator is larger than the denominator, the decimal number is greater than 1. If the opposite is true, the decimal number is less than 1.

To carry out the division longhand, it is easier to divide a whole number in the denominator into the numerator. To do this, move the decimal in both numerator and denominator the same number of places. In effect, you are multiplying both numerator and denominator by the same number, which does not change the value of the fraction.

$$\frac{a}{b} = \frac{a \times c}{b \times c}$$

$$\frac{88.8}{2.44} = \frac{88.8 \times 100}{2.44 \times 100} = \frac{8880}{244} = \underline{\underline{36.4}}$$

Many divisions can be simplified by "cancellation." Cancellation is the elimination of common multipliers in the numerator and denominator. This is possible because a number divided by itself is equal to unity (e.g., $25/25 = 1$). As in multiplication, all units must also be divided. If identical units appear in both numerator and denominator, they can also be cancelled.

$$\frac{a \times \ell}{b \times \ell} = \frac{a}{b}$$

$$\frac{\overset{1}{\cancel{190}} \times 4 \text{ torr}}{\cancel{190} \text{ torr}} = \frac{4}{\underline{\underline{1}}}$$

$$\frac{2500 \text{ cm}^3}{150 \text{ cm}} = \frac{\overset{1}{\cancel{50}} \times 50 \text{ cm} \times \text{cm} \times \text{cm}}{\underset{1}{\cancel{50}} \times 3 \text{ cm}} = \frac{50 \text{ cm}^2}{3} = \underline{\underline{17 \text{ cm}^2}}$$

$$\frac{2800 \text{ miles}}{45 \text{ hr}} = \frac{\cancel{5} \times 560 \text{ miles}}{\cancel{5} \times 9 \text{ hr}} = \frac{62 \text{ miles}}{1 \text{ hr}} = \underline{\underline{62 \text{ miles/hr}}}$$

This is read as 62 miles "per" one hour or simply 62 miles per hour. The word "per" implies a fraction or a ratio with the unit after "per" in the denominator. If a number is not written or read in the denominator with a unit, it is assumed that the number is unity and is known to as many significant figures as the number in the numerator.

Exercises

A-3 Express the following in decimal fraction form:

(a) 892 miles ÷ 41 hr

(e) $\dfrac{1890 \text{ cm}^3}{66 \text{ cm}}$

(b) 982.6 ÷ 0.25

(f) $\dfrac{146 \text{ ft} \cdot \text{hr}}{0.68 \text{ ft}}$

(c) 195 ÷ 2650

(g) $\dfrac{0.8772 \text{ ft}^3}{0.0023 \text{ ft}^2}$

(d) $\dfrac{67.5 \text{ g}}{15.2 \text{ ml}}$

(h) $\dfrac{37.50 \text{ ft}}{0.455 \text{ sec}}$

Answers: (a) 22 miles/hr (b) 3930.4 (c) 0.0736 (d) 4.44 g/ml
(e) 29 cm² (f) 210 hr (g) 380 ft (h) 82.4 ft/sec

4 Multiplication and Division of Fractions

When two or more fractions are multiplied, all numbers *and units* in both numerator and denominator can be combined into one fraction.

Example A-3

$$\frac{3}{5} \times \frac{75}{4} \times \frac{16}{7} = \frac{3 \times 75 \times 16}{\cancel{5} \times 4 \times 7} = \frac{3 \times \overset{15}{\cancel{75}} \times \overset{4}{\cancel{16}}}{\underset{1}{\cancel{5}} \times \underset{1}{\cancel{4}} \times 7} = \frac{180}{7} = \underline{\underline{25.7}}$$

Example A-4

$$\frac{42 \text{ miles}}{\text{hr}} \times \frac{3}{7} \text{ hr} \times \frac{5280 \text{ ft}}{\text{miles}} = \frac{\overset{6}{\cancel{42}} \times 3 \times 5280 \text{ } \cancel{\text{miles}} \times \cancel{\text{hr}} \times \text{ft}}{\underset{1}{\cancel{7}} \text{ } \cancel{\text{hr}} \times \cancel{\text{mile}}} = \underline{\underline{95,000 \text{ ft}}}$$

Example A-5

$$\frac{3}{4} \text{ mol} \times \frac{0.75 \text{ g}}{\text{mol}} \times \frac{1 \text{ ml}}{19.3 \text{ g}} = \frac{3 \times 0.75 \times 1 \times \cancel{\text{mol}} \times \cancel{\text{g}} \times \text{ml}}{4 \times 1 \times 19.3 \times \cancel{\text{mol}} \times \cancel{\text{g}}} = \underline{\underline{0.029 \text{ ml}}}$$

The division of one fraction by another is the same as the multiplication of the numerator by the *reciprocal* of the denominator. The reciprocal of a fraction is simply the fraction in an inverted form (e.g., $\frac{3}{5}$ is the reciprocal of $\frac{5}{3}$).

$$\frac{\dfrac{a}{b}}{c} = a \times \frac{c}{b} \qquad \frac{\dfrac{a}{b}}{\dfrac{c}{d}} = \frac{a}{b} \times \frac{d}{c} = \frac{a \times d}{b \times c}$$

Example A-6

$$\frac{1650}{\dfrac{3}{5}} = 1650 \times \frac{5}{3} = \underline{\underline{2750}}$$

Example A-7

$$\frac{145 \text{ g}}{\dfrac{7.5 \text{ g}}{\text{ml}}} = 145 \text{ } \cancel{\text{g}} \times \frac{1 \text{ ml}}{7.5 \text{ } \cancel{\text{g}}} = \underline{\underline{19.3 \text{ ml}}}$$

Exercises

A-4 Express the following in decimal fraction form. Where units are used, include them in the answer.

(a) $\frac{3}{8} \times \frac{4}{7} \times \frac{21}{20}$

(b) $\frac{250}{273} \times \frac{175}{300} \times (-6)$

(c) $\frac{4}{9} \times (-\frac{5}{8}) \times (-\frac{3}{4})$

(d) $195 \text{ g/ml} \times 47.5 \text{ ml}$

(e) $0.75 \text{ mol} \times 17.3 \text{ g/mol}$

(f) $(3.57 \text{ in.})^2 \times 0.85 \text{ in.} \times \frac{16.4 \text{ cm}^3}{\text{in.}^3}$

(g) $\dfrac{\frac{150}{350}}{\frac{25}{42}}$

(h) $\dfrac{(-\frac{3}{7})}{(-\frac{4}{9})}$

(i) $\dfrac{(-\frac{17}{3})}{\frac{8}{9}}$

(j) $\dfrac{\frac{16}{9} \times \frac{10}{14}}{\frac{5}{6}}$

(k) $\dfrac{75.2 \text{ torr}}{760 \text{ torr/atm}}$

(l) $\dfrac{(55.0 \text{ miles/hr}) \times (5280 \text{ ft/mile}) \times (1 \text{ hr/60 min})}{60 \text{ sec/min}}$

(m) $\dfrac{305 \text{ K} \times (62.4 \text{ liters} \cdot \text{torr/K} \cdot \text{mol}) \times 0.25 \text{ mol}}{650 \text{ torr}}$

Answers: **(a)** 0.225 **(b)** -3.21 **(c)** 0.208 **(d)** 9260 g **(e)** 13 g
(f) 180 cm³ **(g)** 0.72 **(h)** 0.96 **(i)** -6.375 **(j)** 1.52 **(k)** 0.0989 atm
(l) 80.7 ft/sec **(m)** 7.3 liters

5 Roots of Numbers

The square root of a number is expressed as

$$\sqrt{a} \quad \text{or} \quad a^{1/2}$$

It is the number multiplied by itself that gives the number a. For example,

$$\sqrt{4} = 2 \qquad (2 \times 2 = 4)$$
$$\sqrt{9} = 3 \qquad (3 \times 3 = 9)$$

The square roots of numbers that are not even numbers may be computed on a calculator or found in a table. Without these available, an educated approximation can come close to the answer.

Example A-8

What is the square root of 54.0 in.²?

Solution:

We know that

$$\sqrt{49 \text{ in.}^2} = 7 \text{ in.}$$

and

$$\sqrt{64 \text{ in.}^2} = 8 \text{ in.}$$

(Notice that the square roots of units are also expressed.) You can see that the answer is between 7 in. and 8 in. Try 7.5 in.

$$7.5 \text{ in} \times 7.5 \text{ in.} = 56.2 \text{ in.}^2$$

This is close but a little high. Try 7.4 in.

$$7.4 \text{ in.} \times 7.4 \text{ in.} = 54.8 \text{ in.}^2$$

This is still a little high. Try 7.3 in.

$$7.3 \text{ in.} \times 7.3 \text{ in.} = 53.3 \text{ in.}^2$$

This is a little low, but an approximation of $\underline{\underline{7.35}}$ would be on the money.

The cube root of a number is expressed as

$$\sqrt[3]{b} = b^{1/3}$$

It is the number multiplied by itself three times that gives b. For example,

$$\sqrt[3]{27} = 3 \qquad (3 \times 3 \times 3 = 27)$$
$$\sqrt[3]{64} = 4 \qquad (4 \times 4 \times 4 = 64)$$

Exercises

A-5 Find the following roots. If necessary, approximate the answer and check on a calculator or with a table.

(a) $\sqrt{25}$ (c) $\sqrt{144 \text{ ft}^4}$ (e) $\sqrt{7}$ (g) $2^{1/2}$ (i) $100^{1/3}$

(b) $\sqrt{36 \text{ cm}^2}$ (d) $\sqrt{40}$ (f) $110^{1/2}$ (h) $\sqrt[3]{50}$

Answers: (a) 5 (b) 6.0 cm (c) 12.0 ft² (d) 6.32 (e) 2.65 (f) 10.5 (g) 1.41 (h) 3.68 (i) 4.64

6 Construction of Fractions

In the examples of the fractions discussed thus far we have seen that the units of the numerator can be profoundly different from those of the denominator (e.g., miles/hr, g/ml, etc.). In other problems in chemistry we use fractions to express a component part in the numerator to the total in the denominator. In most cases such fractions are expressed without units and in decimal fraction form.

Example A-9

A box of nails contains 985 nails; 415 of these are 6-in. nails, 375 are 3-in. nails, and the rest are roofing nails. What is the fraction of roofing nails?

Solution:

Roofing nails = total − others = 985 − (415 + 375) = 195

$$\frac{\text{Component}}{\text{Total}} = \frac{195}{375 + 415 + 195} = \underline{\underline{0.198}}$$

Example A-10

A mixture contains 4.25 mol of N_2, 2.76 mol of O_2, and 1.75 mol of CO_2. What is the fraction of moles of O_2 present in the mixture? (This fraction is known, not surprisingly, as "the mole fraction." The mole is a unit of quantity, like dozen.)

Solution:

$$\frac{\text{Component}}{\text{Total}} = \frac{2.76}{4.25 + 2.76 + 1.75} = \underline{\underline{0.315}}$$

Exercises

A-6 A grocer has 195 dozen boxes of fruit; 74 dozen boxes are apples, 62 dozen boxes are peaches, and the rest are oranges. What is the fraction of the boxes that are oranges?

A-7 A mixture contains 9.85 mol of gas. A 3.18-mol quantity of the gas is N_2, 4.69 mol is O_2, and the rest is He. What is the mole fraction of He in the mixture?

A-8 The total pressure of a mixture of two gases, N_2 and O_2, is 0.72 atm. The pressure due to O_2 is 0.41 atm. What is the fraction of the pressure due to O_2?

Answers: **A-6** 0.303 **A-7** 0.201 **A-8** 0.57

7 Fractions Expressed as Percent

The decimal fractions that have just been discussed are frequently expressed as percentages. Percent simply means parts "per" 100. Percent is obtained by multiplying a fraction in decimal form by 100.

Example A-11

If 57 out of 180 people at a party are women, what is the percent women?

Solution:

The fraction of women in decimal form is

$$\frac{57}{180} = 0.317$$

The percent women is

$$0.317 \times 100 = 31.7\% \text{ women}$$

The general method used to obtain percent is

$$\frac{\text{Component}}{\text{Total}} \times 100 = \underline{\qquad}\% \text{ of component}$$

To change from percent back to a decimal fraction, divide the percent by 100, which moves the decimal to the left two places.

$$86.2\% = \frac{86.2}{100} = 0.862 \qquad \text{(fraction in decimal form)}$$

Many problems involving percent require the use of algebra. Therefore many of the sample calculations will be illustrated in Appendix B.

Exercises

A-9   Express the following fractions as percent: $\frac{1}{4}$, $\frac{3}{8}$, $\frac{9}{8}$, $\frac{55}{25}$, 0.67, 0.13, 1.75, 0.098.

A-10   A bushel holds 198 apples, of which 27 are green. What is the percent of green apples?

A-11   A basket contains 75 pears, 8 apples, 15 oranges, and 51 grapefruit. What is the percent of each?

Answers: **A-9** $\frac{1}{4} = 25\%$, $\frac{3}{8} = 37.5\%$, $\frac{9}{8} = 112.5\%$, $\frac{55}{25} = 220\%$, $0.67 = 67\%$, $0.13 = 13\%$, $1.75 = 175\%$, $0.098 = 9.8\%$ **A-10** 13.6%, **A-11** 50.3% pears, 5.4% apples, 10.1% oranges, and 34.2% grapefruit

Basic Algebra

B

A Test of Basic Algebra

Following is a test on some of the basic algebra skills that are necessary in chemistry. After you take the test, compare your answers to those listed at the end of the test. If you miss more than one question in any one section, you should work through the review following the test. In any case, you will be referred to this review at specific points in the text when certain calculations are first encountered.

1 Solve the following equations for x. (Isolate all numbers and variables other than x on the right-hand side of the equation.)

(a) $4x + 17 = 73$ 1(a) _____

(b) $\dfrac{y}{3x} + 4 = R$ 1(b) _____

(c) $3x^2 + y = 225 - 2x^2$ 1(c) _____

(d) $\dfrac{1}{2x + 1} = y$ 1(d) _____

2 Write algebraic equations that express the following:
(a) A number x is six more than y.

2(a) _____

(b) A number r is 36 less than the square root of z.

2(b) _____

(c) Four times the product of two numbers s and t is 14% of the sum of two other numbers p and q.

2(c) _____

(d) Forty-three percent of a number k is equal to the reciprocal of the fraction $\frac{2}{3}$.

2(d) _____

3 Write algebraic equations that represent the following (let $x =$ an unknown) and solve.
(a) A large brick weighs 1 lb more than twice as much as a smaller brick. The sum of the weights of the two bricks is 28 lb. What is the weight of each brick?
2(a) Equation _____ Answer _____

(b) A student got a score of 78.9% on a chemistry test that had a total of 180 points. How many points did the student receive?

2(b) Equation _____   Answer _____

(c) There are 751 college students at a rock concert, which is 59.6% of the crowd. What is the size of the crowd?

2(c) Equation _____   Answer _____

(d) There was $9.87 in a cash register, with one more than twice as many dimes as quarters and three less than four times as many pennies as quarters. How many quarters were in the cash register?

2(d) Equation _____   Answer _____

4 Answer the following:

(a) A small car weighs 100 lb more than two-thirds of the weight of a midsize car. The difference in weight between the two cars is 800 lb. What is the weight of the small car?

4(a) _____

(b) A major leaguer during his career got 32 home runs, 25 triples, 55 doubles, and the rest singles. The fraction of his hits that were doubles was 0.119. How many singles did he get?

4(b) _____

(c) An oil refinery held 175 barrels of oil. When refined, each barrel yields 24 gallons of gasoline. If 3120 gallons of gasoline were produced, what percentage of the original barrels of oil were refined?

4(c) _____

(d) A shipment of rabbits originally contained 874 rabbits. On arrival it was found that 12.7% of the rabbits had each produced three more rabbits and that 15.0% had produced two. The others did not multiply. How many rabbits arrived in all?

4(d) _____

5 Write equations for the following. Use k for a constant of proportionality.

(a) A quantity x is directly proportional to y.

5(a) _____

(b) A quantity z is inversely proportional to the square root of r.

5(b) _____

(c) A quantity q is directly proportional to A and inversely proportional to the square of the sum of B and C.

5(c) _____

(d) The cube root of a quantity t is directly proportional to the square of v and inversely proportional to six less than the quantity w.

5(d) _____

Answers: **1(a)** $x = 14$ **1(b)** $x = y/(3R - 12)$ **1(c)** $x = \sqrt{45 - y/5}$
1(d) $x = (1/2y) - (1/2)$ **2(a)** $x = y + 6$ **2(b)** $r = \sqrt{z} - 36$
2(c) $4st = 0.14(p + q)$ **2(d)** $0.43k = \frac{3}{2}$ **3(a)** $x + (2x + 1) = 28.9$ lb,
19 lb   **3(b)** $x = (0.789)(180) = 142$ **3(c)** $0.596(x) = 751, x = 1260$
3(d) $0.25x + 0.10(2x + 1) + 0.01(4x - 3) = 9.87, x = 20$ **4(a)** 1900 lb
4(b) 350   **4(c)** 74.3%   **4(d)** 1469   **5(a)** $x = ky$ **5(b)** $z = k/\sqrt{r}$
5(c) $q = kA/(B + C)^2$ **5(d)** $\sqrt[3]{t} = kv^2/(w - 6)$

B A Review of Basic Algebra

1 Operations on Algebra Equations

Many of the problems of chemistry require the use of basic algebra.
Let's examine the following simple algebra equation:

$$a = (b + c)$$

In any algebraic equation the equality remains valid when *identical* operations are performed on both sides of the equation. The following operations illustrate this principle.

1   A quantity d may be added to or subtracted from both sides of the equation.

$$a + d = (b + c) + d$$
$$a - d = (b + c) - d$$

2   Both sides of the equation may be multiplied or divided by the same quantity d.

$$a \times d = (b + c) \times d$$
$$\frac{a}{d} = \frac{(b + c)}{d}$$

3   Both sides of the equation may be squared, or the square root of both sides of the equation may be taken.

$$a^2 = (b + c)^2$$
$$\sqrt{a} = \sqrt{b + c}$$

4   Both sides of an equation can be inverted

$$\frac{1}{a} = \frac{1}{(b + c)}$$

In addition to operations on both sides of an equation, there are two other points to recall.

1 As in any fraction, identical multipliers in the numerator and the denominator in an algebraic equation may be cancelled.

$$\frac{a \times \cancel{d}}{\cancel{d}} = b + c$$

2 Quantities equal to the same quantity are equal to each other. Thus substitutions for equalities may be made in algebraic equations.

$$a = b + c$$

$$a = 2d + r$$

Therefore, since $a = a$,

$$b + c = 2d + r$$

We can use these basic rules to solve algebraic equations. Usually, we need to isolate one variable on the left-hand side of the equation with all other numbers and variables on the right-hand side of the equation. The following examples illustrate the isolation of a variable on the left-hand side of the equation.

Example B-1

Solve for x in $x + 5 = 92$.

Solution:

Subtract 5 from both sides of the equation.

$$x + 5 \underline{- 5} = 92 \underline{- 5}$$

$$\underline{\underline{x = 87}}$$

In practice, a number or a variable may be moved to the other side of an equation with a change of sign. For example,

$$x \underline{+ 5} = y$$

$$x = y \underline{- 5}$$

Example B-2

Solve for x in $x + y + 8 = z + 6$.

Solution:

$$x = z + 6 \underline{- y - 8}$$
$$= z - y + 6 - 8$$
$$\underline{\underline{= z - y - 2}}$$

Example B-3
Solve for x in

$$\frac{x + 8}{y} = z$$

Solution:
To move the y to the right, multiply both sides of the equation by y. This cancels the y on the left and leaves y in the numerator on the right.

$$\not{y} \cdot \frac{x + 8}{\not{y}} = z \cdot y$$

$$x + 8 = zy$$

$$\underline{\underline{x = zy - 8}}$$

In practice, to move a denominator to the other side of the equation, multiply both sides of the equation by the denominator. To move a numerator, divide both sides by the numerator. For example,

$$\frac{z}{x} = y$$

$$\frac{1}{\not{z}} \cdot \frac{\not{z}}{x} = y \cdot \frac{1}{z}$$

$$\frac{1}{x} = \frac{y}{z}$$

In these problems, the numerical value of x can be found if values for the other variables are known. For example, if $y = -2$ and $z = 8$ in Example B-3,

$$x = zy - 8$$
$$= (-2)(8) - 8 = -16 - 8 = -24$$

Example B-4
Solve for x in

$$\frac{4x + 2}{3 + x} = 7$$

Solution:
Clear of fractions:

$$4x + 2 = 7(3 + x) = 21 + 7x$$

Move $7x$ to the left and 2 to the right:

$$4x - 7x = 21 - 2$$

$$-3x = 19$$

Divide both sides by -3:

$$\frac{-3x}{-3} = \frac{19}{-3}$$

$$x = -\frac{19}{3}$$

Example B-5
Solve for T_2 in

$$\frac{P_1 V_1}{T_1} = \frac{P_2 V_2}{T_2}$$

Solution:
Invert both sides of the equation and switch sides (so that T_2 is on the left).

$$\frac{T_2}{P_2 V_2} = \frac{T_1}{P_1 V_1}$$

Multiply both sides by $P_2 V_2$ to move $P_2 V_2$ to the right-hand side of the equation.

$$\cancel{P_2 V_2} \times \frac{T_2}{\cancel{P_2 V_2}} = \frac{T_1 \cdot P_2 V_2}{P_1 V_1}$$

$$T_2 = \frac{T_1 P_2 V_2}{P_1 V_1}$$

Example B-6
Solve the following equation for y:

$$\frac{2y}{3} + x = 9z + 4$$

Solution:
Multiply both sides of the equation by 3.

$$2y + 3x = 27z + 12$$

$$2y = 27z + 12 - 3x$$

$$y = \frac{27z + 12 - 3x}{2}$$

Example B-7
Solve the following equation for x:

$$x^2 - 49 = y$$

$$x^2 = y + 49$$

$$x = \sqrt{y + 49}$$

Exercises

B-1   Solve for x in $17x = y - 87$.

B-2   Solve for x in

$$\frac{y}{x} + 8 = z + 16$$

B-3   Solve for T in $PV = (\text{wt.}/\text{M.M.})RT$.

B-4   Solve for x in

$$\frac{7x - 3}{6 + 2x} = 3r$$

B-5   Solve for x in $18x - 27 = 2x + 4y - 35$. If $y = 3x$, what is the value of x?

B-6   Solve for x in

$$\frac{x}{4y} + 18 = y + 2$$

B-7   Solve for x in $5x^2 + 12 = x^2 + 37$.

B-8   Solve for r in

$$\frac{80}{2r} + \frac{y}{r} = 11$$

What is the value of r if $y = 14$?

Answers: **B-1** $x = (y - 87)/17$ **B-2** $x = y/(8 + z)$ **B-3** $T = PV \cdot \text{M.M.}/\text{wt.} \cdot R$ **B-4** $x = 3(6r + 1)/(7 - 6r)$ **B-5** $x = (y - 2)/4$. When $y = 3x$, $x = -2$.   **B-6** $x = 4y(y - 16)$ **B-7** $x = \pm 2.5$ **B-8** $r = (40 + y)/11$. When $y = 14$, $r = \frac{54}{11}$.

2 Word Problems

Eventually, a necessary skill in chemistry is the ability to translate word problems into algebra equations and then solve. The key is to assign a variable (usually x) to be equal to a certain quantity and then to treat the variable consistently throughout the equation. Again, examples are the best way to illustrate the problems.

Example B-8

Translate each of the following to an equation.

(a) A number x is 4 larger than another number y.

$$x = y + 4$$

(b) A number z is three-fourths of u.

$$z = \tfrac{3}{4}u$$

(c) The square of a number r is 16.9% of w.

$$r^2 = 0.169w \qquad \text{(Change percent to a decimal fraction)}$$

(d) A number t is equal to 12 plus the square root of q.

$$t = 12 + \sqrt{q}$$

Exercises

Write algebraic equations for the following.

B-9   A number n is 85 smaller than m

B-10   A number y is one-fourth of z.

B-11   Fifteen percent of a number k is equal to the square of another number d.

B-12   A number x is 14 more than the square root of v.

B-13   Four times the sum of two numbers, q and w, is equal to 68.

B-14   Five times the product of two variables, s and t, is equal to 16 less than the square of s.

B-15   Five-ninths of a number C is equal to 32 less than a number F.

Answers: **B-9** $n = m - 85$ **B-10** $y = z/4$ **B-11** $0.15k = d^2$
B-12 $x = \sqrt{v} + 14$ **B-13** $4(q + w) = 68$ **B-14** $5st = s^2 - 16$
B-15 $\tfrac{5}{9}C = F - 32$

We now move from the abstract to the real. In the following examples it is necessary to translate the problem into an algebraic expression as in the previous examples. There are two types of examples that we will use. The first you will certainly recognize, but the second type may be unfamiliar especially if you have just begun the study of chemistry. However, it is *not* important that you understand the units of the chemistry problems at this time. What *is* important is for you to notice that the problems are worked in the same manner regardless of the units.

Example B-9

John is two years more than twice as old as Mary. The sum of their ages is 86. How old is each?

Solution:

Let x = age of Mary. Then $2x + 2$ = age of John.

$$x + (2x + 2) = 86$$

$$3x = 84$$

$$x = \underline{28} \quad \text{(age of Mary)}$$

$$[2(28) + 2] = \underline{58} \quad \text{(age of John)}$$

Example B-10

One mole of SF_6 weighs 30.0 g less than four times the weight of 1 mol of CO_2. The weight of 1 mol of SF_6 plus the weight of 1 mol of CO_2 is 190 g. What is the weight of 1 mol of each?

Solution:

Let x = weight of 1 mol of CO_2. Then $4x - 30$ = weight of 1 mol of SF_6.

$$x + (4x - 30) = 190$$

$$x = \underline{44 \text{ g}} \quad \text{(weight of 1 mol of } CO_2)$$

$$[4(44) - 30] = \underline{146 \text{ g}} \quad \text{(weight of 1 mol of } SF_6)$$

Example B-11

If a grocer has 42.5 dozen boxes of fruit and 16.5% are apples, how many dozen boxes of apples are there? How many dozen boxes are not apples?

Solution:

Let x = dozens of boxes of apples. Then $x = 0.165 \times 42.5$.

$$x = \underline{7.0 \text{ dozen boxes of apples}}$$

Boxes remaining will be $42.5 - x$.

$$42.5 - 7.0 = \underline{35.5 \text{ dozen boxes remaining}}$$

Example B-12

If a 8.75-g quantity of sugar represents 65.7% of a mixture, what is the weight of the mixture?

Solution:

Let x = weight of the mixture. Then

$$0.657x = 8.75$$

$$x = \frac{8.75}{0.657} = \underline{13.3 \text{ g}}$$

Example B-13

In a certain year 14.5% of a rancher's herd of 876 cows had two calves each. The rest had none.

(a) How many calves were born?
(b) How many cows did not have a calf?
(c) How large was the herd including calves?

Solution:

Let x = number of cows having calves. Then $2x$ = number of calves born.

$$x = 0.145 \cdot 876 = 127 \text{ cows having calves}$$

$$2x = 2 \cdot 127 = \underline{354 \text{ calves born}}$$

$$876 - x = 876 - 127 = \underline{749 \text{ cows not having calves}}$$

$$876 + 2x = 876 + 354 = \underline{1230 \text{ total cattle}}$$

Example B-14

This example is directly analogous to the preceding example. A certain compound, AX_2, can break apart (dissociate) into one A and two X's. In a 0.250-mol quantity of AX_2, 6.75% of the moles dissociated. (A mole, abbreviated mol, is a unit of quantity, like dozen or gross.)

(a) How many moles of AX_2 dissociated?
(b) How many moles of AX_2 did not dissociate?
(c) How many total moles of particles (AX_2, A, and X) are present?

Solution:

Let x = number of moles of AX_2 that dissociate. Then

$$x = 0.0675 \cdot 0.250 \text{ mol} = \underline{0.0169 \text{ mol dissociated}}$$

The number of moles of AX_2 that are not dissociated = $0.250 - x$.

$$0.250 - 0.0169 = \underline{0.233 \text{ mol undissociated}}$$

Total moles present

$$= \text{moles } AX_2 \text{ (undissociated)} + \text{moles of A} + \text{moles of X}$$

$$\text{Moles } AX_2 = 0.233$$

$$\text{Moles of A} = x = 0.0169$$

$$\text{Moles of X} = 2x = 0.0338$$

$$\text{Total moles} = 0.233 + 0.0169 + 0.0338 = \underline{0.284 \text{ mol total}}$$

Example B-15

A used car dealer has Fords, Chevrolets, and Plymouths. There are 120 Fords, 152 Chevrolets, and the rest are Plymouths. If the fraction of Fords is 0.31, how many Plymouths are on the lot?

Solution:

Let x = number of Plymouths.

$$\text{Fraction of Fords} = \frac{\text{number of Fords}}{\text{total number of cars}} = 0.31$$

$$\frac{120}{120 + 152 + x} = 0.31$$

$$120 = 0.31(272 + x)$$

$$120 = 84.3 + 0.31x$$

$$x = \underline{\underline{115 \text{ Plymouths}}}$$

Example B-16

There is a 0.605-mol quantity of N_2 present in a mixture of N_2 and O_2. If the mole fraction of N_2 is 0.251, how many moles of O_2 are present?

Solution:

Let $x =$ number of moles of O_2. Then

$$\text{mole fraction } N_2 = \frac{\text{mol of } N_2}{\text{total mol present}} = 0.251$$

$$\frac{0.605}{0.605 + x} = 0.251$$

$$0.605 = 0.251(0.605 + x)$$

$$x = \underline{\underline{1.80 \text{ mol}}}$$

Exercises

In the following exercises, a problem concerning an everyday situation is followed by one or two closely analogous problems concerning a chemistry situation. In both cases the mechanics of the solution are similar. Only the units differ.

B-16 The total length of two boards is 18.4 ft. If one board is 4.0 ft. longer than the other, what is the length of each board?

B-17 An isotope of iodine weighs 10 amu less than two-thirds of the weight of an isotope of thallium. The total weight of the two isotopes is 340 amu. What is the weight of each isotope?

B-18 An isotope of gallium weighs 22 amu more than one-fourth the weight of an isotope of osmium. The difference in the two weights is 122 amu. What is the weight of each?

B-19 In a certain audience, 45.9% were men. If there were 196 people in the audience, how may women were present?

B-20 In the alcohol molecule, 34.8% of the weight is due to oxygen. What is the weight of oxygen in 497 g of alcohol?

B-21   The cost of a hamburger in 1980 is 216% of the cost in 1970. If hamburgers cost $0.75 each in 1970, what do they cost in 1980?

B-22   An unstable isotope of iodine weighs 104% of that of a stable isotope which weighs 126 amu. What is the weight of the unstable isotope?

B-23   In a certain audience, 46.0% are men. If there are 195 men in the audience, how large is the audience?

B-24   If a solution is 23.3% by weight HCl and it contains 14.8 g of HCl, what is the total weight of the solution?

B-25   An oil refinery held 175 barrels of oil. When refined, each barrel yields 24 gallons of gasoline. If 3120 gallons of gasoline were produced, what percentage of the original barrels of oil were refined?

B-26   A solution contained 0.856 mol of a substance A_2X. In solution some of the A_2X's break up into A's and X's (notice that each mole of A_2X yields 2 mol of A). If 0.224 mol of A are present in the solution, what percentage of the moles of A_2X dissociated (broke apart)?

B-27   In Las Vegas, a dealer starts with 264 decks of cards. If 42.8% of the decks were used in an evening, how many jacks (four per deck) were used?

B-28   A solution originally contains a 1.45-mol quantity of a compound A_3X_2. If 31.5% of the A_3X_2 dissociates (three A's and two X's per A_3X_2), how many moles of A are formed? How many moles of X? How many moles of undissociated A_3X_2 remain? How many moles of particles (A's, X's, and A_3X_2's) are present in the solution?

B-29   The fraction of kerosine that can be recovered from a barrel of crude oil is 0.200. After refining a certain amount of oil, 8.9 gallons of kerosine, some gasoline, and 18.6 gallons of other products were produced. How many gallons of gasoline were produced?

B-30   The fraction of moles (mole fraction) of gas A in a mixture is 0.261. If the mixture contains 0.375 mol of gas B and 0.175 mol of gas C as well as gas A, how many moles of gas A are present?

Answers: **B-16**   7.2 ft, 11.2 ft   **B-17**   Thallium, 210 amu; iodine, 130 amu;   **B-18**   gallium, 70 amu; osmium, 192 amu   **B-19**   106 women   **B-20**   173 g   **B-21** $1.62   **B-22**   131 amu   **B-23**   424 people   **B-24**   63.5 g   **B-25**   74.3%   **B-26**   13.1%   **B-27**   452 jacks   **B-28**   1.37 mol of A, 0.194 mol of X, 0.99 mol of A_3X_2, 3.27 mol total   **B-29**   17.0 gallons   **B-30**   0.195 mol

3 Direct and Inverse Proportionalities

There is one other point that should be included in a review on algebra, and that is direct and inverse proportionalities. We will use these terms often in chemistry.

When a quantity is directly proportional to another, it means that an increase in one variable will cause a corresponding increase of the same percent in the other variable. The direct proportionality is shown as

$$A \propto B \qquad (\propto \text{proportionality symbol})$$

which reads "A is directly proportional to B." A proportionality can be easily converted to an algebraic equation by the introduction of a constant (in our examples designated k), called a constant of proportionality. Thus the proportion becomes

$$A = kB$$

or, rearranging,

$$\frac{A}{B} = k$$

Notice that k is not a variable but has a certain numerical value that does not change as do A and B under experimental conditions.

A common, direct proportionality that we will study relates kelvin temperature, T, and volume, V, of a gas at constant pressure. This is written as

$$V \propto T \qquad V = kT \qquad \frac{V}{T} = k$$

(This is known as Charles' law.)

In a hypothetical case, $V = 100$ liters and $T = 200$ K. From this information we can calculate the value of the constant k.

$$\frac{V}{T} = \frac{100 \text{ liters}}{200 \text{ K}} = 0.50 \text{ liters/K}$$

A change in volume or temperature requires a corresponding change in the other *in the same direction*. For example, if the temperature of the gas is changed to 300 K, we can see that a corresponding change in volume is required from the following calculation:

$$\frac{V}{T} = k$$

$$V = kT = 0.50 \text{ liter/\cancel{K}} \times 300 \text{ \cancel{K}} = \underline{\underline{150 \text{ liters}}}$$

When a quantity is inversely proportional to another quantity, an increase in one brings about a corresponding *decrease* in the other. An inverse proportionality between A and B is written as

$$A \propto \frac{1}{B}$$

As before, the proportionality can be written as an equality by introduction of a constant (which has a different value than in the example above).

$$A = \frac{k}{B} \quad \text{or} \quad AB = k$$

A common inverse proportionality that we use relates the volume, V, of a gas to the pressure, P, at a constant temperature. This is written as

$$V \propto \frac{1}{P} \qquad V = \frac{k}{P} \qquad PV = k$$

(This is known as Boyle's law.)

In a hypothetical case, $V = 100$ liters and $P = 1.50$ atm. From this information we can calculate the value of the constant k.

$$PV = k = 1.50 \text{ atm} \times 100 \text{ liters} = 150 \text{ atm} \cdot \text{liters}$$

A change in volume or pressure requires a corresponding change in the other *in the opposite direction.* For example, if the pressure on the gas is changed to 3.00 atm, we can see that a corresponding change in volume is required.

$$PV = k$$

$$V = \frac{k}{P} = \frac{150 \text{ atm} \cdot \text{liters}}{3.00 \text{ atm}} = \underline{\underline{50.0 \text{ liters}}}$$

Quantities can be directly or inversely proportional to the square, square root, or any other function of another variable or number, as illustrated by the following examples.

Example B-17
A quantity C is directly proportional to the square of D. Write an equality for this statement and explain how a change in C affects the value of D.

Solution:
The equation is

$$C = kD^2$$

Notice that a change in D will have a significant effect on the value of C. For example,

$$\text{if } D = 1, \text{ then } C = k$$

$$\text{if } D = 2, \text{ then } C = 4k$$

$$\text{if } D = 3, \text{ then } C = 9k$$

Notice that when the value of D is doubled, the value of C is increased *fourfold.*

Example B-18

A variable X is directly proportional to the square of the variable Y and inversely proportional to the square of another variable Z.

This can be written as two separate equations if it is assumed that Y is constant when Z varies and vice versa.

$$X = k_1Y^2 \qquad (Z \text{ constant})$$

$$X = k_2/Z^2 \qquad (Y \text{ constant})$$

$$(k_1 \text{ and } k_2 \text{ are different constants})$$

This relationship can be combined into one equation when both Y and Z are variables:

$$X = \frac{k_3Y^2}{Z^2}$$

k_3 is a third constant that is a combination of k_1 and k_2.

Exercises

Write equalities for the following relations.

B-31 X is inversely proportional to $Y + Z$.

B-32 $[H_3O^+]$ is inversely proportional to $[OH^-]$.

B-33 $[H_2]$ is directly proportional to the square root of r.

B-34 B is directly proportional to the square of y and the cube of z.

B-35 The pressure P of a gas is directly proportional to the number of moles n, the temperature T, and inversely proportional to the volume V.

Answers: **B-31** $X = k/(Y + Z)$ **B-32** $[H_3O^+] = k/[OH^-]$ **B-33** $[H_2] = k\sqrt{r}$ **B-34** $B = ky^2z^3$ **B-35** $P = knT/V$

C

Scientific Notation

A Test of Scientific Notation

Following is a short test on the manipulation of numbers expressed in scientific notation. Since you should be comfortable with the use of such numbers in calculations before beginning Chapter 8, you are urged to study the review if you miss more than one problem in any one section. Compare your answers to those that follow the test. In the following problems, zeros to the left of a decimal point but to the right of a digit (e.g., 9200) where no decimal point is showing are assumed not to be significant figures (see Section 2-1). Although the numbers used in this appendix do not have units (for simplicity), treat the numbers as measurements so that the answer to a problem shows the proper number of significant figures.

1   Express the following numbers in scientific notation in standard form [one digit to the left of the decimal point in the coefficient (e.g., 7.39×10^6)].

(a)   0.00657   1(a) _____

(b)   14,300   1(b) _____

(c)   0.00030   1(c) _____

(d)   3,457,000,000   1(d) _____

2   Change the following numbers expressed in scientific notation to standard form.

(a) 0.778×10^5 2(a) _____

(b) 145×10^{-8} 2(b) _____

(c) 0.00720×10^{-8} 2(c) _____

(d) 4330×10^{-10} 2(d) _____

3   Carry out the following calculations. Express your answer to the proper decimal place.

(a) $0.049 + (3.7 \times 10^{-2}) + (76.6 \times 10^{-3})$ 3(a) _____

(b) $(1.62 \times 10^6) + (0.78 \times 10^4) - (0.025 \times 10^7)$ 3(b) _____

(c) $(0.363 \times 10^{-6}) + (71.2 \times 10^{-9}) + (619 \times 10^{-12})$ 3(c) _____

(d) $(481 \times 10^6) - (0.113 \times 10^9) + (3.5 \times 10^5)$ 3(d) _____

4 Carry out the following calculations. Express your answer in standard form to the proper number of significant figures.

(a) $(4.8 \times 10^6) \times (17.2 \times 10^{-2})$ **4(a)** _____

(b) $(186 \times 10^{20}) \div (0.044 \times 10^5)$ **4(b)** _____

(c) $[(55.3 \times 10^5) \times (1.5 \times 10^6)] \div (2.2 \times 10^{-5})$ **4(c)** _____

(d) $0.000879 \div 8.22 \times 10^5$ **4(d)** _____

5 Carry out the following calculations. Express the answer in standard form.

(a) $(5.0 \times 10^{-5})^2$ **5(a)** _____

(b) $\sqrt{4.9 \times 10^5}$ **5(b)** _____

(c) $(1.1 \times 10^4)^3$ **5(c)** _____

(d) $\sqrt[3]{1.25 \times 10^{-10}}$ **5(d)** _____

Answers: **1(a)** 6.57×10^{-3} **1(b)** 1.43×10^4 **1(c)** 3.0×10^{-4}
1(d) 3.457×10^9 **2(a)** 7.78×10^4 **2(b)** 1.45×10^{-6} **2(c)** 7.20×10^{-11} **2(d)** 4.33×10^{-7} **3(a)** 0.163 **3(b)** 1.38×10^6 **3(c)** 4.35×10^{-7}
3(d) 3.68×10^8 **4(a)** 8.3×10^5 **4(b)** 4.2×10^{18} **4(c)** 3.8×10^{17}
4(d) 1.07×10^{-9} **5(a)** 2.5×10^{-9} **5(b)** 7.0×10^2 **5(c)** 1.3×10^{12}
5(d) 5.00×10^{-4}

B A Review of Scientific Notation

Although this topic was first introduced in Section 2-2 in this text, we will focus on a review of the mathematical manipulation of numbers expressed in scientific notation in this appendix. Specifically, addition, multiplication, division, and taking the roots of numbers expressed in scientific notation are covered.

As mentioned in Chapter 2, scientific notation makes use of powers of 10 to simplify awkward numbers that employ more than two or three zeros that are not significant figures. The exponent of 10 simply indicates how many times we should multiply or divide a number (called the coefficient) by 10 to produce the actual number. For example, $8.9 \times 10^3 = 8.9$ (the coefficient), multiplied by 10 *three* times, or

$$8.9 \times 10 \times 10 \times 10 = 8900$$

Also, $4.7 \times 10^{-3} = 4.7$ (the coefficient), divided by 10 *three* times, or

$$\frac{4.7}{10 \times 10 \times 10} = 0.0047$$

1 Expressing Numbers in Standard Scientific Notation

The method for expressing numbers in scientific notation was explained in Section 2-2. However, in order to simplify a number or to express it in the standard form with one digit to the left of the decimal point in the coefficient, it is often necessary to change a number already expressed in scientific notation. Since this is often done in a hurry, a large number of errors may result. Thus, it is worthwhile to practice moving the decimal point of numbers expressed in scientific notation.

Example C-1
Change the following numbers to the standard form:
(a) 489×10^4 (b) 0.00489×10^8

Procedure:
All you need to remember is to raise the power of 10 one unit for each place the decimal point is moved to the left and lower the power of 10 one unit for each place that the decimal point is moved to the right in the coefficient.

Solution:

$$489 \times 10^4 = (4\,8\,9) \times 10^4 = 4.89 \times 10^{4+2} = \underline{\underline{4.89 \times 10^6}}$$

$$0.00489 \times 10^8 = (0.0\,0\,4\,8\,9) \times 10^8 = 4.89 \times 10^{8-3} = \underline{\underline{4.89 \times 10^5}}$$

As an aid to remembering whether you should raise or lower the exponent as you move the decimal point, it is suggested that you write (or at least imagine) the coefficient on a slant. For each place that you move the decimal point *up*, add one to the exponent. For each place that you move the decimal point *down*, subtract one from the exponent. Notice that the exponent moves up or down with the decimal point.

Example C-2
Change the following numbers to the standard form in scientific notation:
(a) 4223×10^{-7} (b) 0.00076×10^{18}

Solution:

$$4223 \times 10^{-7} = \begin{bmatrix} 4^{+3} \\ 2^{+2} \\ 2^{+1} \\ 3 \end{bmatrix} \times 10^{-7} = 4.223 \times 10^{-7+3} = 4.223 \times 10^{-4}$$

$$0.00076 \times 10^{18} = \begin{bmatrix} 0 \\ . \\ 0^{-1} \\ 0^{-2} \\ 0^{-3} \\ 7^{-4} \\ 6 \end{bmatrix} \times 10^{18} = 7.6 \times 10^{18-4} = \underline{\underline{7.6 \times 10^{14}}}$$

Exercises

C-1 Change the following numbers to standard scientific notation with one digit to the left of the decimal point in the coefficient.

    (a) 787×10^{-6} **(d)** 0.0037×10^{9}

    (b) 43.8×10^{-1} **(e)** 49.3×10^{15}

    (c) 0.015×10^{-16} **(f)** 6678×10^{-16}

C-2 Change the following numbers to a number with two digits to the left of the decimal point in the coefficient.

    (a) 9554×10^{4} **(d)** 116.5×10^{4}

    (b) 1.6×10^{-5} **(e)** 0.023×10^{-1}

    (c) 1×10^{6} **(f)** 0.005×10^{23}

Answers: **C-1(a)** 7.87×10^{-4} **C-1(b)** 4.38 **C-1(c)** 1.5×10^{-18}
C-1(d) 3.7×10^{6} **C-1(e)** 4.93×10^{16} **C-1(f)** 6.678×10^{-13}
C-2(a) 95.54×10^{6} **C-2(b)** 16×10^{-6} **C-2(c)** 10×10^{5}
C-2(d) 11.65×10^{5} **C-2(e)** 23×10^{-4} **C-2(f)** 50×10^{19}

2 Addition and Subtraction

Addition or subtraction of numbers in scientific notation can be accomplished only when all coefficients have the same exponent of 10. When all the exponents are the same, the coefficients are added and then multiplied by the power of 10. The correct number of places to the right of the decimal point must be shown as discussed in Section 2-1.

Example C-3

Add the following:

$$3.67 \times 10^{-4}, 4.879 \times 10^{-4}, \text{ and } 18.2 \times 10^{-4}$$

Solution:

$$
\begin{array}{r}
3.67 \times 10^{-4} \\
4.879 \times 10^{-4} \\
18.2 \times 10^{-4} \\
\hline
26.749 \times 10^{-4} = \underline{\underline{26.7 \times 10^{-4}}}
\end{array}
$$

Example C-4

Add the following:

$$320.4 \times 10^{3}, 1.2 \times 10^{5}, \text{ and } 0.0615 \times 10^{7}$$

Solution:
Before adding, change all three numbers to the same exponent of 10.

$$
\begin{aligned}
320.4 \times 10^3 &= 3.204 \times 10^5 \\
1.2 \times 10^5 &= 1.2 \quad \times 10^5 \\
0.0615 \times 10^7 &= 6.15 \quad \times 10^5 \\
\hline
&\;\; 10.554 \times 10^5 = 10.6 \times 10^5 = 1.06 \times 10^6
\end{aligned}
$$

Exercises

C-3   Add the following numbers. Express the answer to the proper decimal place.

     (a) $152 + (8.635 \times 10^2) + (0.021 \times 10^3)$

     (b) $(10.32 \times 10^5) + (1.1 \times 10^5) + (0.4 \times 10^5)$

     (c) $(1.007 \times 10^{-8}) + (118 \times 10^{-11}) + (0.1141 \times 10^{-6})$

     (d) $(0.0082) + (2.6 \times 10^{-4}) + (159 \times 10^{-4})$

C-4   Carry out the following calculations. Express your answer to the proper decimal place.

     (a) $(18.75 \times 10^{-6}) - (13.8 \times 10^{-8}) + (1.0 \times 10^{-5})$

     (b) $(1.52 \times 10^{-11}) + (17.7 \times 10^{-12}) - (7.5 \times 10^{-15})$

     (c) $(481 \times 10^6) - (0.113 \times 10^9) + (8.5 \times 10^5)$

     (d) $(0.363 \times 10^{-6}) + (71.2 \times 10^{-9}) + (519 \times 10^{-12})$

Answers: **C-3(a)** 1.036×10^3 **C-3(b)** 1.18×10^6 **C-3(c)** 1.254×10^{-7} **C-3(d)** 2.44×10^{-2} **C-4(a)** 2.9×10^{-5} **C-4(b)** 3.29×10^{-11} **C-4(c)** 3.69×10^8 **C-4(d)** 4.35×10^{-7}

3 Multiplication and Division

When numbers expressed in scientific notation are multiplied, the exponents of 10 are *added*. When the numbers are divided, the exponent of 10 in the denominator (the divisor) is subtracted from the exponent of 10 in the numerator (the dividend).

Example C-5
Carry out the following calculation:

$$(4.75 \times 10^6) \times (3.2 \times 10^5)$$

Solution:
In the first step, group the coefficients and the powers of 10. Carry out each step separately.

$$= (4.75 \times 3.2) \times (10^6 \times 10^5)$$

$$= 15.200 \times 10^{6+5} = 15 \times 10^{11} = \underline{\underline{1.5 \times 10^{12}}}$$

Example C-6

Carry out the following calculation:

$$(1.62 \times 10^{-8}) \div (8.55 \times 10^{-3})$$

Solution:

$$= \frac{1.62 \times 10^{-8}}{8.55 \times 10^{-3}} = \frac{1.62}{8.55} \times \frac{10^{-8}}{10^{-3}} = 0.189 \times 10^{-8-(-3)}$$

$$= 0.189 \times 10^{-5} = \underline{\underline{1.89 \times 10^{-6}}}$$

Exercises

C-5 Carry out the following calculations. Express your answer to the proper number of significant figures with one digit to the left of the decimal point.

(a) $(7.8 \times 10^{-6}) \times (1.12 \times 10^{-2})$

(b) $(0.511 \times 10^{-3}) \times (891 \times 10^{-8})$

(c) $(156 \times 10^{-12}) \times (0.010 \times 10^4)$

(d) $(16 \times 10^9) \times (0.112 \times 10^{-3})$

(e) $(2.35 \times 10^3) \times (0.3 \times 10^5) \times (3.75 \times 10^2)$

(f) $(6.02 \times 10^{23}) \times (0.0100)$

C-6 Follow the instructions in Prob. C-5.

(a) $(14.6 \times 10^8) \div (2.2 \times 10^8)$

(b) $(6.02 \times 10^{23}) \div (3.01 \times 10^{20})$

(c) $(0.885 \times 10^{-7}) \div (16.5 \times 10^3)$

(d) $(0.0221 \times 10^3) \div (0.57 \times 10^{18})$

(e) $238 \div (6.02 \times 10^{23})$

C-7 Follow the instructions in Prob. C-5.

(a) $[(8.70 \times 10^6) \times (3.1 \times 10^8)] \div (5 \times 10^{-3})$

(b) $(47.9 \times 10^{-6}) \div [(0.87 \times 10^6) \times (1.4 \times 10^2)]$

(c) $1 \div [(3 \times 10^6) \times (4 \times 10^{10})]$

(d) $1.00 \times 10^{-14} \div [(6.5 \times 10^5) \times (0.32 \times 10^{-5})]$

(e) $[(147 \times 10^{-6}) \div (154 \times 10^{-6})] \div (3.0 \times 10^{12})$

Answers: **C-5(a)** 8.7×10^{-8} **C-5(b)** 4.55×10^{-9} **C-5(c)** 1.6×10^{-8}
C-5(d) 1.8×10^{6} **C-5(e)** 3×10^{10} **C-5(f)** 6.02×10^{21} **C-6(a)** 6.6
C-6(b) 2.00×10^{3} **C-6(c)** 5.36×10^{-12} **C-6(d)** 3.9×10^{-17}
C-6(e) 3.95×10^{-22} **C-7(a)** 5×10^{17} **C-7(b)** 3.9×10^{-13} **C-7(c)** $8 \times$
10^{-18} **C-7(d)** 4.8×10^{-15} **C-7(e)** 3.2×10^{-13}

4 Powers and Roots

When a number expressed in scientific notation is raised to a power, the coefficient is raised to the power and the exponent of 10 is *multiplied* by the power.

Example C-7
Carry out the following calculation:

$$(3.2 \times 10^{3})^{2} = (3.2)^{2} \times 10^{3 \times 2}$$
$$= 10.24 \times 10^{6} = \underline{1.0 \times 10^{7}}$$

$$[(10^{3})^{2} = 10^{3} \times 10^{3} = 10 \times 10 \times 10 \times 10 \times 10 \times 10 = 10^{6}]$$

Example C-8
Carry out the following calculation:

$$(1.5 \times 10^{-3})^{3} = (1.5)^{3} \times 10^{-3 \times 3}$$
$$= \underline{3.4 \times 10^{-9}}$$

When the root of a number expressed in scientific notation is taken, take the root of the coefficient and divide the exponent of 10 by the root. Before taking a root, however, adjust the number so that the exponent of 10 divided by the root produces a whole number. Adjust the exponent in the direction that will leave the coefficient as a number greater than 1 (to avoid mistakes).

Example C-9
Carry out the following calculation:

$$\sqrt{2.9 \times 10^{5}}$$

Solution:
First adjust the number so that the exponent of 10 is divisible by 2.

$$\sqrt{2.9 \times 10^{5}} = \sqrt{29 \times 10^{4}} = \sqrt{29} \times \sqrt{10^{4}} = \sqrt{29} \times 10^{4/2}$$
$$= 5.4 \times 10^{2} = 540$$

Example C-10
Carry out the following calculation:

$$\sqrt[3]{6.9 \times 10^{-8}}$$

Solution:

Adjust the number so that the exponent of 10 is divisible by 3.

$$\sqrt[3]{6.9 \times 10^{-8}} = \sqrt[3]{69 \times 10^{-9}} = \sqrt[3]{69} \times \sqrt[3]{10^{-9}}$$
$$= 4.1 \times 10^{-9/3} = 4.1 \times 10^{-3}$$

Exercises

C-8 Carry out the following operations.

(a) $(6.6 \times 10^4)^2$

(b) $(0.7 \times 10^6)^3$

(c) $(1200 \times 10^{-5})^2$ (It will be easier if you change the number to $1.2 \times 10^?$ first. It will be easier to square.)

(d) $(0.035 \times 10^{-3})^3$

(e) $(0.7 \times 10^7)^4$

C-9 Take the following roots. Approximate the answer if necessary.

(a) $\sqrt{36 \times 10^4}$ (d) $\sqrt[3]{1.6 \times 10^5}$

(b) $\sqrt[3]{27 \times 10^{12}}$ (e) $\sqrt{81 \times 10^{-7}}$

(c) $\sqrt{64 \times 10^9}$ (f) $\sqrt{180 \times 10^{10}}$

Answers: **C-8(a)** 4.4×10^9 **C-8(b)** 3×10^{17} **C-8(c)** 1.4×10^{-4}
C-8(d) 4.3×10^{-14} **C-8(e)** 2×10^{27} **C-9(a)** 6.0×10^2 **C-9(b)** 3.0×10^4
C-9(c) 2.5×10^5 **C-9(d)** 54 **C-9(e)** 2.8×10^{-3} **C-9(f)** 1.3×10^6

D

Problem Solving by the Unit-Factor Method

A large number of the problems in basic chemistry fit into the category of *conversion* problems. Whether it is from grams to milligrams, grams to pounds, moles to number of molecules, volume of a gas to moles, or weight of solute to volume of solution, these problems can all be solved by the same method. Once the method is understood, the actual solution becomes routine. The problems have the same basic feature; a quantity is given in one unit of measurement and you are asked to convert this to another unit of measurement. To solve such problems we use the unit-factor method, which employs one or more conversion factors.

The basic procedure for a one-step conversion is

$$[\text{Given}] \text{ old unit} \times \left[\begin{array}{c} \text{conversion} \\ \text{factor} \end{array} \right] \frac{\text{new unit}}{\text{old unit}} = [\text{requested}] \text{ new unit}$$

If you are asked to convert 24 in. to feet, the answer comes easily because of your familiarity with the units and the relationship of inches to feet. To obtain the answer of 2 ft, most of us probably aren't even aware of what we did mathematically. In this appendix, however, we will carefully show what we do with units as well as numbers even for the simple conversions. By getting into the habit of including units in the calculation, the more complex problems can be simplified.

Three types of units are employed in the examples that follow:

1 Familiar English units
$\begin{cases} \text{Weight—ounces, pounds} \\ \text{Distance—feet, yards, miles} \\ \text{Volume—quarts, gallons} \\ \text{Quantity—dozen, gross} \\ \text{Time—hours, minutes} \end{cases}$

2 Alien units
$\begin{cases} \text{Weight—gerbs} \\ \text{Distance—frims} \\ \text{Volume—moks} \\ \text{Quantity—nums, rims} \\ \text{Time—bleems} \end{cases}$

3 Chemical units (metric or SI)
$\begin{cases} \text{Weight—grams} \\ \text{Distance—meters, kilometers} \\ \text{Volume—milliliters, liters} \\ \text{Quantity—moles} \\ \text{Time—hours, minutes} \\ \text{And others} \end{cases}$

There is a reason for this approach. Most of the math you need in general chemistry is quite simple and was reviewed in the previous appendixes. Often, however, the units and the awkward numbers cause confusion. It is the goal of this appendix to end the confusion.

First we will work examples and exercises in familiar English units, where you probably have a feeling for the correct answer before doing the actual calculation. By using the unit-factor method in these problems we are in effect perfecting the use of a "decoder" which can be used with less familiar systems. Next we travel to an alien planet with our "decoder" to work problems with units that have no meaning to us. In these problems we certainly have no feeling for the approximate answer to a calculation, so there is no way to work the problems without using the "decoder." The unit-factor method leads us confidently to the answers in this strange world.

Finally, we return to earth to work examples with the units used in chemistry. To the beginning student these units may also appear alien. However, until these units become familiar, we may rely on the "decoder" to lead us to the correct answers.

In summary, don't be overly concerned at this point if some of the units used in chemistry are unfamiliar. Our calculations in the alien and chemical units is like flying an airplane on instruments. The experienced pilot trusts the instruments to lead safely to the destination when the ground can't be seen. Likewise, you can trust the units to lead safely to the answer when you are not sure whether to multiply or divide. Understandably, pilots prefer to see where they are going with their own eyes. Likewise, you will eventually wish to see where you are going by becoming familiar with the units and calculations.

To be comfortable with the unit-factor method, there are three essential requirements: (1) to understand the origin and function of the conversion factor, (2) to understand the manipulation of units, and finally, (3) to carry out the calculation itself correctly. We will take up these three points individually.

A Conversion Factors

Conversion factors originate from the relationship of one unit of measurement to another. The relationships originate in two ways:

1 From an *exact* definition (e.g., 1000 m = 1 km, 12 in. = 1 ft, 60 min = 1 hr), or

2 From a measurement (e.g., 1.00 in. is equal to 2.54 cm, 1.00 ml weighs 4.64 g, 1.00 lb costs $1.75)

These equalities or equivalency relationships can also be expressed as a fraction or *factor form*. For example, 12 in. = 1 ft can be expressed as

$$\frac{12 \text{ in.}}{1 \text{ ft}} \quad \text{or} \quad \frac{1 \text{ ft}}{12 \text{ in.}}$$

and 1.00 ml weighs 4.64 g as

$$\frac{1.00 \text{ ml}}{4.64 \text{ g}} \quad \text{or} \quad \frac{4.64 \text{ g}}{1.00 \text{ ml}}$$

These are sometimes called "unity factors" because they relate a quantity to exactly "one" of the other. When unity factors from a measurement are shown, it is understood that the "one" is known to as many significant figures as the other quantity. Also, in cases where the one is in the denominator, it is often not shown or read. Thus the relationship between ml and g shown above is simplified in this text as

$$\frac{1 \text{ ml}}{4.64 \text{ g}} \quad \text{or} \quad \frac{4.64 \text{ g}}{\text{ml}}$$

The following examples illustrate the construction of unity factors originating from a measurement. In calculations it is convenient (but certainly not necessary) to express conversion factors as unity factors.

Example D-1

If 0.250 miles is found to equal 1320 ft, how many feet are there per mile?

Solution:

The problem tells us that 1320 ft = 0.250 mile. We put the relationship in factor form and carry out the math.

$$\frac{1320 \text{ ft}}{0.250 \text{ mile}} = \underline{\underline{5280 \text{ ft/mile}}}$$

Example D-2

A substance from the alien planet weighs 488 gerbs and has a volume of 14.2 moks. How many gerbs are there per mok?

Solution:

$$488 \text{ gerbs} = 14.2 \text{ moks}$$

$$\frac{488 \text{ gerbs}}{14.2 \text{ moks}} = \underline{\underline{34.4 \text{ gerbs/mok}}}$$

Example D-3

A 0.785-mole (abbreviated mol) quantity of a compound has a weight of 175 g. What is the weight in grams per mole?

Solution:

$$175 \text{ g} = 0.785 \text{ mol}$$

$$\frac{175 \text{ g}}{0.785 \text{ mol}} = \underline{\underline{223 \text{ g/mol}}}$$

Exercises

In the following problems, include the units of the answer with the number where appropriate.

D-1 The unit price of groceries is sometimes listed in cost per ounce. Which of the two has the smallest cost per ounce: 16 oz of baked beans costing $1.45, or 26 oz costing $2.10?

D-2 An 82.3-dozen quantity of oranges weighs 247 lb. What is the weight per dozen oranges? How many dozen oranges are there per pound?

D-3 If two men can dig an 8-ft trench in 2.5 hr, what is the number of feet dug per man per hour? (The units will be ft/man · hr.)

D-4 There are 1480 years in 3.25 bleems. How many bleems are there per year?

D-5 There are 8.61×10^5 objects in 0.574 nums. How many objects are there per num?

D-6 A gas has a volume of 146 liters and a weight of 48.5 g. What is the weight per liter? What is the volume per gram?

D-7 There are 8.85×10^5 coulombs in 9.17 faradays. How many coulombs are there per faraday?

D-8 A quantity of 6.02×10^{23} atoms of tungsten weighs 184 g. What is the weight per atom?

Answers: **D-1** The 16-oz can costs $0.091/oz, the 26-oz can costs $0.081/oz **D-2** 3.00 lb/doz, 0.333 doz/lb **D-3** 1.6 ft/man · hr **D-4** 2.20×10^{-3} bleems/year **D-5** 1.50×10^6 objects/num **D-6** 0.332 g/liter, 3.01 liters/g **D-7** 9.65×10^4 coulombs/faraday **D-8** 3.06×10^{-22} g/atom.

B Manipulation of Units

In the unit-factor method it is essential to multiply, divide, or cancel units as well as numbers. Although this topic was discussed in Appendix A, it is important enough to be reemphasized in this section. The manipulation of units is illustrated by the following examples. We are interested only in the units of the answer at this time. Later we will include the numerical value.

Example D-4

What are the unit or units of the answers implied by the following calculations?

$$\frac{ft}{mile} \times \frac{mile}{min} \times \frac{min}{hr}$$

Cancel units that occur in both numerator and denominator.

$$\frac{ft}{\cancel{mile}} \times \frac{\cancel{mile}}{\cancel{min}} \times \frac{\cancel{min}}{hr} = \frac{ft}{hr}$$

Example D-5

$$\frac{frim^3/mok \times 1/bleem}{frim^2/mok}$$

Rearrange into one quotient and cancel (see Appendix A for manipulation of fractions).

$$\frac{frim^{\overset{1}{\cancel{3}}} \times \cancel{mok}}{\cancel{frim^2} \times \cancel{mok} \times bleem} = \frac{frim}{bleem}$$

Example D-6

$$\frac{(liter \cdot atm/K \cdot mol) \times K}{atm}$$

Rearrange into one quotient and cancel.

$$\frac{liter \times \cancel{atm} \times \cancel{K}}{\cancel{K} \times mol \times \cancel{atm}} = \frac{liter}{mol}$$

Exercises

In the following exercises, write the units of the answer implied from the calculations.

D-9 $\dfrac{ft}{mile} \times lb \times mile$

D-10 $\dfrac{(mile/hr) \times (ft/mile) \times (hr/min)}{sec/min}$

D-11 $\dfrac{(rim/num) \times (1/bleem)}{1/num}$

D-12 $\dfrac{(gerb/frim^3) \times mok \times frim}{mok/frim}$

D-13 $\dfrac{g \times (liter \cdot torr/K \cdot mol) \times K}{torr \times liter}$

D-14 $\dfrac{1/atm \times liter}{mol \times (liter \cdot atm/K \cdot mol)}$

D-15 $\dfrac{coulomb \times mol/F}{(mol/g) \times coulomb}$

Answers: **D-9** ft · lb **D-10** ft/sec **D-11** rim/bleem **D-12** gerb/frim **D-13** g/mol **D-14** K/liter^2 · atm^2 **D-15** g/F.

The main lesson to be learned from these examples and exercises is to be sure that the units in the calculation lead to the units desired in the answer. For example, if the answer should be ml but the units come out to be g^2/ml, the units indicate that you have made a mistake. Be careful in your manipulation of units to avoid *forcing* a cancellation, such as

$$\frac{g}{\cancel{ml}} \times \frac{1}{\cancel{ml}} = g \qquad \text{(Wrong)}$$

$$\frac{g}{\cancel{ml}} \times \cancel{ml} = g \qquad \text{(Right)}$$

C One-Step Conversions

Before we work some examples, a word of warning is in order. In working problems in the English system, this procedure should not become so mechanical that you lose a feeling for the approximate answer. For example, if a car travels 50 miles/hr and you are asked how long it takes to travel 160 miles, you know that the answer is more than 3 hr. If you multiply instead of divide, you should immediately sense that the answer of 8000 hr is ridiculous even without help from the units. *Think about your answer.* It is understood at this point, however, that you may not be able to approximate the answer when working with chemical units. Eventually you will. For example, if the weight of one tiny atom is requested, the answer is very small ($\sim 10^{-23}$ g). An answer of 10 g or even 10^{23} g (about the weight of the moon) is out of the money and should be so recognized. Our purpose here is to set you at ease about the calculation itself. When you study the particular topic, you should answer the very important question: "Does this answer make sense?"

In the following problems we will list what is given and requested, a shorthand procedure, the conversion factor properly set up to follow the procedure, and finally the solution. This effort pays off for the more complex problems later.

Remember that the conversion factor is set up so that the old unit (which is given) is in the denominator and cancels. The new unit (which is requested) is in the numerator and remains in the answer. This assumes a simple one-step conversion as shown in the following examples.

Example D-7

If a car travels 55 miles/hr, how many miles does it travel in 8.5 hr?

Given:

$$8.5 \text{ hr}$$

Requested:

$$\underline{\quad ? \quad} \text{ miles}$$

Procedure:

$$\text{hr} \rightarrow \text{miles}$$

Conversion factor:

$$55 \text{ miles/hr}$$

Solution:

$$8.5 \ \cancel{\text{hr}} \times \frac{55 \text{ miles}}{\cancel{\text{hr}}} = \underline{\underline{470 \text{ miles}}}$$

Example D-8

If the same car as in Example D-7 travels 785 miles, how many hours does it take?

Given:

$$785 \text{ miles}$$

Requested:

$$\underline{\quad ? \quad} \text{ hr}$$

Procedure:

$$\text{miles} \rightarrow \text{hr}$$

Conversion factor:

$$\frac{1 \text{ hr}}{55 \text{ miles}}$$

Solution:

$$785 \ \cancel{\text{miles}} \times \frac{1 \text{ hr}}{55 \ \cancel{\text{miles}}} = \underline{\underline{14 \text{ hr}}}$$

Notice in Example D-7 that a small number (8.5) is converted to a large number (470). In such a case, the conversion factor is always larger than unity. The opposite is true in Example D-8, where a large number (785) is converted to a small number (14). Notice that for such a conversion the factor is less than unity. A simple check to make sure that the conversion factor changes the magnitude of the number in the proper direction saves many mistakes.

Example D-9

A certain type of chain costs $0.55/ft. What does it cost per mile? (5280 ft = 1 mile).

Given:

$$\$0.55/\text{ft}$$

Requested:

$$\frac{\$}{\text{mile}}$$

Procedure:

$$\frac{\$}{\text{ft}} \rightarrow \frac{\$}{\text{mile}}$$

Conversion factor:

$$5280 \text{ ft/mile}$$

In this case, the unit of the denominator is to be changed. The factor should therefore have the given unit in the *numerator* and the requested unit in the *denominator*.

Solution:

$$\frac{\$0.55}{\text{ft}} \times \frac{5280\,\text{ft}}{\text{mile}} = \$2904/\text{mile}$$

One popular exception to the proper use of significant figures is where money is involved. Traditionally, price is expressed to the nearest cent regardless of how it is obtained. An argument to a salesperson in the above example that the price should only be $2900 because of significant figures would not go very far.

Example D-10

A section of cable weighs 18.5 gerbs/frim. How many frims are there in 97.2 gerbs?

Given:

$$97.2 \text{ gerbs}$$

Requested:

$$\underline{\qquad ? \qquad} \text{ frims}$$

Procedure:

$$\text{gerbs} \rightarrow \text{frims}$$

Conversion factor:

$$\frac{1 \text{ frim}}{18.5 \text{ gerb}}$$

Solution:

$$97.2 \text{ \sout{gerb}} \times \frac{1 \text{ frim}}{18.5 \text{ \sout{gerb}}} = \underline{\underline{5.25 \text{ frim}}}$$

Example D-11

A space ship can travel 1.85×10^6 frims in 0.240 bleems. How far can it travel in 0.525 bleem?

Given:

$$0.525 \text{ bleem}$$

Requested:

$$\underline{\qquad ? \qquad} \text{ frims}$$

Procedure:

$$\text{bleems} \rightarrow \text{frims}$$

Conversion factor:

$$\frac{1.85 \times 10^6 \text{ frims}}{0.240 \text{ bleem}}$$

Solution:

$$0.525 \text{ \sout{bleem}} \times \frac{1.85 \times 10^6 \text{ frims}}{0.240 \text{ \sout{bleem}}} = \underline{\underline{4.05 \times 10^6 \text{ frims}}}$$

Example D-12

What is the weight of a 3.25-mol quantity of a substance if it weighs 80.0 g/mol?

Given:

$$3.25 \text{ moles}$$

Requested:

$$\underline{\qquad ? \qquad} \text{ g}$$

Procedure:

$$\text{mol} \rightarrow \text{g}$$

Conversion factor:

$$80.0 \text{ g/mol}$$

Solution:

$$3.25 \ \cancel{mol} \times \frac{80.0 \text{ g}}{\cancel{mol}} = \underline{\underline{260 \text{ g}}}$$

Example D-13

If there are 6.02×10^{23} molecules/mol, how many moles does 5.65×10^{24} molecules represent?

Given:

$$5.65 \times 10^{24} \text{ molecules}$$

Requested:

$$\underline{\quad ? \quad} \text{ mol}$$

Procedure:

$$\text{molecules} \rightarrow \text{mol}$$

Conversion factor:

$$\frac{1 \text{ mol}}{6.02 \times 10^{23} \text{ molecules}}$$

Solution:

$$5.65 \times 10^{24} \ \cancel{\text{molecules}} \times \frac{1 \text{ mol}}{6.02 \times 10^{23} \ \cancel{\text{molecules}}} = \underline{\underline{9.39 \text{ mol}}}$$

Exercises

Practice one-step conversions on the following problems.

D-16 If the unit price of chili is $0.18/oz, what is the cost of a 12-oz can of chili? How many ounces can you buy at this rate for $1.75?

D-17 If a bullet travels 1860 ft/sec, how far does it travel in 4.50 sec?

D-18 If there are 4.55 lb of corn per dozen ears, how many dozen ears of corn are in 265 lb?

D-19 There are 1.50×10^6 commas per num (remember, a num is a unit of quantity, like dozen, gross, or mole). How many nums is 83×10^8 commas? How many commas are in 0.420 nums?

D-20 A resident of an alien planet can lose 0.475 gerbs in 2.15 bleems when on a diet. How many gerbs can be lost in 0.880 bleem? How many bleems would it take to lose 187 gerbs?

D-21   If 18.0 g of a gas has a volume of 22.4 liters under certain conditions, what is the weight of the gas that would have a volume 1.35 liters?

D-22   A pressure of 1 atm is equal to 760 torr. How many torr are there in 0.66 atm?

D-23   There are 6.02×10^{23} molecules/mole. How many molecules are present in 0.212 mol?

D-24   Using the relationship from the previous problem, how many molecules are there per gram of a particular substance if 1 mol weighs 18.0 g?

D-25   There are 96,500 coulombs/faraday. How many faradays are equivalent to 4820 coulombs?

Answers: **D-16** $2.16,   9.72   oz   **D-17**   8370   ft   **D-18**   58.2   doz
D-19 5.5×10^3 nums, 6.30×10^5 commas   **D-20**   0.194 gerbs, 846 bleems
D-21   1.08   g   **D-22**   500   torr   **D-23** 1.28×10^{23} molecules
D-24 3.34×10^{22} molecules/g   **D-25**   0.0499 faradays

D Two-Step or More Conversions

Whether you know it or not, you have already worked some fairly sophisticated problems in chemistry in the preceding exercises. Hopefully they didn't cause major problems, so we are now ready to work problems where one direct relationship is not available between units of what's given and the units of the answer. In these cases, you must carefully outline a plan to make the conversion in two or more steps. The general procedure of a simple two-step conversion is as follows:

$$[\text{Given}]\,\cancel{\text{old unit}} \times \left[\frac{\text{conversion}}{\text{factor (a)}}\right] \frac{\cancel{\text{unit X}}}{\cancel{\text{old unit}}}$$

$$\times \left[\frac{\text{conversion}}{\text{factor (b)}}\right] \frac{\text{new unit}}{\cancel{\text{unit X}}} = [\text{answer}]\ \text{new unit}$$

Remember that a conversion factor is needed for each step in the problem. Be sure that the conversion factor for each step leaves a new unit in the numerator and the old unit to be cancelled in the denominator. For example, assume that a procedure is planned as follows for a problem where lb is given and $ is requested. The procedure is

$$\text{lb} \xrightarrow{(a)} \text{doz} \xrightarrow{(b)} \$$$

Step (a) converts lb to doz.

$$\boxed{\cancel{lb} \times \frac{doz}{\cancel{lb}}} = \underline{} \text{ doz}$$

Step (b) converts doz to $.

$$lbs \times \boxed{\frac{\cancel{doz}}{lbs} \times \frac{\$}{\cancel{doz}}} = \$\underline{}$$

In the rest of this appendix, many of the relationships used are obtained from Tables 2-1 and 2-2.

Example D-14

If a train travels 75.0 miles/hr, how many miles does it travel in 416 min?

Given:

$$416 \text{ min}$$

Requested:

$$\underline{?} \text{ miles}$$

Procedure:

$$\text{min} \xrightarrow{\text{(a)}} \text{hr} \xrightarrow{\text{(b)}} \text{miles}$$

Conversion factors:

$$\text{(a)} \frac{1 \text{ hr}}{60 \text{ min}} \qquad \text{(b)} \frac{75.0 \text{ miles}}{\text{hr}}$$

Solution:

$$416 \cancel{\text{ min}} \times \frac{1 \cancel{\text{ hr}}}{60 \cancel{\text{ min}}} \times \frac{75.0 \text{ miles}}{\cancel{\text{hr}}} = \underline{\underline{520 \text{ miles}}}$$

Example D-15

One gallon of gas costs $0.98 (in 1979). What is the cost of 48.0 liters (the volume unit used in Mexico and Canada)?

Given:

$$48.0 \text{ liters}$$

Requested:

$$\$\underline{?}$$

Procedure:

$$\text{liters} \xrightarrow{\text{(a)}} \text{qts} \xrightarrow{\text{(b)}} \text{gal} \xrightarrow{\text{(c)}} \$$$

Conversion factors:

$$\text{(a)} \ \frac{1.06 \text{ qt}}{\text{liter}} \qquad \text{(b)} \ \frac{1 \text{ gal}}{4 \text{ qt}} \qquad \text{(c)} \ \frac{\$0.98}{\text{gal}}$$

Solution:

$$48.0 \ \cancel{\text{liters}} \times \frac{1.06 \ \cancel{\text{qt}}}{\cancel{\text{liter}}} \times \frac{1 \ \cancel{\text{gal}}}{4 \ \cancel{\text{qt}}} \times \frac{\$0.98}{\cancel{\text{gal}}} = \underline{\underline{\$12.47}}$$

Example D-16

If two men can dig an 8-ft trench in 2.5 hr, how long a trench can six men dig in 18 hr if all the men are working at the same rate?

Given:

$$6 \text{ men} \qquad 18 \text{ hr}$$

Requested:

$$\underline{\quad ? \quad} \text{ ft}$$

Procedure:

$$\text{men, hr} \xrightarrow{\text{(a)}} \text{ft}$$

Conversion factor:

$$\frac{8 \text{ ft}}{2 \text{ men} \cdot 2.5 \text{ hr}}$$

Solution:

$$6 \ \cancel{\text{men}} \times 18 \ \cancel{\text{hr}} \times \frac{8 \text{ ft}}{2 \ \cancel{\text{men}} \cdot 2.5 \ \cancel{\text{hr}}} = \underline{\underline{173 \text{ ft}}}$$

Example D-17

The speed limit of a space ship is 55.0 frims/millibleem (10^3 millibleems per bleem). How many frims can a space ship travel in 0.205 bleem?

Given:

$$0.205 \text{ bleem}$$

Requested:

$$\underline{\quad ? \quad} \text{ frims}$$

Procedure:

$$\text{bleems} \xrightarrow{\text{(a)}} \text{millibleems} \xrightarrow{\text{(b)}} \text{frims}$$

Conversion factors:

$$\text{(a)} \ \frac{10^3 \text{ millibleems}}{\text{bleem}} \qquad \text{(b)} \ \frac{55.0 \text{ frims}}{\text{millibleem}}$$

Solution:

$$0.205 \text{ bleem} \times \frac{10^3 \text{ millibleems}}{\text{bleem}} \times \frac{55.0 \text{ frims}}{\text{millibleem}} = \underline{\underline{11,300 \text{ frims}}}$$

Example D-18

There are 1.50×10^6 numecules in a num. If the weight of 1 num (the "num mass") of a substance is 24.0 gerbs, what is the weight in gerbs of 7.50×10^5 numecules of the substance?

Given:

$$7.50 \times 10^5 \text{ numecules}$$

Requested:

$$\underline{\quad ? \quad} \text{ gerbs}$$

Procedure:

$$\text{numecules} \xrightarrow{\text{(a)}} \text{nums} \xrightarrow{\text{(b)}} \text{gerbs}$$

Conversion factors:

$$\text{(a)} \ \frac{1 \text{ num}}{1.50 \times 10^6 \text{ numecule}} \qquad \text{(b)} \ \frac{24.0 \text{ gerbs}}{\text{num}}$$

Solution:

$$7.50 \times 10^5 \text{ numecule} \times \frac{1 \text{ num}}{1.50 \times 10^6 \text{ numecule}} \times \frac{24.0 \text{ gerbs}}{\text{num}} = \underline{\underline{12.0 \text{ gerbs}}}$$

Example D-19

Using the information given in Example D-18, determine how many numecules there are in 175 gerbs.

Given:

$$175 \text{ gerbs}$$

Requested:

$$\underline{\quad ? \quad} \text{ numecules}$$

Procedure:

$$\text{gerbs} \xrightarrow{\text{(a)}} \text{nums} \xrightarrow{\text{(b)}} \text{molecules}$$

Conversion factors:

$$\text{(a)} \ \frac{1 \text{ num}}{24.0 \text{ gerbs}} \qquad \text{(b)} \ \frac{1.50 \times 10^6 \text{ numecules}}{\text{num}}$$

Solution:

$$175 \text{ gerbs} \times \frac{1 \text{ num}}{24.0 \text{ gerbs}} \times \frac{1.50 \times 10^6 \text{ numecules}}{\text{num}} = \underline{\underline{1.09 \times 10^7 \text{ numecules}}}$$

Example D-20
Gold weighs 19.3 g/ml. What is the volume (in milliliters) of exactly 1 lb of gold?

Given:

$$1.00 \text{ lb}$$

Requested:

$$\underline{\qquad ? \qquad} \text{ ml}$$

Procedure:

$$\text{lb} \xrightarrow{\text{(a)}} \text{g} \xrightarrow{\text{(b)}} \text{ml}$$

Conversion factors:

$$\text{(a)} \ \frac{454 \text{ g}}{\text{lb}} \qquad \text{(b)} \ \frac{1 \text{ ml}}{19.3 \text{ g}}$$

Solution:

$$1.00 \ \cancel{\text{lb}} \times \frac{454 \ \cancel{\text{g}}}{\cancel{\text{lb}}} \times \frac{1 \text{ ml}}{19.3 \ \cancel{\text{g}}} = \underline{\underline{23.5 \text{ ml}}}$$

Example D-21
There are 6.02×10^{23} molecules in 1 mol. If the weight of 1 mol (the "molar mass") of carbon dioxide is 44.0 g. What is the weight of 8.46×10^{22} molecules of carbon dioxide?

Given:

$$8.46 \times 10^{22} \text{ molecules}$$

Requested:

$$\underline{\qquad ? \qquad} \text{ g}$$

Procedure:

$$\text{molecules} \xrightarrow{\text{(a)}} \text{mol} \xrightarrow{\text{(b)}} \text{g}$$

Conversion factors:

$$\text{(a)} \ \frac{1 \text{ mol}}{6.02 \times 10^{23} \text{ molecules}} \qquad \text{(b)} \ \frac{44.0 \text{ g}}{\text{mol}}$$

Solution:

$$8.46 \times 10^{22} \ \cancel{\text{molecules}} \times \frac{1 \ \cancel{\text{mol}}}{6.02 \times 10^{23} \ \cancel{\text{molecules}}} \times \frac{44.0 \text{ g}}{\cancel{\text{mol}}} = 6.18 \text{ g}$$

Example D-22
Using the information given in Example D-21, determine how many molecules are in 890 g of carbon dioxide.

Given:

$$890 \text{ g}$$

Requested:

$$\underline{\quad ? \quad} \text{ molecules}$$

Procedure:

$$g \xrightarrow{\text{(a)}} \text{mol} \xrightarrow{\text{(b)}} \text{molecules}$$

Conversion factors:

$$\text{(a)} \frac{1 \text{ mol}}{44.0 \text{ g}} \qquad \text{(b)} \frac{6.02 \times 10^{23} \text{ molecules}}{\text{mol}}$$

Solution:

$$890 \text{ g} \times \frac{1 \text{ mol}}{44.0 \text{ g}} \times \frac{6.02 \times 10^{23} \text{ molecules}}{\text{mol}} = \underline{\underline{1.22 \times 10^{25} \text{ molecules}}}$$

Exercises

The following problems require two or more conversions to obtain the answer.

D-26 If a train travels at a speed of 85 miles/hr, how many hours does it take to travel 17,000 ft? How many yards can it travel in 37 min? (1760 yards = 1 mile.)

D-27 A certain size of nail costs $1.25/lb. What is the cost of 3.25 kg of nails?

D-28 Two men can push a car 110 yards in 15 min. How many minutes would it take three men to push a car 250 yards at the same rate?

D-29 If a car travels 55.0 miles/hr, what is its speed in feet per second?

D-30 Low-tar cigarettes contain 11.0 mg of tar per cigarette. If all of the tar gets into the lungs, how many packages of cigarettes (20 cigarettes per package) would have to be smoked to produce 0.500 lb of tar? If a person smoked two packages per day, how many years would it take to accumulate 0.500 lb of tar?

D-31 If a car gets 24.5 miles to the gallon of gasoline and gasoline costs $0.98/gallon, what would it cost to drive 350 km?

D-32 A Martian fruit weighs 0.820 gerbs/rim. What is the volume (in moks) of 425 rims if there are 115 gerbs per mok?

D-33 If there are 1.50×10^6 numecules/num, how many numecules are in 1 rim if there are 135 rims/num?

D-34 If numecules weigh 118 gerbs/num, how many gerbs are in 4.50×10^8 numecules (1.50×10^6 numecules/num)? How many numecules are in 45.5 gerbs?

D-35 A liquid solution from the alien planet contains 0.250 nums of a substance per mok. If the substance weighs 164 gerbs/num, how many gerbs of the substance are in 152 moks of solution?

D-36 A liquid solution on earth contains 0.250 mol of sulfuric acid per liter. How many grams of sulfuric acid are in 6.75 liters of solution if the sulfuric acid weighs 98.0 g/mol? What is the volume (in liters) occupied by 15.5 g of sulfuric acid?

D-37 If there are 6.02×10^{23} molecules/mol, how many molecules of sulfuric acid are in the two liquid solutions in Problem D-36?

D-38 How many grams of silver are deposited by 46,200 coulombs of electricity if 1 faraday is equal to 96,500 coulombs and 1 faraday deposits 108 g of silver?

D-39 If 1 mol of a gas occupies 22.4 liters under certain conditions, how many liters will 75.0 g of this gas occupy if the gas weighs 60.0 g/mol?

D-40 One atmosphere pressure is equal to 760.0 torr. What is 355.0 torr expressed in kilopascals (kPa) if 1 atm is equal to 101.4 kPa?

D-41 There are 3 mol of hydrogen atoms per mole of phosphoric acid. How many moles of hydrogen atoms are in 16.0 g of phosphoric acid if 1 mol of phosphoric acid weighs 98.0 g?

Answers: **D-26** 0.038 hr, 9.2×10^4 yards **D-27** $8.94 **D-28** 23 min **D-29** 80.7 ft/sec **D-30** 1030 packages, 1.41 years **D-31** $8.69 **D-32** 3.03 moks **D-33** 1.11×10^4 numecules/rim **D-34** 3.54×10^4 gerbs, 5.78×10^5 numecules **D-35** 6230 gerbs **D-36** 165 g, 0.633 liter **D-37** 1.02×10^{24} molecules, 9.52×10^{22} molecules **D-38** 51.7 g **D-39** 28.0 liters **D-40** 47.36 kPa **D-41** 0.490 mol of H atoms

Logarithms

In certain areas of science the expression of numbers in scientific notation can become tedious and awkward. Such an area in chemistry concerns the concentration of acids and bases in water. A simpler way to express the number is by use of common logarithms. With common logarithms (or just logs), both the coefficient and the exponent of 10 are expressed as one simple number.

A common log of a number is the power to which 10 must be raised to give that number. For example, the number 100 is equal to 10^2. Two is therefore the log of 100, since that is the power that 10 must be raised to give 100. Other simple examples are

$$
\begin{array}{llll}
1 = 10^0 & \log 1 = 0 & 0.1 = 10^{-1} & \log 0.1 = -1 \\
10 = 10^1 & \log 10 = 1 & 0.01 = 10^{-2} & \log 0.01 = -2 \\
100 = 10^2 & \log 100 = 2 & 0.001 = 10^{-3} & \log 0.001 = -3 \\
1000 = 10^3 & \log 1000 = 3 & 0.0001 = 10^{-4} & \log 0.0001 = -4
\end{array}
$$

A Logs of Numbers Between 1 and 10

Notice that $\log 1 = 0$ and $\log 10 = 1$. The log of numbers between 1 and 10 will therefore be between 0 and 1. The log of these numbers can be conveniently found in a *log table* as found on page 402. This is a four-place log table, which suits our purpose in this text. Two-, three-, and five-place log tables are available which give fewer or more significant figures, respectively. A four-place log table provides the log of a three-significant-figure number directly from the table. In addition, the log of a four-significant-figure number can be calculated by interpolation between numbers in the table. The log of a number is found in the body of the table. To find the log of a number between 1 and 10, locate the first two digits in the left-hand column and then move across the table to locate the third at the top of the table. For each significant figure in the original number, include one place to the right of the decimal point.

Example E-1

What is the log of (a) 5.82, (b) 3.78, and (c) 1.11?

Solution:
The answer is between 0 and 1 and is expressed as

$$0. \text{ log}$$

log 5.82 = 0.765	(58 down, 2 across)
log 3.78 = 0.578	(37 down, 8 across)
log 1.11 = 0.045	(11 down, 1 across)

Exercises

E-1 Find logarithms of the following numbers:
   (a) 9.87 (b) 7.65 (c) 2.550 (d) 1.08

Answers: (a) 0.994 (b) 0.884 (c) 0.4065 (d) 0.033

B Logs of Numbers Less Than 1 or Greater Than 10

To find the log of a number less than 1 or greater than 10, follow these steps:

1 Write the number in standard scientific notation (one digit to the left of the decimal point in the coefficient).

2 Look up the log of the coefficient and write as described in Section A.

3 The value of the log of the coefficient is added to the value of the exponent of 10. That is a result of the general rule

$$\log(A \times B) = \log A + \log B$$

$$\log(A \times 10^B) = \log A + \log 10^B = (\log A) + B$$

The sum of the two numbers (the exponent plus the log of the coefficient) has two parts. *The number to the right of the decimal point is called the* **mantissa.** As mentioned before, the mantissa is expressed to the same number of decimal places as there are significant figures in the coefficient. *The whole number to the left of the decimal point is called the* **characteristic.** For example,

$$\log(2.65 \times 10^6) = \log 2.65 + \log 10^6 = 0.423 + 6$$
$$= \boxed{6} . \boxed{423}$$

Characteristic   Mantissa

Example E-2
What is the log of
(a) 4.76×10^3

$$\log(4.76 \times 10^3) = \log 4.76 + \log 10^3$$
$$= 0.678 + 3 = \underline{3.678}$$

(b) 15.8

$$\begin{aligned} \log 15.8 &= \log(1.58 \times 10^1) \\ &= \log 1.58 + \log 10^1 \\ &= 0.199 + 1 = 1.199 \end{aligned}$$

(c) 0.0666

$$\begin{aligned} \log 0.0666 &= \log(6.66 \times 10^{-2}) \\ &= \log 6.66 + \log 10^{-2} \\ &= 0.824 + (-2) \\ &= 0.824 - 2 = \underline{-1.176} \end{aligned}$$

Notice that the logs of numbers less than 1 have *negative values*.

(d) 87.1×10^{-8}

$$\begin{aligned} \log(8.71 \times 10^{-7}) &= \log 8.71 + \log 10^{-7} \\ &= 0.940 + (-7) \\ &= 0.940 - 7 = \underline{-6.060} \end{aligned}$$

Exercises

E-2 Find the logs of the following numbers.
  (a) 7.43 (b) 11.8 (c) 0.875 (d) 0.0878 (e) 85.2 (f) 1780
  (g) 7.33×10^4 (h) 7.33×10^{-4} (i) 4.11×10^{-12} (j) 4.55
  (k) 0.0577×10^8 (l) 32.8×10^{-5} (m) 164×10^6 (n) 0.601×10^{-7}

E-3 pH is defined as the negative of the log of a number (e.g., $-\log 10^{-2} = -(-2) = 2$). Find $-\log$ of the following:
  (a) 8.55×10^{-6} (b) 3.42×10^{-11} (c) 71.9×10^{-10} (d) 0.217×10^{-3} (e) 1.15

Answers: **E-2(a)** 0.871 **E-2(b)** 1.072 **E-2(c)** -0.058 **E-2(d)** -1.056
E-2(e) 1.930   **E-2(f)** 3.250   **E-2(g)** 4.865   **E-2(h)** -3.135
E-2(i) -11.386 **E-2(j)** 0.658   **E-2(k)** 6.761   **E-2(l)** -3.484
E-2(m) 8.215   **E-2(n)** -7.221 **E-3(a)** 5.068   **E-3(b)** 10.466
E-3(c) 8.143 **E-3(d)** 3.664 **E-3(e)** -0.061

C The Antilog of a Positive Logarithm

To find the antilog of a log, follow the opposite procedure from taking the log. We are asking the question, "What is the number whose log is a certain value?" For example, what is the antilog (x) of 2?

$$\log x = 2$$

$$x = \underline{\underline{10^2}} \quad \text{since } \log 10^2 = 2$$

When the log is between 0 and 1, the number is between 1 and 10. (This is just the reverse of the statement in Section A.) To find the antilog, locate the given log in the *body* of the log table. Find the two numbers on the left margin and the third on the top which corresponds to this log. Write the three numbers you found with one digit to the left of the decimal point.

Example E-3
What is the antilog of
(a) 0.847

$$\text{Antilog } 0.847 = \underline{\underline{7.03}}$$

(b) 0.021

$$\text{Antilog } 0.021 = \underline{\underline{1.05}}$$

(c) 0.956

$$\text{Antilog } 0.956 = \underline{\underline{9.04}}$$

If the log is a number greater than 1, divide the number into two parts, the mantissa which lies to the right of the decimal and the characteristic which lies to the left. For example,

$$2.567 = 0.567 + 2$$

$$11.118 = 0.118 + 11$$

Take the antilog of the mantissa and express the number as explained above. This becomes the coefficient of the number. The characteristic then becomes the exponent of 10.

Example E-4
Find the antilog of (a) 4.655 and (b) 3.897.
(a) $4.655 = 0.655 + 4$

$$\text{Antilog } 0.655 = 4.52$$

$$\text{Antilog } 4 = 10^4$$

$$\text{Answer} = \underline{\underline{4.52 \times 10^4}}$$

(b) $3.897 = 0.897 + 3$

$$\text{Antilog } 0.897 = 7.89$$

$$\text{Antilog } 3 = 10^3$$

$$\text{Answer} = \underline{\underline{7.89 \times 10^3}}$$

D The Antilog of a Negative Logarithm

A negative log must be adjusted so that the mantissa is positive without changing the actual value of the log itself. For example, -0.182 is the same number as $(+0.818 - 1)$, and -3.289 is the same number as $(+0.711 - 4)$. To do this, first separate and write the mantissa and the characteristic as described before. *Add* 1 to the mantissa and *subtract* 1 from the characteristic. Notice that if you add and subtract the same number from a quantity, you have not actually changed the number (e.g., $15 + 1 - 1 = 15$).

	Separate	Add 1 to mantissa	Subtract 1 from characteristic	
$-3.289 =$	$-0.289 - 3 =$	$(-0.289 + 1) -$	$3 - 1$	$= +0.711 - 4$
$-0.182 =$	$-0.182 - 0 =$	$(-0.182 + 1) -$	$0 - 1$	$= +0.818 - 1$

Locate the mantissa in the body of the log table as described before and locate the corresponding numbers in the left-hand column and the top. The negative characteristic becomes the negative exponent of 10.

Example E-5
Find the following antilogs.
(a) -0.013

$$-0.013 = (-0.013 + 1) - 1 = +0.987 - 1$$

$$\text{Antilog } 0.987 = 9.71$$

$$\text{Antilog } (-1) = 10^{-1}$$

$$\text{Answer} = 9.71 \times 10^{-1} = \underline{\underline{0.971}}$$

(b) -4.548

$$-4.548 = -0.548 - 4 = (-0.548 + 1) - 4 - 1 = +0.452 - 5$$

$$\text{Antilog } 0.452 = 2.83$$

$$\text{Antilog } (-5) = 10^{-5}$$

$$\text{Answer} = \underline{\underline{2.83 \times 10^{-5}}}$$

Exercises

E-4 Find the antilog of **(a)** 0.813 **(b)** 3.459 **(c)** -0.699 **(d)** -5.783 **(e)** 10.945 **(f)** -2.600 **(g)** -0.086 **(h)** 8.401 **(i)** 0.345 **(j)** -5.996.

Answers: **E-4(a)** 6.50   **E-4(b)** 2.88×10^3 **E-4(c)** 0.200
E-4(d) 1.65×10^{-6} **E-4(e)** 8.81×10^{10} **E-4(f)** 2.51×10^{-3}
E-4(g) 0.820 **E-4(h)** 2.52×10^8 **E-4(i)** 2.21 **E-4(j)** 1.01×10^{-6}

Logarithms

	0	1	2	3	4	5	6	7	8	9
10	0000	0043	0086	0128	0170	0212	0253	0294	0334	0374
11	0414	0453	0492	0531	0569	0607	0645	0682	0719	0755
12	0792	0828	0864	0899	0934	0969	1004	1038	1072	1106
13	1139	1173	1206	1239	1271	1303	1335	1367	1399	1430
14	1461	1492	1523	1553	1584	1614	1644	1673	1703	1732
15	1761	1790	1818	1847	1875	1903	1931	1959	1987	2014
16	2041	2068	2095	2122	2148	2175	2201	2227	2253	2279
17	2304	2330	2355	2380	2405	2430	2455	2480	2504	2529
18	2553	2577	2601	2625	2648	2672	2695	2718	2742	2765
19	2788	2810	2833	2856	2878	2900	2923	2945	2967	2989
20	3010	3032	3054	3075	3096	3118	3139	3160	3181	3201
21	3222	3243	3263	3284	3304	3324	3345	3365	3385	3404
22	3424	3444	3464	3483	3502	3522	3541	3560	3579	3598
23	3617	3636	3655	3674	3692	3711	3729	3747	3766	3784
24	3802	3820	3838	3856	3874	3892	3909	3927	3945	3962
25	3979	3997	4014	4031	4048	4065	4082	4099	4116	4133
26	4150	4166	4183	4200	4216	4232	4249	4265	4281	4298
27	4314	4330	4346	4362	4378	4393	4409	4425	4440	4456
28	4472	4487	4502	4518	4533	4548	4564	4579	4594	4609
29	4624	4639	4654	4669	4683	4698	4713	4728	4742	4757
30	4771	4786	4800	4814	4829	4843	4857	4871	4886	4900
31	4914	4928	4942	4955	4969	4983	4997	5011	5024	5038
32	5051	5065	5079	5092	5105	5119	5132	5145	5159	5172
33	5185	5198	5211	5224	5237	5250	5263	5276	5289	5302
34	5315	5328	5340	5353	5366	5378	5391	5403	5416	5428
35	5441	5453	5465	5478	5490	5502	5514	5527	5539	5551
36	5563	5575	5587	5599	5611	5623	5635	5647	5658	5670
37	5682	5694	5705	5717	5729	5740	5752	5763	5775	5786
38	5798	5809	5821	5832	5843	5855	5866	5877	5888	5899
39	5911	5922	5933	5944	5955	5966	5977	5988	5999	6010
40	6021	6031	6042	6053	6064	6075	6085	6096	6107	6117
41	6128	6138	6149	6160	6170	6180	6191	6201	6212	6222
42	6232	6243	6253	6263	6274	6284	6294	6304	6314	6325
43	6335	6345	6355	6365	6375	6385	6395	6405	6415	6425
44	6435	6444	6454	6464	6474	6484	6493	6503	6513	6522
45	6532	6542	6551	6561	6571	6580	6590	6599	6609	6618
46	6628	6637	6646	6656	6665	6675	6684	6693	6702	6712
47	6721	6730	6739	6749	6758	6767	6776	6785	6794	6803
48	6812	6821	6830	6839	6848	6857	6866	6875	6884	6893
49	6902	6911	6920	6928	6937	6946	6955	6964	6972	6981
50	6990	6998	7007	7016	7024	7033	7042	7050	7059	7067
51	7076	7084	7093	7101	7110	7118	7126	7135	7143	7152
52	7160	7168	7177	7185	7193	7202	7210	7218	7226	7235
53	7243	7251	7259	7267	7275	7284	7292	7300	7308	7316
54	7324	7332	7340	7348	7356	7364	7372	7380	7388	7396

Logarithms

	0	1	2	3	4	5	6	7	8	9
55	7404	7412	7419	7427	7435	7443	7451	7459	7466	7474
56	7482	7490	7497	7505	7513	7520	7528	7536	7543	7551
57	7559	7566	7574	7582	7589	7597	7604	7612	7619	7627
58	7634	7642	7649	7657	7664	7672	7679	7686	7694	7701
59	7709	7716	7723	7731	7738	7745	7752	7760	7767	7774
60	7782	7789	7796	7803	7810	7818	7825	7832	7839	7846
61	7853	7860	7868	7875	7882	7889	7896	7903	7910	7917
62	7924	7931	7938	7945	7952	7959	7966	7973	7980	7987
63	7993	8000	8007	8014	8021	8028	8035	8041	8048	8055
64	8062	8069	8075	8082	8089	8096	8102	8109	8116	8122
65	8129	8136	8142	8149	8156	8162	8169	8176	8182	8189
66	8195	8202	8209	8215	8222	8228	8235	8241	8248	8254
67	8261	8267	8274	8280	8287	8293	8299	8306	8312	8319
68	8325	8331	8338	8344	8351	8357	8363	8370	8376	8382
69	8388	8395	8401	8407	8414	8420	8426	8432	8439	8445
70	8451	8457	8463	8470	8476	8482	8488	8494	8500	8506
71	8513	8519	8525	8531	8537	8543	8549	8555	8561	8567
72	8573	8579	8585	8591	8597	8603	8609	8615	8621	8627
73	8633	8639	8645	8651	8657	8663	8669	8675	8681	8686
74	8692	8698	8704	8710	8716	8722	8727	8733	8739	8745
75	8751	8756	8762	8768	8774	8779	8785	8791	8797	8802
76	8808	8814	8820	8825	8831	8837	8842	8848	8854	8859
77	8865	8871	8876	8882	8887	8893	8899	8904	8910	8915
78	8921	8927	8932	8938	8943	8949	8954	8960	8965	8971
79	8976	8982	8987	8993	8998	9004	9009	9015	9020	9025
80	9031	9036	9042	9047	9053	9058	9063	9069	9074	9079
81	9085	9090	9096	9101	9106	9112	9117	9122	9128	9133
82	9138	9143	9149	9154	9159	9165	9170	9175	9180	9186
83	9191	9196	9201	9206	9212	9217	9222	9227	9232	9238
84	9243	9248	9253	9258	9263	9269	9274	9279	9284	9289
85	9294	9299	9304	9309	9315	9320	9325	9330	9335	9340
86	9345	9350	9355	9360	9365	9370	9375	9380	9385	9390
87	9395	9400	9405	9410	9415	9420	9425	9430	9435	9440
88	9445	9450	9455	9460	9465	9469	9474	9479	9484	9489
89	9494	9499	9504	9509	9513	9518	9523	9528	9533	9538
90	9542	9547	9552	9557	9562	9566	9571	9576	9581	9586
91	9590	9595	9600	9605	9609	9614	9619	9624	9628	9633
92	9638	9643	9647	9652	9657	9661	9666	9671	9675	9680
93	9685	9689	9694	9699	9703	9708	9713	9717	9722	9727
94	9731	9736	9741	9745	9750	9754	9759	9763	9768	9773
95	9777	9782	9786	9791	9795	9800	9805	9809	9814	9818
96	9823	9827	9832	9836	9841	9845	9850	9854	9859	9863
97	9868	9872	9877	9881	9886	9890	9894	9899	9903	9908
98	9912	9917	9921	9926	9930	9934	9939	9943	9948	9952
99	9956	9961	9965	9969	9974	9978	9983	9987	9991	9996

F Graphs

A Direct Linear Relationships

How can as few as two experimental measurements be used to predict the results of other measurements? The answer is by a graph. Assuming that the two measurements are accurate and are the result of a consistent linear relationship between two measurable properties, a graph can multiply the two results into many. For example, the following experimental measurements were recorded on a sample of gas at constant pressure:

Experiment 1 At a temperature of 350°C, the volume was 12.5 liters.

Experiment 2 At a temperature of 250°C, the volume was 10.5 liters.

Actually, it is always risky to assume that only two experiments provide an accurate graph or give all of the needed information, but for simplicity we will assume that it is valid in this case. We will check the graph with an additional experiment later.

To record this information on a graph, we first need some graph paper with regularly spaced divisions. On the vertical axis (called the ordinate or y axis) we put temperature. Pick divisions on the graph paper that spread out the range of temperatures as much as possible. In our example the range of temperatures is between 250°C and 350°C or 100 degrees. The origin (where the two axes intersect) need not start at zero. We then plot the volume on the horizontal axis (called the abscissa or x axis). The range of volumes is between 12.5 liters and 10.5 liters or 2.0 liters.

The graph is shown in Figure F-1. an ⊙ is marked at the location of the

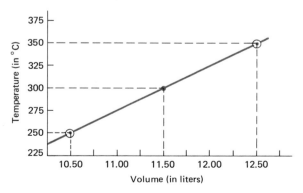

Figure F-1

two points from the experiments and a straight line has been drawn between them.

We will now check the accuracy of our graph. A third experiment tells us that at $T = 300°C, V = 11.5$ liters. Refer back to the graph. Locate $300°C$ on the ordinate and trace that line in to where it intersects the vertical line representing the volume of 11.5 liters. Since the point of intersection is right on the line, we have confirmation of the accuracy of the original plot. *Normally, four or five measurements are required to provide enough points on the graph so that a line can be drawn through or near as many points as possible.*

A straight-line graph can be expressed by the linear algebraic equation

$$y = mx + b$$

where

y = a value on the ordinate
x = the corresponding value on the abscissa
b = the point on the y axis or ordinate where the straight line intersects the axis
m = the slope of the line

The slope of a line is the ratio of the change on the ordinate to the change on the abscissa. It is determined by choosing two widely spaced points on the line $(x_1, y_1$ and $x_2, y_2)$. The slope is the difference in y divided by the difference in x.

$$m = \frac{y_2 - y_1}{x_2 - x_1}$$

The larger the slope *for a particular graph*, the steeper the straight line. For the graph shown in Figure F-1, if we pick point 1 to be $x_1 = 10.50, y_1 = 250$ and point 2 to be $x_2 = 12.50, y_2 = 350$, then

$$m = \frac{350 - 250}{12.50 - 10.50} = \frac{100}{2.00} = 50.0$$

The value for b cannot be determined directly from Figure F-1. The value for b is found by determining the point on the y axis where the line intersects when $x = 0$. If we plotted the data on such a graph, we would find that $b = -273°C$. Therefore, the linear equation for this line is

$$y = 50.0x - 273$$

From this equation we can determine the value for y (the temperature) by substitution of any value for x (the volume).

Exercises

F-1 A sample of water was heated at a constant rate and its temperature was recorded. The following results were obtained.

Experiment	Temp (°C)	Time (min)
1	10.0	0
2	20.0	4.5
3	30.0	7.5
4	55.0	17.8
5	85.0	30.0

(a) Construct a graph that includes all of the above information with temperature on the ordinate and time on the abscissa.

(b) Calculate the slope of the line and the linear equation $y = mx + b$ for the line.

(c) From the graph find the temperature at 11.0 min and compare it to the value calculated from the equation.

(d) From the graph find the time when the temperature is 65.0°C and compare it to the value calculated from the equation.

Answers: **F-1(a)** Draw the line touching or coming close to as many points as possible. Notice that the straight line does not go through all points. **F-1(b)** $m = 2.5, y = 2.5x + 10$ **F-1(c)** From the equation, when $x = 11.0$, $y = 2.5 \cdot 11.0 + 10$, so $y = 37.5$°C. The graph agrees. **F-1(d)** From the equation, when $y = 65$, $65 = 2.5x + 10$, and $x = 22.0$ min. The graph agrees.

B Nonlinear Relationships

Thus far we have been dealing with direct linear relationships that produce straight lines when graphed. Other relationships, such as an inverse relationship (see Appendix B), produce curves when plotted on a graph. This is illustrated by the following example.

In the following experiments the pressure on a volume of gas was varied and the volume was measured at constant temperature.

Experiment	Pressure (torr)	Volume (liters)
1	500	15.2
2	600	12.6
3	760	10.0
4	900	8.44
5	1200	6.40
6	1600	4.80

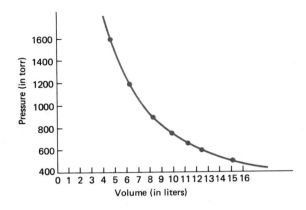

Figure F-2 An inverse relationship.

This information has been graphed in Figure F-2.

Notice that the line has a gradually changing slope, starting with a very steep slope on the left and going to a small slope on the right of the graph. We can interpret this. In the region of the steep slope the increase in pressure causes little decrease in volume compared to the part of the curve with the small slope to the right. Notice that this is a plot of an inverse relationship. That is, the higher the pressure, the lower the volume. If we regraph the information and include much higher pressures and the corresponding volumes as shown in Figure F-3, we see that *the curve approaches the y axis but never actually touches it,* no matter how high the pressure. Such a curve is said to approach the axis **asymptotically.** In our example, the curve would eventually appear to change to a vertical line parallel to the *y* axis.

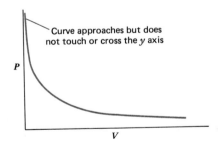

Figure F-3

Exercises

F-2 Construct a graph from the following experimental information. Plot the time on the abscissa and the concentration on the ordinate. Describe what this graph tells you about the change in concentration of HI as a function (how it varies with) the time.

Experiment	Time (min)	Concentration of HI (mol/liter)
1	0.00	2.50×10^{-2}
2	5.0	1.50×10^{-2}
3	10.0	0.90×10^{-2}
4	15.0	0.65×10^{-2}
5	20.0	0.55×10^{-2}
6	35.0	0.50×10^{-2}

(a) From the graph, find the concentration of HI at 12.0 min; at 25 min.

(b) From the graph, find the time corresponding to when the concentration of HI is 2.00×10^{-2}; 0.75×10^{-2}.

Answers: **F-2** The concentration decreases rapidly at the start (the curve has a large slope) but then changes little after about 25 min. At that point the curve appears to change to a straight line parallel to the x axis. (a) At 12.0 min, the concentration is 0.80×10^{-2} mol/liter; at 25 min, the concentration is about 0.52×10^{-2} mol/liter. (b) When the concentration is 2.00×10^{-2} mol/liter, the time is 2.4 min; when the concentration is 0.75×10^{-2} mol/liter, the time is 13.0 min.

Glossary*

G

A

Accuracy. How close the value of a measurement is to the true value. (2)

Acid. A substance that increases the H_3O^+ ion concentration in water. In the Brønsted-Lowry definition, an acid is a proton donor. (12)

Acid anhydride. A nonmetal oxide that reacts with water to form an oxyacid. (12)

Acidic solution. An aqueous solution that has a hydronium ion (H_3O^+) concentration greater than $10^{-7} M$ (pH less than 7) at 25°C. (15)

Acid-salt. An ionic compound whose anion contains one or more acidic hydrogens. (12)

Actinide. One of 14 elements (Th through Lr) in which the $5f$ subshell is filling. (5)

Alcohol. An organic compound containing at least one hydroxyl group (i.e., R—OH). (16)

Aldehyde. An organic compound containing a carbonyl group bound to a hydrogen and one

hydrocarbon group (i.e., $R-\overset{\displaystyle O}{\overset{\displaystyle \|}{C}}-H$). (16)

Alkali metal. An element in group IA of the periodic table. (5)

Alkaline Earth metal. An element in group IIA of the periodic table. (5)

Alkane. A hydrocarbon containing only single bonds (a saturated hydrocarbon). The general formula is C_nH_{2n+2} for open-chain and C_nH_{2n} for cyclic alkanes. (16)

Alkene. A hydrocarbon containing at least one double bond (an unsaturated hydrocarbon). The general formula for open-chain alkenes is C_nH_{2n}. (16)

Alkyl group. A substituent which is an alkane minus a hydrogen atom (e.g., $-C_2H_5$). (16)

Alkylation. The formation of large hydrocarbon molecules from small ones in the refining process. (16)

Alkyne. A hydrocarbon containing a triple bond. The general formula for open-chain alkynes is C_nH_{2n-2}. (16)

Allotropes. Different forms of the same element (e.g., graphite and diamond are allotropes of carbon). (5)

Alpha (α) particle. A helium nucleus that is emitted from a large radioactive nucleus. (3)

Amide. A derivative of a carboxylic acid in which an amine group is substituted for the hydroxyl

group of the acid (i.e., $R-\overset{\displaystyle O}{\overset{\displaystyle \|}{C}}-NH_2$). (16)

Amine. An organic compound containing a nitrogen with single bonds to a hydrocarbon group and a total of two other hydrocarbon groups or hydrogens (i.e., $R-NH_2$). (16)

Amphoteric. A substance that can act as an acid when a base is present or as a base when an acid is present. (12)

Anion. A negatively charged ion. (3)

Anode. The electrode at which oxidation takes place. (13)

Aromatic compound. A cyclic hydrocarbon containing alternating single and double bonds between adjacent carbon atoms. (16)

Atom. The smallest particle of an element that can exist and have the properties of that element. It is the smallest unit of an element that takes part in chemical reactions. (3)

Atomic mass unit (amu). A mass or weight that is exactly one-twelfth of the mass of an atom of ^{12}C, which is defined as exactly 12 amu. One amu is equal to 1.6605×10^{-24} g. (3)

Atomic number. The number of protons in the nucleus of the atom of an element. (3)

Atomic radius. The distance from the nucleus of an atom to the outermost electrons. (5)

Atomic weight. The weight of a certain isotope compared to ^{12}C, which is defined as exactly 12 amu. For an element it is the (weighted) average of the atomic weights of all of the isotopes present. (3)

Aufbau principle. Electrons fill the lowest available energy levels in an atom first. (4)

---

* The number in parentheses at the end of each entry refers to the chapter in which the entry is first discussed.

Autoionization. The ionization of one covalent molecule to form a cation and an anion. (15)

Avogadro's law. Equal volumes of gases at the same temperature and pressure contain equal numbers of molecules. (10)

Avogadro's number. The number of objects or particles in 1 mol, which is 6.02×10^{23}. (8)

B

Balanced equation. A chemical equation that has the same numbers of like atoms and the same total charge on both sides of the equation. (9)

Barometer. A closed-end glass tube filled with mercury which is used to measure atmospheric pressure. (10)

Base. A substance that increases the OH^- ion concentration in water. In the Brønsted-Lowry definition, a base is a proton acceptor. (12)

Base anhydride. A metal oxide that reacts with water to form an ionic hydroxide. (12)

Basic solution. An aqueous solution that has a hydroxide ion (OH^-) concentration greater than $10^{-7} M$ (pH greater than 7) at 25°C. (15)

Battery. A combination of one or more voltaic cells. (13)

Beta (β) particle. A high-energy electron emitted from a radioactive nucleus. (3)

Binary compounds. Molecules or formula units containing two different elements. (6)

Boyle's law. The volume of a quantity of gas is inversely proportional to the pressure at constant temperature. (10)

Buffer solution. An aqueous solution containing a weak acid and a salt of its conjugate base (or a weak base and a salt of its conjugate acid). A buffer solution resists changes in pH. (15)

Buoyancy. The ability of a solid or liquid to float in a certain liquid. (2)

C

Carbonyl group. A functional group made up of a carbon with a double bond to an oxygen (i.e.,

$$\overset{\displaystyle O}{\underset{\displaystyle \|}{}}$$
$-C-$). It is the functional group of aldehydes and ketones. (16)

Carboxylic acid. Organic compounds containing

a carboxyl group (i.e., $-\overset{\displaystyle O}{\overset{\|}{C}}-OH$). (16)

Catalyst. A substance that is not consumed in a reaction but whose presence increases the rates of both the forward and the reverse reaction. (14)

Cathode. The electrode at which reduction takes place in an electrolytic or voltaic cell. (13)

Cation. A positively charged ion. (3)

Celsius scale (°C). A temperature scale with 100 equal divisions between the freezing point of water (defined as 0°C) and the normal boiling point of water (defined as 100°C). (2)

Chain reaction. A self-sustaining reaction. In a nuclear chain reaction, one reacting neutron produces an average of three other neutrons which in turn cause reactions. (3)

Charles' law. The volume of a quantity of gas is directly proportional to the kelvin temperature at constant pressure. (10)

Chemical bond. The electrostatic (ionic) or covalent forces that hold atoms or ions together in a compound. (3)

Chemical change. The change of substances into other substances through a chemical reaction. (1)

Chemical equation. The representation of a chemical reaction by means of chemical symbols and coefficients. (8)

Chemical kinetics. The study of the interrelationships of reaction rates, equilibrium, and the mechanism of a reaction. (14)

Chemical property. The property of a substance that concerns its ability or tendency to change into other substances by a chemical reaction. (1)

Chemistry. The branch of science dealing with the nature and composition of matter and the changes it undergoes. (1)

Collision theory. A theory suggesting that chemical reactions result from collisions of reactant molecules having proper orientation and sufficient energy. (14)

Combined gas law. A combination of Boyle's and Charles' laws (i.e., $PV/T =$ a constant). (10)

Combustion. The rapid chemical reaction of a substance with oxygen, releasing heat and light. (1)

Compound. A pure substance composed of two or more different elements. (1)

Concentration. The amount of solute present in a certain amount of solvent or solution. (11)

Conjugate acid (base). The conjugate acid of a base is formed by the addition of one H^+ to the base. The conjugate base of an acid is formed by the loss of one H^+ from the acid. (12)

Conversion factor. An equality or equivalency relationship expressed in factor form. It is used to convert units of one measurement to another. (2)

Covalent bond. Atoms held together in an ion or molecule by sharing one to three pairs of electrons. (6)

Cracking. The formation of small hydrocarbon molecules from large ones in the refining process. (16)

Crystal. A solid substance made up of ions or molecules in an orderly and regular geometric arrangement. (3)

D

Dalton's law of partial pressure. The total pressure exerted by a mixture of gases is the sum of the partial pressures of the component gases. (10)

Daniell cell. A voltaic cell made up of zinc and copper electrodes each immersed in solutions of their ions connected by a salt bridge. (13)

Density. The ratio of the mass of a substance (usually in grams) to its volume (usually in milliliters or cubic centimeters). In other words, it is mass per unit volume. (2)

Derivative. A compound produced by the substitution of a group on the molecules of the original or parent compound. (16)

Diamagnetic. An atom or molecule with all electrons paired. (4)

Dipole. The existence of a partial charge separation in a chemical bond. This produces a polar covalent bond. (6)

Discrete spectra. The definite wavelengths of light (spectral lines) that are emitted by hot, gaseous atoms. (4)

Dissociation. The formation of ions in solution from the breakup of ionic or polar covalent compounds in solution. (12)

Distillation. The process of separation of a solution into its components by vaporization followed by condensation of the more volatile component. (1)

Double bond. A chemical bond in which two pairs of electrons are shared. (6)

Double displacement reaction. A reaction involving an exchange of ions in a solution that leads to the formation of a precipitate or a covalent compound. (9)

Dry cell. A voltaic cell composed of zinc and graphite electrodes immersed in an aqueous slurry of NH_4Cl and MnO_2. (13)

Dynamic equilibrium. The state at which opposing changes occur at equal rates. In a chemical reaction it is the point at which the rate of the forward reaction equals the rate of the reverse reaction. (12)

E

Electrode. A conducting surface at which reactions take place in a voltaic or electrolytic cell. (13)

Electrolysis. The process of forcing electrical energy through an electrolytic cell, thereby producing a certain chemical reaction. (13)

Electrolyte. A substance that forms ions when dissolved in water. The solution is then a conductor of electricity. (11)

Electrolytic cell. An electrochemical cell that converts electrical energy into chemical energy. (13)

Electron. A subatomic particle with a negative charge of minus one. It exists outside of the nucleus with a very small mass compared to the entire atom. (3)

Electron affinity. The energy released when a gaseous atom adds an electron to form a gaseous ion. (5)

Electron notation. The arrangement of electrons in an atom according to shell and subshell. (4)

Electronegativity. The tendency of an atom in a bond to attract electrons to itself. (6)

Electrostatic forces. Forces of attraction between centers of opposite charge and forces of repulsion between centers of the same charge. (3)

Element. A pure substance that is one of the 106 basic forms of nature. Elements cannot be separated into other substances by ordinary chemical means. (1)

Empirical formula. The simplest whole-number ratio of the atoms of the elements that make up a compound. (8)

Endothermic reaction. A chemical reaction that absorbs energy from the surroundings. (1)

Energy. The ability to do work. It can take the

form of light, heat, mechanical, chemical, electrical, or nuclear energy. (1)

Equilibrium. See *Dynamic equilibrium.*

Equilibrium constant. A number that defines the position of equilibrium for a particular reaction at a certain temperature. (14)

Equilibrium constant expression. The fraction containing the concentrations of products in the numerator and reactants in the denominator each raised to the power corresponding to the respective coefficients in the balanced equation. (14)

Ester. A derivative of a carboxylic acid where a hydrocarbon group (R′) is substituted for the hydrogen on the oxygen (i.e., R—C(=O)—O—R′). (16)

Ether. An organic compound containing an oxygen bonded to two hydrocarbon groups (i.e., R—O—R′). (16)

Exact number. A number that results from a definition or an actual count. (2)

Excited state. The presence of an electron in an atom in an energy state higher than the lowest available state. (4)

Exothermic reaction. A chemical reaction that releases energy to the surroundings. (1)

Extrapolate. To extend a line on a graph beyond the experimental points. (10)

F

Fahrenheit scale (°F). A temperature scale with 180 equal divisions between the freezing point of water (32°F) and the normal boiling point of water (212°F). (2)

Family. See *Group.*

Fission. The splitting of a large, unstable nucleus into two smaller nuclei of similar size with a production of energy. (3)

Formula. The symbols of the elements and the number of atoms of each element that make up a compound. (3)

Formula weight. The total weight of all of the atoms in a molecule (or formula unit for an ionic compound) expressed in atomic mass units. (8)

Functional group. The part of a molecule in an organic compond where most of the chemical reactions occur. (16)

Fusion (nuclear). The combination of two small nuclei with the production of energy. (3)

G

Gamma ray. A high-energy form of light emitted from a radioactive nucleus. (3)

Gas. One of the three physical states of matter. A gas does not have a definite shape or volume, so it fills the container uniformly. (1)

Gay-Lussac's law. The pressure of a gas is directly proportional to the kelvin temperature at constant volume. (10)

Graham's law of diffusion. The rate of diffusion of a gas is inversely proportional to the square root of its formula weight. (10)

Ground state. The lowest possible energy state available to an electron in an atom. (4)

Group. All of the elements in a vertical column in the periodic table. (4)

H

Half-life. The time required for one-half of a given sample of an isotope to undergo radioactive decay. (3)

Half-reaction. The balanced oxidation or reduction reaction written separately. (13)

Halogen. An element from group VIIA of the periodic table. (5)

Heterogeneous matter. A nonuniform mixture of substances consisting of two or more phases with definite boundaries between the phases. (1)

Homogeneous matter. Matter existing in one phase or physical state with the same properties and composition throughout. Homogeneous matter may be either a solution or a pure substance. (1)

Homologous series. A series of related compounds in which one member differs from the next by a constant amount. (16)

Hund's rule. Electrons occupy separate orbitals of the same energy level with their spins parallel. (4)

Hydration. The formation of electrostatic interactions between polar water molecules and solute ions or molecules. (11)

Hydrocarbon. An organic compound containing only carbon and hydrogen. (16)

Hydrogen bond. An electrostatic (dipole–dipole) interaction between a hydrogen on one molecule and the negative dipole of another. (11)

Hydrolysis. The reaction of ions with water to produce an acidic or basic solution. (15)

Hydrometer. A device used to measure directly the density or specific gravity of a liquid. (2)

Hydronium ion. The H_3O^+ ion, which is a commonly used representation of the H^+ ion in aqueous solution. (12)

Hydroxyl group. The —OH group as found in an organic compound. It is the functional group of alcohols. (16)

I

Ideal gas. A hypothetical gas whose molecules are considered to have no volume or interactions with each other. An ideal gas obeys the ideal gas law at all temperatures and pressures. (10)

Ideal gas law. $PV = nRT$. (10)

Ion. An electrically charged atom or group of atoms. (3)

Ion product of water. The equilibrium constant expression for the autoionization of water. (15)

Ionic bond. The ion–ion electrostatic interactions between oppositely charged ions. (6)

Ionization. The process of forming an ion or ions from a molecule or atom. (6)

Ionization constant. The equilibrium constant for the ionization (dissociation) of a weak acid (K_a) or a weak base (K_b). (15)

Ionization energy. The energy required to remove an electron from a gaseous atom to form a gaseous ion. (5)

Irreversible reaction. A reaction that goes 100% to the right. (14)

Isomers. Compounds with the same molecular formula but different molecular structures. (16)

Isotopes. Atoms with different mass numbers but the same atomic number. (3)

K

Kelvin scale (K). The temperature scale on which zero kelvin is known as absolute zero, which is the lowest possible temperature. The kelvin scale relates to the celsius scale by K = °C + 273. (10)

Ketone. An organic compound containing a carbonyl group bonded to two hydrocarbon groups

$$\text{(i.e., R—}\overset{\displaystyle O}{\overset{\displaystyle \|}{C}}\text{—R')}. \text{ (16)}$$

Kinetic energy. Mechanical energy due to motion. K.E. $= \frac{1}{2}$ [mass × (velocity)^2]. (1)

Kinetic theory of gases. The assumptions of the nature of a gas that explain the gas laws and other observations. (10)

L

Lanthanide. One of 14 elements (Ce through Lu) in which the $4f$ subshell is filling. (5)

Lattice. The regular three-dimensional arrangement of ions or molecules in a solid crystal. (6)

Le Châteleir's principle. When stress is applied to a system at equilibrium, the system reacts in such a way so as to counteract the stress. (14)

Lead–acid battery. A rechargeable voltaic battery composed of lead and lead dioxide electrodes in a sulfuric acid solution. (13)

Lewis structure. The representation of an atom, molecule, or ion showing all outer s and p electrons as dots and/or dashes. (6)

Limiting reactant. The reactant that is completely consumed in a reaction and thereby determines the theoretical yield. (9)

Liquid. One of the three physical states of matter. Liquids have a definite volume but not a definite shape. They flow and take the shape of the bottom of the container. (1)

M

Mass. The quantity of matter that a substance contains. (2)

Mass number. The number of nucleons in an isotope. (3)

Matter. Anything that has mass and occupies space. (1)

Measurement. The quantity, dimensions, or extent of something, usually in comparison to a standard. (2)

Mechanism. The means by which reactant molecules collide or otherwise behave so as eventually to form products. (14)

Metalloids. Elements with properties intermediate between metals and nonmetals. (5)

Metals. Elements that can lose one or more electrons easily to form positive ions. Physically, they are solids (except Hg) and conduct heat and electricity. (5)

Metric system. A system of measurement based on multiples of 10. (2)

Molar mass. The weight in grams of 1 mol of an element or compound. (8)

Molar volume. The volume of 1 mol of a gas measured at STP, which is 22.4 liters. (10)

Molarity. A unit of concentration in solution which refers to the number of moles of solute per liter of solution. (11)

Mole. A quantity of 6.02×10^{23} objects or particles. (8)

Molecular formula. The formula of a compound showing the exact number of atoms of each element in the compound. (8)

Molecular weight. See *Formula weight*.

Molecule. A group of two or more atoms held together by chemical bonds. (3)

Monomer. Single molecules that can be joined together in a chain to form a repeating structure called a polymer. (16)

N

Net ionic equation. The balanced ionic equation written without spectator ions. (11)

Neutralization. The reaction of an acid and a base to form a salt and water. (12)

Neutron. A neutral subatomic particle that exists in the nucleus of the atom. It has a mass of approximately 1 amu. (3)

Noble gas. An element in group O in the periodic table. (5)

Nonelectrolyte. A covalent compound that dissolves in water without formation of ions. A solution of a nonelectrolyte does not conduct electricity. (11)

Nonmetals. Elements in the upper right-hand corner of the periodic table. They are characterized by high ionization energies. (5)

Nonpolar bond. A covalent bond in which the electrons are shared equally between the two atoms. (6)

Nuclear equation. A symbolization of the changes of a nucleus or nuclei into other nuclei and particles. (3)

Nuclear reactor. A device that generates electrical energy from a controlled nuclear fission reaction. (3)

Nucleons. The protons and neutrons that make up the nucleus of the atom. (3)

Nucleus. The small, dense interior of the atom containing the nucleons. (3)

O

Octet rule. The tendency of an atom of the representative elements to achieve access to eight outer electrons by gaining or losing electrons or by sharing electrons with other atoms. (6)

Orbital. A region of space within a certain energy level which can be occupied by one or two electrons. The orbitals of a certain subshell have different orientations in space. (4)

Organic chemistry. The branch of chemistry that deals with the compounds of carbon. (16)

Oxidation. The loss of electrons which results in an increase in oxidation state. (13)

Oxidation state. The charge that the atoms of an element would have if the electrons in each bond are assigned to the more electronegative element. (7)

Oxidation–reduction (redox) reaction. A chemical reaction involving an exchange of electrons between two substances. (13)

Oxidizing agent. A substance that accepts electrons from the substance oxidized. (13)

Oxyacid. An acid derived from hydrogen plus an oxyanion. (7)

Oxyanion. An anion containing oxygen and at least one other element. (7)

P

Paramagnetic. An atom or a molecule with one or more unpaired electrons. (4)

Partial pressure. The pressure due to one component in a mixture of gases. (10)

Pauli exclusion principle. No two electrons in an atom can have the same set of four quantum numbers. (4)

Percent by weight. A unit of concentration in which the weight of solute is expressed as a percent of the weight of the solution. (11)

Percent composition. The weight of each element in a compound expressed as a percent of the formula weight. (8)

Percent yield. The ratio of the actual yield to the theoretical yield times 100. (9)

Period. The elements between inert gases in the periodic table. (5)

Periodic table. An arrangement of elements in order of increasing atomic number. Elements with the same number of outer electrons are arranged in vertical columns. (4)

pH. An expression of the H_3O^+ concentration in an aqueous solution as the negative of the log of the H_3O^+ molarity. (15)

Phase. A physically distinct state; either solid, liquid, or gas. (1)

Physical change. A change that does not alter the chemical composition of a substance. Usually a change in physical state. (1)

Physical property. A property of a substance that can be observed without changing the substance into some other substance. (1)

Physical science. A science concerned with the natural laws of matter other than those concerned with life or the earth. (2)

pOH. An expression of the OH^- concentration in an aqueous solution as the negative of the log of the OH^- molarity. (15)

Polar covalent bond. A covalent bond in which the pair of electrons resides closer to one atom, establishing a separation of charge (a dipole). (6)

Polyatomic ion. An ion containing more than one atom. (6)

Polymer. A high-molecular-weight substance made from repeating units of an alkene or from combinations of other molecules. (16)

Potential energy. Energy due to position. (1)

Precipitate. A solid compound that is formed in solution and settles to the bottom of the container. (11)

Precision. The exactness of a measurement. It also refers to the reproducibility of a measurement. (2)

Pressure. Force per unit area. (10)

Property. A description of a unique, observable characteristic or trait. (1)

Proton. A positively charged subatomic particle that exists in the nucleus of an atom. The mass of a proton is approximately 1 amu. (3)

Pure substance. Matter that has definite and unchanging composition and properties. Pure substances are either elements or compounds. (1)

Q

Quantum numbers. The four numbers that describe the energy, the orbital, and the spin of an electron in an atom. (4)

R

Radiation. Particles (α or β) or high-energy light rays (γ or X) that cause ionization of matter. (3)

Radioactive decay series. A series of elements formed from the successive emission of alpha and beta particles starting from a long-lived isotope and ending with a stable isotope. (3)

Radioactivity. The emission of energy or particles from an unstable nucleus. (3)

Radionuclide. A radioactive isotope. (3)

Rate law. An equality that relates the rate of a reaction to the concentration of reactants or any other substance that affects the rate of the reaction. (14)

Reactants. The compounds or elements that undergo a chemical reaction. (9)

Redox reaction. See *Oxidation–reduction reaction.*

Reducing agent. A substance that donates the electrons to the substance reduced. (13)

Reduction. The gain of electrons resulting in a decrease in oxidation state. (13)

Reforming. The rearrangement of long-chain hydrocarbons to branched hydrocarbons and/or the removal of hydrogen from hydrocarbons. (16)

Representative elements. Groups IA through VIIA in the periodic table. (5)

Resonance structures. The representation of all possible Lewis structures of a molecule or ion having the same number of electrons and arrangement of atoms. (6)

Reversible reaction. A reaction that proceeds in both directions under the conditions of the reaction and reaches a point of equilibrium. (14)

S

Salt. An ionic compound composed of a positive ion from a base and a negative ion from an acid. (12)

Saturated hydrocarbon. A hydrocarbon containing only single bonds (an alkane). (16)

Scientific notation. A number in which nonsignificant zeros in a very large or small number are expressed as powers of 10. (2)

Shell (orbit). The principal energy state for electrons in an atom. It corresponds to the distance of the electrons from the nucleus. (4)

Shielding. The effect that one electron has on another as to the nuclear charge it feels. (5)

Significant figure. A number that is known to be correct in a measurement. The last significant figure in a measurement may be estimated, however. (2)

SI units. The international system of units of measurement. (2)

Single displacement reaction. A reaction in which a metal substitutes for another metal ion in an ionic compound. (9)

Solid. One of the three physical states of matter. A solid is characterized by a definite volume and a definite shape. (1)

Solubility. The maximum amount of a solute that dissolves in a given amount of solvent at a certain temperature. (11)

Solute. A substance that dissolves in a solvent to form a homogeneous solution. (11)

Solution. A homogeneous mixture of substances with uniform properties and composition. (1)

Solvent. A medium that disperses another substance called a solute to form a homogeneous mixture. The solvent and the solution exist in the same physical state. (11)

Specific gravity. The ratio of the density of a substance to the density of water. (2)

Spectator ion. An ion that is not directly involved in an aqueous reaction and does not appear in the net ionic equation. (11)

Spectrum. The separate color components of a beam of light. (4)

Stock method. The naming of metal ions in a compound by writing the name of the metal followed by the oxidation state in Roman numerals. (7)

Stoichiometry. The quantitative relationships among reactants and products in a chemical reaction. (9)

STP. Standard temperature and pressure, which is 1 atm (760 torr) pressure and 0°C (273 K). (10)

Subshell. One of the four (s, p, d, or f) subdivisions of a shell in an atom. Each subshell has a different shape. (4)

Substance. A distinct form of matter. (1)

Substituent. An atom or group of atoms that can substitute for a hydrogen in an organic compound. (16)

Symbol. The letter or letters used to represent an element. (1)

T

Temperature. A measure of the intensity of heat (the average kinetic energy) of a substance. (2)

Theoretical yield. The amount of a product formed when at least one of the reactants is completely consumed. (9)

Thermometer. A device used to measure temperature. (2)

Torr. A unit of gas pressure equivalent to millimeters of mercury. (10)

Total ionic equation. A reaction in aqueous solution involving ionic compounds written so that all ions are shown independently in the equation. (11)

Transition element. Elements in which the d subshell is filling. These elements are designated as the B elements in the periodic table. (5)

Transmutation. The changing of one element into another by decay or a nuclear reaction. (3)

Triple bond. A chemical bond in which three pairs of electrons are being shared. (6)

U

Unity factor. A conversion factor that relates a quantity in one unit to "one" of another unit. (2)

Unsaturated hydrocarbon. A hydrocarbon that contains at least one double or triple bond. (16)

V

Volatility. The tendency of a compound to vaporize. (16)

Voltaic cell. A device that converts the chemical energy from a spontaneous redox reaction directly to electrical energy. (13)

W

Wavelength (λ). The distance between two equivalent points on a wave. (4)

Weak electrolyte. A substance that dissolves in water to form a limited concentration of ions. The solution is a weak conductor of electricity. (11)

Weight. The measure of the attraction of gravity for a substance. (2)

Answers to Problems

Chapter 1

1-1 **(a)** Homogeneous   **(b)** Heterogeneous   **(c)** Heterogeneous
(d) Homogeneous   **(e)** Homogeneous   **(f)** Homogeneous
(g) Heterogeneous **(h)** Heterogeneous **(i)** Homogeneous

1-2 **(a)** Compound   **(b)** Element   **(c)** Element   **(d)** Compound
(e) Element

1-3 **(a)** F **(b)** Ne **(c)** Sn **(d)** Mg **(e)** Mn **(f)** W

1-4 **(a)** Iron   **(b)** Bismuth   **(c)** Europium   **(d)** Uranium
(e) Curium **(f)** Mercury **(g)** Silver **(h)** Antimony

1-5 **(a)** Physical **(b)** Chemical **(c)** Physical **(d)** Chemical **(e)**
Chemical **(f)** Physical **(g)** Physical **(h)** Chemical

1-6 **(a)** Chemical   **(b)** Physical   **(c)** Physical   **(d)** Chemical
(e) Physical

Chapter 2

2-1 **(a)** Three **(b)** Two **(c)** Three **(d)** One **(e)** Four **(f)** Two
(g) Two **(h)** Three

2-3 **(a)** 16.0   **(b)** 1.01   **(c)** 0.665   **(d)** 4880   **(e)** 87,600
(f) 0.0272 **(g)** 301

2-5 **(a)** 188 **(b)** 12.90 **(c)** 2300 **(d)** 48 **(e)** 0.84

2-7 **(a)** 120 cm^2 **(b)** 394 ft · lb **(c)** 2 cm **(d)** 2.3 in.

2-9 **(a)** 1.57×10^2 **(b)** 1.57×10^{-1} **(c)** 3.00×10^{-2} **(d)** 4.0×10^7
(e) 3.49×10^{-2} **(f)** 3.2×10^4 **(g)** 3.2×10^{10} **(h)** 7.71×10^{-4}
(i) 2.34×10^3

2-10 **(a)** 0.000476   **(b)** 6550   **(c)** 0.00788   **(d)** 48,900   **(e)** 4.75
(f) 0.0000034

2-12 **(a)** 720 cm, 7.2 m, 7.2×10^{-3} km **(b)** 56,400 mm, 5640 cm,
0.0564 km **(c)** 2.50×10^5 mm, 2.50×10^4 cm, 250 m

2-13 (a) 8.9 g, 8.9×10^{-3} kg (b) 25,700 mg, 0.0257 kg (c) 1.25×10^6 mg, 1250 g

(*Note:* Your answer may vary from those below in the last significant figure depending on the conversion factor used.)

2-15 (a) 4.8 miles, 26,000 ft, 7.8 km (b) 2380 ft, 725 m, 0.725 km (c) 1.70 miles, 2740 m, 2.74 km (d) 4.21 miles, 22,200 ft, 6780 m

2-16 (a) 27.1 qt, 25.6 liters (b) 170 gal, 630 liters (c) 2.08×10^3 gal, 8.33×10^3 qt

2-18 $28.0 \, \cancel{m} \times \dfrac{100 \, \cancel{cm}}{\cancel{m}} \times \dfrac{1 \, \cancel{in.}}{2.54 \, \cancel{cm}} \times \dfrac{1 \, \cancel{ft}}{12 \, \cancel{in.}} \times \dfrac{1 \, yd}{3 \, \cancel{ft}} = 30.6$ yd

The team should look for a new punter.

2-20 14.6 gal

2-22 89 km/hr

2-23 $10,044 or $10,032 depending on which weight conversion factor you use. (When it comes to money, the price is usually not rounded off.)

2-25 2100 sec or 0.58 hr

2-26 $0.262/liter; $20.97

2-28 $25.95; $14.10

2-29 $500 \, \cancel{aspirin} \times \dfrac{0.324 \, \cancel{g}}{\cancel{aspirin}} \times \dfrac{1 \, lb}{454 \, \cancel{g}} = 0.357$ lb

2-31 $28.20

2-33 2.60 g/ml

2-35 670 g

2-36 625 ml

2-37 0.951 g/ml; 4790 ml; pumice floats on water but sinks in ethyl alcohol.

2-39 2080 g

2-41   1 liter of H_2O weighs 1000 g, 1 liter of gasoline weighs 670 g

2-43 $\left(\dfrac{2.54 \text{ cm}}{\text{in.}}\right)^3 = \dfrac{16.4 \text{ cm}^3}{\text{in.}^3} = \dfrac{16.4 \text{ ml}}{\text{in.}^3} \quad \left(\dfrac{12 \text{ in.}}{\text{ft}}\right)^3 = \dfrac{1728 \text{ in.}^3}{\text{ft}^3}$

$\dfrac{1.00 \text{ g}}{\text{ml}} \times \dfrac{1.00 \text{ lb}}{454 \text{ g}} \times \dfrac{16.4 \text{ ml}}{\text{in.}^3} \times \dfrac{1728 \text{ in.}^3}{\text{ft}^3} = 62.4 \text{ lb/ft}^3$

2-44 2.0×10^5 lb (100 tons)

2-45   572°F

2-46   24°C

2-48 -38°F

2-49   Since °F = °C, then $\frac{9}{5}$C $+ 32 = \frac{5}{9}$(F $- 32) = \frac{5}{9}$(C $- 32)$. Therefore °C $= -40$.

Chapter 3

3-1   (a) Six carbons, four hydrogens, and two chlorines (b) Two carbons, six hydrogens, and one oxygen (c) One copper, one sulfur, 18 hydrogens, and 13 oxygens

3-2   (a) SO_2 (c) H_2SO_4 (e) $(NH_4)_3PO_4$

3-3   (a) 21 protons, 24 neutrons, 21 electrons (b) 92 protons, 143 neutrons, 92 electrons (d) 38 protons, 52 neutrons, 36 electrons (e) 34 protons, 45 neutrons, 36 electrons

3-4   (a) Re: at. no. 75, at. wt. 186 (b) Co: at. no. 27, at. wt. 58.9 (c) Br: at. no. 35, at. wt. 79.9 (d) Si: at. no. 14, at. wt. 28.1

3-5   28.09

3-6   Let X = decimal fraction of ^{35}Cl and Y = decimal fraction of ^{37}Cl. Since there are only two isotopes present, $X + Y = 1$, $Y = 1 - X$.

$(X \times 35) + (Y \times 37) = 35.5$

$(X \times 35) + [(1 - X) \times 37] = 35.5$

$X = 0.75 \qquad (75\% \ ^{35}Cl)$

$Y = 0.25 \qquad (25\% \ ^{37}Cl)$

3-8   (a) $_{-1}^{0}e$ (b) $_{38}^{90}Sr$ (c) $_{90}^{231}Th$ (d) $_{20}^{41}Ca$ (e) $_{81}^{210}Tl$

3-9   (a) $^{230}_{90}\text{Th} \rightarrow\ ^{226}_{88}\text{Ra} +\ ^{4}_{2}\text{He}$

        (b) $^{214}_{84}\text{Po} \rightarrow\ ^{210}_{82}\text{Pb} +\ ^{4}_{2}\text{He}$

        (c) $^{210}_{84}\text{Po} \rightarrow\ ^{210}_{85}\text{At} +\ ^{0}_{-1}e$

3-10   (a) $^{90}_{39}Y$ (b)   50 years

3-12   About 11,500 years old

3-13 $^{234}_{90}\text{Th},\ ^{234}_{91}\text{Pa},\ ^{234}_{92}\text{U},\ ^{230}_{90}\text{Th},\ ^{226}_{88}\text{Ra},\ ^{222}_{86}\text{Rn},\ ^{218}_{84}\text{Po},\ ^{214}_{82}\text{Pb},\ ^{214}_{83}\text{Bi},\ ^{210}_{81}\text{Tl},$
$^{210}_{82}\text{Pb},\ ^{210}_{83}\text{Bi},\ ^{210}_{84}\text{Po},\ ^{206}_{82}\text{Pb}$

3-14   (a) $^{87}_{34}\text{Se}$ (b) $4^{1}_{0}n$

3-15   (a) $^{35}_{16}\text{S}$ (b) $^{2}_{1}\text{H}$ (d) $8_{-1}^{0}e$ (e) $^{254}_{102}\text{No}$ (g) $4^{1}_{0}n$

3-16 $^{239}_{93}\text{Np},\ _{-1}^{0}e$

Chapter 4

4-1   When $n = 4$, $l = 0, 1, 2,$ and 3; $l = 0$ (s subshell), $l = 1$ (p subshell), $l = 2$ (d subshell), $l = 3$ (f subshell)

4-2   Five subshells; the g subshell holds 18 electrons.

4-4 $2d$, $3f$, and $1p$ do not exist.

4-5   S $[\text{Ne}]3s^2 3p^4$; Zn $[\text{Ar}]4s^2\ 3d^{10}$; Y $[\text{Kr}]5s^2 4d^1$

4-6   (a)   F   (b)   Ga   (c)   Ba   (d)   Gd

4-7   Element number 121, assuming normal filling

4-8 $[\text{Rn}]7s^2 5f^{14}6d^{10}7p^2$; it will be in group IVA under lead.

4-10   (a) $5s$ (b) $5p$ (c) $6s$ (d) $5s$

4-11   (b)   Te

4-12 $m_l = 3, 2, 1, 0, -1, -2, -3$; seven orbitals

4-14   Nine orbitals (one s, three p, and five d orbitals)

4-15   (a) $2s$ (b) $3d$ (c) $5p$ (d) $4d$ (e) $5f$ (f) $4d$

       In order of increasing energy: $2s$, $3d$, $4d = 4d$, $5p$, $5f$

4-16 **(b)** because m_l is impossible, and **(c)**, because l is impossible

4-18 Both electrons would have the same spin in the same orbital, which violates the Pauli exclusion principle.

4-19 IIB, none; VB, three; VIA, two; VIB, 6; Pm, five

4-20 This is a $4p$ orbital, shaped like the $2p$ orbitals shown in the text.

Chapter 5

5-1 **(a)** Metal **(b)** Metal **(c)** Metal **(d)** Nonmetal **(e)** Non-metal

5-2 Te

5-3 32 in 7th period, 50 in 8th period

5-4 **(a)** Transition metal **(b)** Representative element **(c)** Inner transition metal **(d)** Transition metal

5-5 IA because it has a ns^1 notation and VIIA because it is the element before a noble gas.

5-6 **(e)**, **(d)**, **(f)**, **(c)**, **(b)**, **(a)**

5-7 **(a)**, **(b)**, **(d)**, **(c)**

5-8 **(a)** As **(b)** Ru **(c)** Ba

5-9 The outer electron in Hf is in a shell higher in energy than Zr. This alone would make Hf a larger atom. However, in between Zr and Hf lie several subshells including the long $4f$ subshell (Ce through Lu). The filling of these subshells, especially the $4f$, causes a gradual contraction that offsets the higher shell for Hf.

5-10 **(a)** V **(b)** Cl **(c)** Mg **(d)** Fe **(e)** B

5-11 Ga^+, 578.8 kJ; Ga^{2+}, 2558 kJ; Ga^{3+}, 5521 kJ; Ga^{4+}, 11,700 kJ. The fourth electron must be removed from an inner shell.

5-13 **(a)**, **(d)**, and **(g)**. [(b) removes electron from inner shell; (c) nonmetals do not form positive ions easily; (e) same; (f) too high a charge.]

5-14 **(a)** Six in the second, 14 in the fourth **(b)** 14, 28 **(c)** 46

$[42]6s^2 5d^1 4f^1)$ **(d)** elements 11 and 17 **(e)** element 12, element 12 **(f)** element 7, element 7, element 7 **(g)** 16^+, 13^-, 1^-, 15^+

Chapter 6

6-1 **(a)** $\overset{\cdot}{\underset{\cdot}{Ca}}$ · **(b)** ·$\overset{\cdot}{\underset{\cdot}{Sb}}$· **(c)** ·$\overset{\cdot}{Sn}$· **(d)** ·$\overset{\cdot\cdot}{\underset{\cdot\cdot}{I}}$:

6-2 **(a)** group IIIA **(b)** group VA **(c)** group IIA

6-3 S^-, Cr^{2+}, In^+

6-4 Se^{2-}, Br^-, Rb^+, Sr^{2+}, Y^{3+}

6-5 Cs_2S, Cs_3N; $BaBr_2$, BaS, Ba_3N_2; $InBr_3$, In_2S_3, InN

6-6 **(a)** O is $2-$, so Cr is $3+$ **(b)** F is $1-$, so Fe is $3+$ **(c)** S is $2-$, so Mn is $2+$ **(d)** O is $2-$, so Co is $2+$

6-7 Mg and S, Cr and Cl

6-8 The word "molecule" implies discrete $BaCl_2$ units (one Ba attached to two Cl's). In fact the ions in an ionic compound such as $BaCl_2$ are not attached to just one or two other ions but are surrounded by ions of opposite charge. The formula in this case just gives the ratio of ions in the compound.

6-9 **(a)** H—$\overset{\displaystyle H}{\underset{\displaystyle H}{C}}$—$\overset{\displaystyle H}{\underset{\displaystyle H}{C}}$—H **(b)** H—$\overset{\cdot\cdot}{\underset{\cdot\cdot}{O}}$—$\overset{\cdot\cdot}{\underset{\cdot\cdot}{O}}$—H **(c)** :$\overset{\cdot\cdot}{\underset{\cdot\cdot}{F}}$—$\overset{\cdot\cdot}{N}$—$\overset{\cdot\cdot}{\underset{\cdot\cdot}{F}}$:

with :$\overset{\cdot\cdot}{\underset{\cdot\cdot}{F}}$: below N

6-10 **(1)(a)** 17 **(b)** 31 **(c)** 51_3 **(d)** 97 **(e)** 7(13) **(f)** $10(13)_2$
(g) 67_2 **(h)** 3_26
(2) On Zerk, six electrons fill the outer s and p orbitals to make a noble gas structure. Therefore we have a "sextet" rule on Zerk.

(a) 1—$\overset{\cdot\cdot}{7}$: **(b)** 3^+1:$^-$ (ionic) **(c)** 1—5—1 **(d)** 9^+:$\overset{\cdot\cdot}{7}$:$^-$ (ionic)

with 1 below the 5

(e) :$\overset{\cdot\cdot}{7}$—1$\overset{\cdot\cdot}{3}$: **(f)** 10^{2+}(:$\overset{\cdot\cdot}{1}3$:$^-$)$_2$ (ionic)

(g) :$\overset{\cdot\cdot}{7}$—$\overset{\cdot\cdot}{6}$—$\overset{\cdot\cdot}{7}$: **(h)** $(3^+)_2$:$\overset{\cdot\cdot}{6}$:$^{2-}$ (ionic)

6-11 **(a)** SO_4^{2-} **(b)** ClO_3^- **(c)** SO_3^{2-} **(d)** PO_4^{3-}

6-12   **(a)** :C≡O: **(b)** 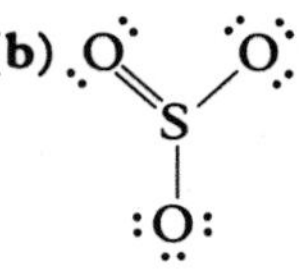 structure   **(c)** K⁺[:C≡N:]⁻

6-12 (b) O, O bonded to S with :O: below

(d)

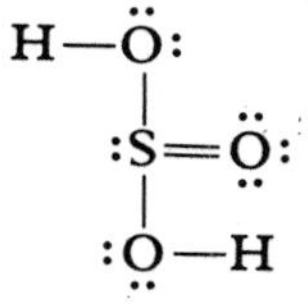

H—O:
    |
:S=O:
    |
:O—H

6-13   **(a)** N=O=N   **(b)** Ca²⁺ [O—N—O]⁻ ₂

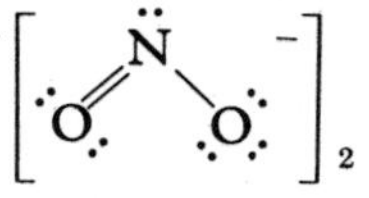

(c) 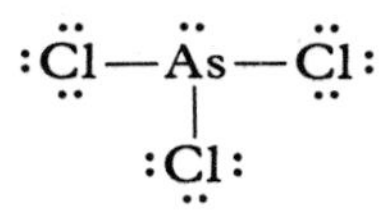 :Cl—As—Cl:
       |
     :Cl:

(d) 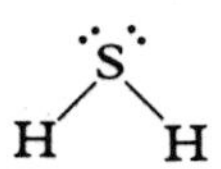 S with H H

(e) 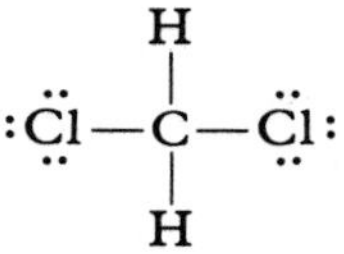
H
|
:Cl—C—Cl:
|
H

(f) 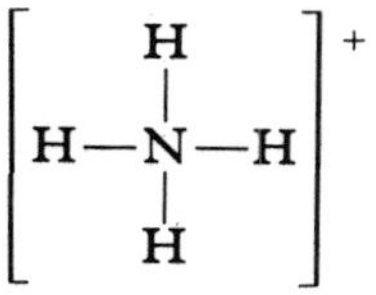
[H
   |
H—N—H
   |
   H   ]⁺

6-14   **(a)** :Cl—O—Cl: **(b)** 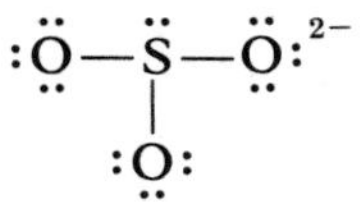 :O—S—O:²⁻
                     |
                    :O:

(c) 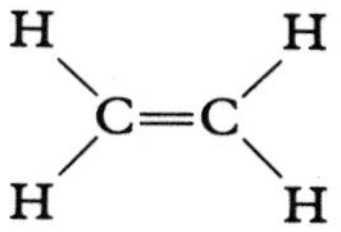
H   H
  \   /
   C=C
  /   \
H   H

(d) 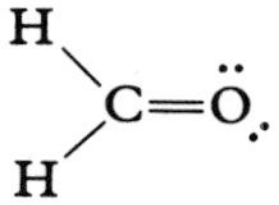
H
 \
  C=O
 /
H

(e) 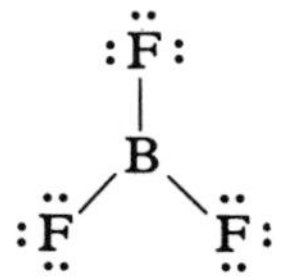
:F:
|
B
:F   F:

(f) :N≡O:⁺

6-17   **(a)**

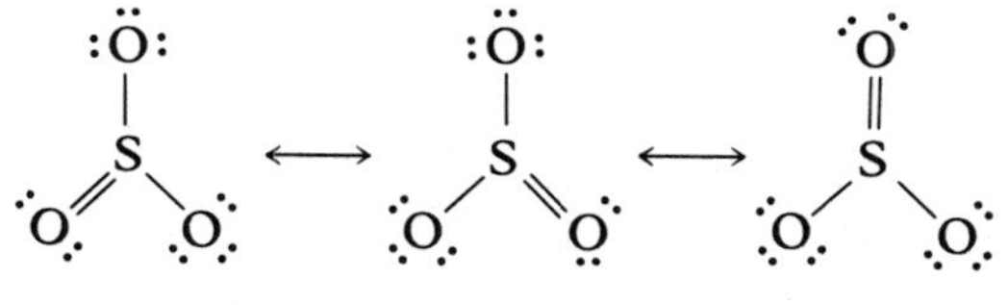

:O: :O: O
|
S   ⟷   S   ⟷   S
O   O   O   O   O   O

(c) 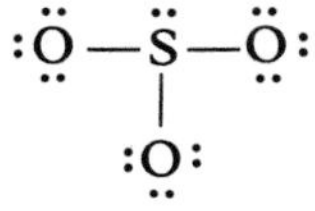 :O—S—O: (only one structure)
      |
    :O:

6-18   I—I and N—N

6-19 $K \overset{+}{\underset{-}{\rightleftharpoons}} O, Al \overset{+}{\underset{-}{\rightleftharpoons}} F, Ca \overset{+}{\underset{-}{\rightleftharpoons}} Cl, Al \overset{+}{\underset{-}{\rightleftharpoons}} Cl = I \overset{+}{\underset{-}{\rightleftharpoons}} F, N \overset{+}{\underset{-}{\rightleftharpoons}} H,$

$I \overset{+}{\underset{-}{\rightleftharpoons}} B = O \overset{+}{\underset{-}{\rightleftharpoons}} N, Se \overset{+}{\underset{-}{\rightleftharpoons}} Br, B \overset{+}{\underset{-}{\rightleftharpoons}} H$

The difference in electronegativity predicts that the K—O, Al—F, and the Ca—Cl bonds are primarily ionic.

6-20   **(a)** Nonpolar (equal dipoles cancel) **(b)** Polar (unequal dipoles) **(c)** Nonpolar (equal dipoles cancel) **(f)** Polar (unequal dipoles do not cancel) **(g)** Nonpolar (equal dipoles cancel)

Chapter 7

7-1   **(a)** Pb $+4$, O -2, **(b)** P $+5$, O -2 **(c)** C -1, H $+1$ **(d)** N -2, H $+1$ **(e)** Li $+1$, H -1 **(f)** B $+3$, Cl -1 **(g)** Rb $+1$, Se -2 **(h)** Bi $+3$, S -2 **(i)** Cl $+4$, O -2

7-2   **(a)** P $+5$ **(b)** C $+3$ **(c)** Cl $+7$ **(d)** Cr $+6$ **(e)** S $+6$ **(f)** N $+5$ **(g)** Mn $+7$

7-3   **(a)** Lithium fluoride **(b)** Tin(IV) oxide **(c)** Calcium selenide **(d)** Manganese(VII) oxide

7-4 SnO_2, Mn_2O_7, and Bi_2O_5

7-5   **(a)** Cu_2S **(b)** V_2O_3 **(c)** K_3P

7-6   **(a)** Chromium(II) sulfate **(b)** Aluminum sulfite **(c)** Iron(II) cyanide **(d)** Rubidium hydrogen carbonate

7-7   **(a)** Sodium chloride **(b)** Sodium hydrogen carbonate **(c)** Calcium carbonate **(d)** Sodium hydroxide **(e)** Sodium nitrate **(f)** Ammonium chloride **(g)** Aluminum oxide **(h)** Calcium hydroxide **(i)** Potassium hydroxide

7-8   **(a)** $Mg(MnO_4)_2$ **(b)** $Co(CN)_2$ **(c)** $Sr(OH)_2$ **(d)** Tl_2SO_3 **(e)** $In(HSO_4)_3$ **(f)** $Ba(IO_4)_2$

7-9   **(a)** Si **(b)** I **(c)** B **(d)** Kr **(e)** H

7-10   **(a)** Carbon disulfide **(b)** Boron trifluoride **(c)** Tetraphosphorus decoxide **(d)** Dibromine trioxide **(e)** Sulfur trioxide

7-11   **(a)** P_4O_6 **(b)** CBr_4 **(c)** IF_3 **(d)** Cl_2O_7

7-12 (a) Hydrochloric acid (b) Nitric acid (c) Hypochlorous acid
(d) Permanganic acid

7-13 (a) $H_2C_2O_4$ (b) HNO_2 (c) $H_2Cr_2O_7$ (d) H_3PO_4

7-14 H_3AsO_3, arsenous acid; H_3AsO_4, arsenic acid

Chapter 8

8-1 79.4 g

8-3 $50.0 \text{ g of Al} \times \dfrac{1.00 \text{ mol}}{27.0 \text{ g of Al}} = 1.85$ mol of Al

$50.0 \text{ g of Fe} \times \dfrac{1.00 \text{ mol}}{55.8 \text{ g of Fe}} = 0.896$ mol of Fe

There are more moles of atoms (or more atoms) in 50.0 g of Al.

8-4 $20.0 \text{ g of Ni} \times \dfrac{1.00 \text{ mol}}{58.7 \text{ g of Ni}} = 0.341$ mol of Ni

$2.85 \times 10^{23} \text{ atoms} \times \dfrac{1.00 \text{ mol of Ni}}{6.02 \times 10^{23} \text{ atoms}} = 0.473$ mol of Ni

The 2.85×10^{23} atoms contain more Ni.

8-5 (a) 0.468 mol of P, 2.82×10^{23} atoms (b) 150 g of Rb, 1.05×10^{24}
atoms (c) 27.0 g, 1.00 mol (d) 5.00 mol, Ge (e) 7.96×10^{-23} g,
1.66×10^{-24} mol

8-6 (a) 63.5 g (b) 16 g (c) 40.1 g

8-7 (a) 107 amu (b) 80.1 amu (c) 108 amu (d) 98.1 amu
(e) 106 amu

8-8 C 5.10 mol, H 15.3 mol, O 2.55 mol; 23.0 mol total; C 61.2 g,
H 15.4 g, O 40.8 g; 117.4 g total

8-10 $1.20 \times 10^{22} \text{ molecules} \times \dfrac{1.00 \text{ mol of } O_2}{6.02 \times 10^{23} \text{ molecules}}$

$= 0.0199$ mol of O_2 molecules

$0.0199 \text{ mol of } O_2 \times \dfrac{2 \text{ mol of O atoms}}{\text{mol of } O_2} = 0.0398$ mol of O atoms

$0.0199 \text{ mol of } O_2 \times \dfrac{32.0 \text{ g}}{\text{mol of } O_2} = 0.637$ g of O_2

Same weight of 0 atoms

8-11   **(b)**   189 g, 6.32×10^{24} molecules   **(c)**   0.339 g, 5.00×10^{-3} mol   **(d)**   0.219 mol, 1.32×10^{23} molecules   **(e)**   0.0209 g, 7.22×10^{19} molecules   **(f)**   598 g, 7.48 mol   **(g)** 7.62×10^{-3} mol, 4.59×10^{21} molecules

8-12   **(a)**   C 52.2%, H 13.0%, O 34.8%   **(b)**   C 85.7%, H 14.3%   **(c)** 85.7%, H 14.3%

8-13   Na 12.1%, B 11.3%, O 71.4%, H 5.27%

8-15   N 25.89%, O 74.11%

8-16 N_2O_3

8-18 KO_2

8-20 $1.20 \; \cancel{\text{g of CO}_2} \times \dfrac{1.00 \; \cancel{\text{mol of CO}_2}}{44.0 \; \cancel{\text{g of CO}_2}} \times \dfrac{1 \; \text{mol of C}}{1 \; \cancel{\text{mol of CO}_2}}$

$$= 0.0273 \; \text{mol of C}$$

$0.489 \; \cancel{\text{g of H}_2\text{O}} \times \dfrac{1.00 \; \cancel{\text{mol of H}_2\text{O}}}{18.0 \; \cancel{\text{g of H}_2\text{O}}} \times \dfrac{2 \; \text{mol of H}}{1 \; \cancel{\text{mol of H}_2\text{O}}}$

$$= 0.0543 \; \text{mol of H}$$

$\dfrac{0.0273}{0.0273} = 1 \qquad \dfrac{0.0543}{0.0273} = 2 \qquad \underline{CH_2}$

8-21 $C_9H_{12}Cl_{12}$

8-23 $B_2C_2O_6H_8$

Chapter 9

9-1   **(a)** $2Na(s) + 2H_2O \rightarrow H_2(g) + 2NaOH(aq)$

       (b) $2KClO_3(s) \rightarrow 2KCl(s) + 3O_2(g)$

       (c) $NaCl(aq) + AgNO_3(aq) \rightarrow AgCl(s) + NaNO_3(aq)$

       (d) $2H_3PO_4(aq) + 3Ca(OH)_2(aq) \rightarrow Ca_3(PO_4)_2(aq) + 6H_2O$

9-2   **(a)** $CaCO_3 \rightarrow CaO + CO_2$

       (b) $4Na + O_2 \rightarrow 2Na_2O$

       (c) $H_2SO_4 + 2NaOH \rightarrow Na_2SO_4 + 2H_2O$

       (d) $2H_2O_2 \rightarrow 2H_2O + O_2$

       (e) $Si_2H_6 + 8H_2O \rightarrow 2Si(OH)_4 + 7H_2$

(f) $2Al + 2H_3PO_4 \rightarrow 2AlPO_4 + 3H_2$

(g) $Ca(OH)_2 + 2HCl \rightarrow CaCl_2 + 2H_2O$

(h) $Na_2NH + 2H_2O \rightarrow NH_3 + 2NaOH$

9-3 **(a)** Single replacement **(b)** Decomposition **(c)** Double displacement **(d)** Double displacement **(e)** Combustion **(f)** This reaction does not fall neatly into one of the classifications mentioned. It could be classified as a single displacement where the SO_2 replaces the H_2O in $Ca(OH)_2$.

9-5 **(a)** $\dfrac{1 \text{ mol of Mg}}{2 \text{ mol of HCl}}, \dfrac{1 \text{ mol of Mg}}{1 \text{ mol of MgCl}_2}, \dfrac{1 \text{ mol of Mg}}{1 \text{ mol of H}_2}, \dfrac{2 \text{ mol of HCl}}{1 \text{ mol of H}_2},$

$\dfrac{2 \text{ mol of HCl}}{1 \text{ mol of MgCl}_2},$ **(c)** $\dfrac{2 \text{ mol of C}_4\text{H}_{10}}{13 \text{ mol of O}_2}, \dfrac{2 \text{ mol of C}_4\text{H}_{10}}{8 \text{ mol of CO}_2},$

$\dfrac{2 \text{ mol of C}_4\text{H}_{10}}{10 \text{ mol of H}_2\text{O}}, \dfrac{13 \text{ mol of O}_2}{8 \text{ mol of CO}_2}, \dfrac{13 \text{ mol of O}_2}{10 \text{ mol of H}_2\text{O}}$

9-6 **(a)** 0.400 mol of H_2 **(b)** 0.0400 mol of H_2O

9-7 **(a)** 1.35 mol of CO_2, 1.80 mol of H_2O, 2.25 mol of O_2 **(b)** 4.80 g of H_2O **(c)** 1.10 g of C_3H_8

9-8 **(a)** $5.45 \text{ mol of CO}_2 \times \dfrac{3 \text{ mol of O}_2}{2 \text{ mol of CO}_2} \times \dfrac{32.0 \text{ g of O}_2}{\text{mol of O}_2}$

$$= 262 \text{ g of O}_2$$

(b) $4.58 \times 10^{-4} \text{ mol of C}_2\text{H}_6\text{O} \times \dfrac{3 \text{ mol of H}_2\text{O}}{1 \text{ mol of C}_2\text{H}_6\text{O}}$

$$\times \dfrac{18.0 \text{ g of H}_2\text{O}}{\text{mol of H}_2\text{O}} = 0.0247 \text{ g of H}_2\text{O}$$

(c) $125 \text{ g of C}_2\text{H}_6\text{O} \times \dfrac{1 \text{ mol of C}_2\text{H}_6\text{O}}{46.0 \text{ g of C}_2\text{H}_6\text{O}} \times \dfrac{3 \text{ mol of H}_2\text{O}}{1 \text{ mol of C}_2\text{H}_6\text{O}}$

$$\times \dfrac{18.0 \text{ g of H}_2\text{O}}{\text{mol of H}_2\text{O}} = 147 \text{ g of H}_2\text{O}$$

(d) $50.0 \text{ g of H}_2\text{O} \times \dfrac{1 \text{ mol of H}_2\text{O}}{18.0 \text{ g of H}_2\text{O}} \times \dfrac{2 \text{ mol of CO}_2}{3 \text{ mol of H}_2\text{O}}$

$$\times \dfrac{44.0 \text{ g of CO}_2}{\text{mol of CO}_2} = 81.5 \text{ g of CO}_2$$

(e) $8.54 \times 10^{25} \text{ molecules} \times \dfrac{1 \text{ mol of O}_2}{6.02 \times 10^{23} \text{ molecules}}$

$$\times \dfrac{1 \text{ mol of C}_2\text{H}_6\text{O}}{3 \text{ mol of O}_2} \times \dfrac{46.0 \text{ g of C}_2\text{H}_6\text{O}}{\text{mol of C}_2\text{H}_6\text{O}} = 2180 \text{ g of C}_2\text{H}_6\text{O}$$

9-9   47.2 g of N_2

9-11 $125 \text{ g of Fe}_2\text{O}_3 \times \dfrac{1.00 \text{ mol of Fe}_2\text{O}_3}{160 \text{ g of Fe}_2\text{O}_3} \times \dfrac{2 \text{ mol of Fe}_3\text{O}_4}{3 \text{ mol of Fe}_2\text{O}_3}$

$\times \dfrac{3 \text{ mol of FeO}}{1 \text{ mol of Fe}_3\text{O}_4} \times \dfrac{1 \text{ mol of Fe}}{1 \text{ mol of FeO}} \times \dfrac{55.85 \text{ g of Fe}}{\text{mol of Fe}} = 87.3 \text{ g of Fe}$

9-12   30.0 g of SO_3 (theoretical yield); 70.7% yield

9-14   86.4%

9-15   If 86.4% is converted to CO_2, 13.6% is converted to CO. Thus 0.136×57.0 g = 7.75 g of C_8H_{18} is converted to CO. Notice that 1 mol of C_8H_{18} would form 8 mol of CO (because of the eight carbons in C_8H_{18}). Thus,

$7.75 \text{ g of C}_8\text{H}_{18} \times \dfrac{1.00 \text{ mol of C}_8\text{H}_{18}}{114 \text{ g of C}_8\text{H}_{18}} \times \dfrac{8 \text{ mol of CO}}{1 \text{ mol of C}_8\text{H}_{18}}$

$\times \dfrac{28.0 \text{ g of CO}}{\text{mol of CO}} = 15.2 \text{ g of CO}$

9-17   40.0 g of O_2 forms 0.833 mol of N_2, 1.50 mol of NH_3 forms 0.750 mol of N_2. Therefore, NH_3 is the limiting reactant.

9-18 $20.0 \text{ g of AgNO}_3 \times \dfrac{1.00 \text{ mol of AgNO}_3}{170 \text{ g of AgNO}_3}$

$\times \dfrac{2 \text{ mol of AgCl}}{2 \text{ mol of AgNO}_3} = 0.118 \text{ mol of AgCl}$

$10.0 \text{ g of CaCl}_2 \times \dfrac{1.00 \text{ mol of CaCl}_2}{111 \text{ g of CaCl}_2}$

$\times \dfrac{2 \text{ mol of AgCl}}{1 \text{ mol of CaCl}_2} = 0.180 \text{ mol of AgCl}$

Since $AgNO_3$ produces the smallest yield, it is the limiting reactant.

$0.118 \text{ mol of AgCl} \times \dfrac{143 \text{ g of AgCl}}{\text{mol of AgCl}} = 16.9 \text{ g of AgCl}$

9-20   3.53 g of H_2O

9-21 NH_3 is the limiting reactant. The theoretical yield based on NH_3 is 141 g of NO. The percent yield is 28.4%.

Chapter 10

10-1 (a) 2.17 atm (b) 0.0266 torr (c) 9560 torr (d) 0.0557 atm

10-2 For a column of Hg that is 1 cm² in area and 76.0 cm high, weight = 76.0 cm × 1 cm² × 13.6 g/cm³ = 1030 g. For water, volume × density = weight, volume = height × area, height × 1 cm² × 1.00 g/cm³ = 1030 g, h = 1030 cm.

$$1030 \ \text{cm} \times \frac{1 \ \text{in.}}{2.54 \ \text{cm}} \times \frac{1 \ \text{ft}}{12 \ \text{in.}} = 33.8 \ \text{ft}$$

10-4 10.2 liters

10-6 67.9 torr

10-7 $\dfrac{V_{\text{final}}}{V_{\text{initial}}} = \dfrac{1}{15} = \dfrac{P_{\text{initial}}}{P_{\text{final}}}; P_{\text{final}} = 14.2$ atm

10-9 1.94 liters

10-11 77°C

10-12 T_2 = 341 K or 68°C

10-14 2.94 atm

10-15 191 K or −82°C

10-16 596 K or 323°C

10-18 1.24 atm

10-19 76.8 liters

10-20 88.5 K or −185°C

10-23 1270 miles/hr

10-24 SF_6, SO_2, $N_2O \approx CO_2$, N_2, H_2

10-26 $\dfrac{r_{235U}}{r_{238U}} = \sqrt{\dfrac{352}{349}} = 1.004; r_{235U} = 1.004 r_{238U}$

10-27 $X_{SO_2} = 0.293; X_{O_2} = 0.606; X_{SO_3} = 0.101$

10-29 $X_{O_2} = 0.107; X_{CO_2} = 0.412; X_{N_2} = 0.185; X_{CO} = 0.296$

10-30 $P_{N_2} = 756$ torr, $P_{O_2} = 84$ torr, $P_{SO_2} = 210$ torr

10-33 $X_A = 0.314$

10-34 $P_A = P_T X_A;$ $\qquad\qquad P_T = 0.455 \times 0.175 = 0.630$ atm;

$\qquad X_A = \dfrac{0.455 \text{ atm}}{0.630 \text{ atm}} = 0.722$

10-36 $P_{N_2} = 300$ torr; $P_{O_2} = \dfrac{85 \text{ torr} \times 4 \text{ liters}}{2 \text{ liters}} = 170$ torr;

$\qquad P_{CO_2} = \dfrac{450 \text{ torr} \times 1 \text{ liter}}{2 \text{ liters}} = 225$ torr; $P_T = 300 + 170 + 225 =$

$\qquad \underline{695 \text{ torr}}$

10-37 $P_A = P_T X_A = P_T \dfrac{n_A}{n_A + n_B};$ $0.355 = 0.655 \dfrac{2}{2 + n_B};$ $n_B = \underline{1.69 \text{ mol}}$

10-38   1.70 liters

10-39 $\dfrac{V_1}{V_2} = \dfrac{n_1}{n_2};$ $n_2 = 5.47 \times 10^{-3}$ mol (original gas + added N_2); $n_{N_2} =$

$\qquad n_{total} - n_{original} = 5.47 \times 10^{-3} - 2.50 \times 10^{-3} = 2.97 \times 10^{-3}$ mol of

$\qquad N_2;$ 2.97×10^{-3} mol of $N_2 \times \dfrac{28.0 \text{ g of } N_2}{\text{mol of } N_2} = \underline{0.0832 \text{ g of } N_2}$

10-41   98°C

10-43   13.0 g of NH_3

10-45   Weight of He (4.2×10^6 g) 9.2×10^3 lb, weight of air (3.0×10^7 g) 6.6×10^4 lb, lifting power = 57,000 lb; lifting power with $H_2 = 61,000$ lb. H_2 is combustible whereas He is not. In 1937 the German airship *Hindenburg* (which used H_2) was destroyed by fire with great loss of life.

10-46   Molar mass = 41.9 g; empirical formula is CH_2; molecular formula is C_3H_6

10-48   34.0 g/mol

10-50 1.39×10^{-3} g of NO_2

10-51   25.8 liter of CO_2

10-53   **(a)** $85.0 \text{ g of } \cancel{C_4H_{10}} \times \dfrac{1.00 \text{ mol of } \cancel{C_4H_{10}}}{58.0 \text{ g of } \cancel{C_4H_{10}}} \times \dfrac{8 \text{ mol of } \cancel{CO_2}}{2 \text{ mol of } \cancel{C_4H_{10}}}$

$$\times \dfrac{22.4 \text{ liters}}{\cancel{\text{mol of } CO_2}} = 131 \text{ liters}$$

        (b)   96.3 liters of O_2

        (c)   124 liters of CO_2

10-54 1.72×10^5 g (380 lb)

Chapter 11

11-1   **(a)** $K_2Cr_2O_7 \rightarrow 2K^+(aq) + Cr_2O_7{}^{2-}(aq)$

        (b) $Li_2SO_4 \rightarrow 2Li^+(aq) + SO_4{}^{2-}(aq)$

        (e) $2(NH_4)_2S \rightarrow 4NH_4{}^+(aq) + 2S^{2-}(aq)$

        (f) $4Ba(OH)_2 \rightarrow 4Ba^{2+}(aq) + 8OH^-(aq)$

        (h) $2Sr(C_2H_3O_2)_2 \rightarrow 2Sr^{2+}(aq) + 4C_2H_3O_2{}^-(aq)$

11-2   **(a)** $2K^+(aq) + S^{2-}(aq) + Pb^{2+}(aq) + 2NO_3{}^-(aq) \rightarrow$
$$PbS(s) + 2K^+(aq) + 2NO_3{}^-(aq)$$

           $S^{2-}(aq) + Pb^{2+}(aq) \rightarrow PbS(s)$

        (b) $2NH_4{}^+(aq) + CO_3{}^{2-}(aq) + Ca^{2+}(aq) + 2Cl^-(aq) \rightarrow$
$$CaCO_3(s) + 2NH_4{}^+(aq) + 2Cl^-(aq)$$

           $CO_3{}^{2-}(aq) + Ca^{2+}(aq) \rightarrow CaCO_3(s)$

        (c) $2Ag^+(aq) + 2ClO_4{}^-(aq) + 2K^+(aq) + CrO_4{}^{2-}(aq) \rightarrow$
$$Ag_2CrO_4(s) \times 2K^+(aq) + 2ClO_4{}^-(aq)$$

           $2Ag^+(aq) + CrO_4{}^{2-}(aq) \rightarrow Ag_2CrO_4(s)$

11-3 $PbSO_4$, $MgSO_3$, NiS, Hg_2Br_2, Ag_2O

11-4   **(a)** $2KI(aq) + Pb(C_2H_3O_2)_2(aq) \rightarrow PbI_2(s) + 2KC_2H_3O_2(aq)$

        (b)   No reaction

        (c)   No reaction

        (d) $MgS(aq) + Hg_2(NO_3)_2(aq) \rightarrow Hg_2S(s) + Mg(NO_3)_2(aq)$

11-5   **(a)** $2K^+(aq) + 2I^-(aq) + Pb^{2+}(aq) + 2C_2H_3O_2{}^-(aq) \rightarrow$
$$PbI_2(s) + 2K^+(aq) + 2C_2H_3O_2{}^-(aq)$$

(d) $Mg^{2+}(aq) + S^{2-}(aq) + Hg_2^{2+}(aq) + 2NO_3^-(aq) \rightarrow$
$$Hg_2S(s) + Mg^{2+}(aq) + 2NO_3^-(aq)$$

11-6   **(a)** $Pb^{2+}(aq) + 2I^-(aq) \rightarrow PbI_2(s)$

      (d) $Hg_2^{2+}(aq) + S^{2-}(aq) \rightarrow Hg_2S(s)$

11-7   **(a)** $CuCl_2(aq) + Na_2CO_3(aq) \rightarrow CuCO_3(s) + 2NaCl(aq)$
          Filter the solid $CuCO_3$.

      (b) $(NH_4)_2SO_3(aq) + Pb(NO_3)_2(aq) \rightarrow PbSO_3(s) + 2NH_4NO_3(aq)$
          Filter the solid $PbSO_3$.

      (c) $2KI(aq) + Hg_2(NO_3)_2(aq) \rightarrow Hg_2I_2(s) + 2KNO_3(aq)$
          Filter the solid Hg_2I_2.

      (d) $NH_4Cl(aq) + AgNO_3(aq) \rightarrow AgCl(s) + NH_4NO_3(aq)$
          Filter off the solid $AgCl$; the desired product remains after the
          water is boiled off.

11-8   1.49%

11-10   0.375 mol of NaOH

11-12   0.846 mol

11-13   **(a)**   0.873 M **(b)**   1.40 M **(c)**   12.4 liters   **(d)**   41.3 g

       (e)   0.294 M   **(f)**   0.024 mol

11-14 $1.17 \times 10^{-3} M$

11-15 $[Ba^{2+}] = 0.166$, $[OH^-] = 0.332$

11-16   1.84 M

11-19   0.833 liter

11-21   0.72 liter

11-22 $0.250 \; \cancel{liter} \times \dfrac{0.200 \; \cancel{mol}}{\cancel{liter}} \times \dfrac{60.0 \; \cancel{g}}{\cancel{mol}} \times \dfrac{1.00 \; ml}{1.05 \; \cancel{g}} = 2.86 \; ml$

11-24 $n = (0.150 \; \cancel{liter} \times 0.250 \; mol/\cancel{liter})$
$$+ (0.450 \; \cancel{liter} \times 0.375 \; mol/\cancel{liter}) = 0.2062 \; mol$$
$V = 0.150 \; liter + 0.450 \; liter = 0.600 \; liter$
$M = n/V = 0.344$

11-25 $\left(0.500 \cancel{\text{liter}} \times 0.250 \dfrac{\text{mol}}{\cancel{\text{liter}}}\right) \times \dfrac{1 \cancel{\text{mol of Cr(OH)}_3}}{3 \cancel{\text{mol of KOH}}} \times$

$$\times \dfrac{103 \text{ g of Cr(OH)}_3}{\cancel{\text{mol of Cr(OH)}_3}} = 4.29 \text{ g of Cr(OH)}_3$$

11-26 NaOH is the limiting reactant and produces 1.75 g of $Mg(OH)_2$.

11-28 **(c)** $0.650 \times 0.100 = 0.0650$ mol of H_2S

$$0.0650 \cancel{\text{mol of H}_2\text{S}} \times \dfrac{2 \text{ mol of NO}}{3 \cancel{\text{mol of H}_2\text{S}}} = 0.0433 \text{ mol of NO}$$

$$V = \dfrac{nRT}{P} = \dfrac{0.0433 \cancel{\text{mol}} \times \dfrac{62.4 \text{ liter} \cdot \cancel{\text{torr}}}{\cancel{\text{mol}} \cdot \cancel{K}} \times 300 \cancel{K}}{720 \cancel{\text{torr}}}$$

$$= 1.13 \text{ liter}$$

Chapter 12

12-1 **(a)** HNO_3, nitric acid **(b)** HNO_2, nitrous acid **(c)** $HClO_3$, chloric acid **(d)** H_2SO_3, sulfurous acid

12-2 **(a)** CsOH, cesium hydroxide **(b)** $Sr(OH)_2$, strontium hydroxide **(c)** $Al(OH)_3$, aluminum hydroxide **(d)** $Mn(OH)_3$, manganese(III) hydroxide

12-3 **(a)** Acid **(b)** Salt **(c)** Acid **(d)** Base **(e)** Acid-salt

12-4 **(a)** $HC_2H_3O_2 + KOH \rightarrow KC_2H_3O_2 + H_2O$

(b) $2HI + Ca(OH)_2 \rightarrow CaI_2 + 2H_2O$

12-5 **(b)** $2H^+(aq) + 2I^-(aq) + Ca^{2+}(aq) + 2OH^-(aq) \rightarrow$
$$Ca^{2+}(aq) + 2I^-(aq) + 2H_2O$$

$2H^+(aq) + 2OH^-(aq) \rightarrow 2H_2O$

12-6 **(a)** $2HBr + Ca(OH)_2 \rightarrow CaBr_2 + 2H_2O$

(d) $2H_3PO_4 + 3Mg(OH)_2 \rightarrow Mg_3(PO_4)_2 + 6H_2O$

12-8 $Ca(OH)_2(aq) + 2H_3PO_4(aq) \rightarrow Ca(H_2PO_4)_2(aq) + 2H_2O$

12-9 $[H_3O^+] = 0.55$

12-10 $[H_3O^+] = 0.030 \times 0.55 = 0.016$

12-12 From first ionization, $[H_3O^+] = 0.354$.

From second ionization. $[H_3O^+] = 0.25 \times 0.354 = 0.088$. Total $[H_3O^+] = 0.442$.

12-13 (a) NO_3^- (b) HSO_4^- (c) PO_4^{3-}

12-14 (a) $NH_3CH_3^+$ (b) $H_2PO_4^-$ (c) HNO_3

12-15 (a) $H_2SO_3 + H_2O \rightarrow HSO_3^- + H_3O^+$
$\qquad A_1 \quad\; B_2 \qquad\quad B_1 \qquad\quad A_2$

(b) $HClO + H_2O \rightarrow ClO^- + H_3O^+$
$\qquad A_1 \quad\; B_2 \qquad\qquad B_1 \qquad\; A_2$

(c) $HBr + H_2O \rightarrow Br^- + H_3O^+$
$\qquad A_1 \quad B_2 \qquad\quad B_1 \qquad\; A_2$

(d) $HSO_3^- + H_2O \rightarrow SO_3^{2-} + H_3O^+$
$\qquad A_1 \qquad B_2 \qquad\; B_1 \qquad\; A_2$

12-16 (a) $NH_3 + H_2O \longrightarrow NH_4^+ + OH^-$
$\qquad B_1 \qquad\quad A_1 \qquad\qquad$
$\qquad\qquad A_2 \qquad\qquad\quad B_2$

(b) $N_2H_4 + H_2O \longrightarrow N_2H_5^+ + OH^-$
$\qquad B_1 \qquad\quad A_1 \qquad\qquad$
$\qquad\qquad A_2 \qquad\qquad\quad B_2$

(c) $HS^- + H_2O \longrightarrow H_2S + OH^-$
$\qquad B_1 \qquad\quad A_1 \qquad\qquad$
$\qquad\qquad A_2 \qquad\qquad\quad B_2$

12-17 (a) $HNO_3 + H_2O \rightarrow H_3O^+ + NO_3^-$

(b) $NO_2^- + H_3O^+ \rightleftharpoons HNO_2 + H_2O$

(c) N.R.

(d) $HOCl + H_2O \rightleftharpoons H_3O^+ + OCl^-$

12-18 Brønsted-Lowry base: $HS^- + H_3O^+ \rightleftharpoons H_2S + H_2O$

Brønsted-Lowry acid: $HS^- + OH^- \rightleftharpoons S^{2-} + H_2O$

12-19 $CaC_2 + 2H_2O \rightarrow C_2H_2(g) + Ca^{2+} + 2OH^-$

12-22 $CO_3^{2-} + H_3O^+ \rightleftharpoons HCO_3^- + H_2O$

$HCO_3^- + H_3O^+ \rightleftharpoons H_2CO_3 + H_2O$

$H_2CO_3 \rightarrow H_2O + CO_2(g)$

12-23 **(a)** $Sr(OH)_2$ **(b)** H_2SeO_4 **(c)** H_3PO_4 **(d)** $CsOH$ **(e)** HNO_2
(f) $HClO_3$

12-24 $CO_2(g) + LiOH(aq) \rightarrow LiHCO_3(aq)$

12-25 $2LiNO_3(s)$

Chapter 13

13-1 **(a)** Yes **(b)** No **(c)** Yes **(d)** No

13-2 **(a)** Oxidation **(b)** Reduction **(e)** Reduction

13-3 **(b)** CH_4; CO_2; O_2; CO_2 and H_2O; O_2; CH_4

(c) Fe^{2+}; Fe^{3+}; MnO_4^-; Mn^{2+}; MnO_4^-; Fe^{2+}

13-4 **(a)**
$$\overset{\displaystyle \overset{-5e \times 4 = -20e^-}{\boxed{}}}{\underset{-3}{}\underset{+2}{}}$$
$4NH_3 + 5O_2 \longrightarrow 4NO + 6H_2O$

$$\underset{\underset{+4e^- \times 5 = +20e^-}{\boxed{}}}{0 \qquad -2 \qquad -2}$$

(b)
$$\overset{\displaystyle \overset{-4e^- \times 1 = -4e^-}{\boxed{}}}{\underset{0}{}\underset{+4}{}}$$
$Sn + 4HNO_3 \longrightarrow SnO_2 + 4NO_2 + 2H_2O$

$$\underset{\underset{+1e^- \times 4 = +4e^-}{\boxed{}}}{+5 \qquad\qquad +4}$$

13-5 **(a)** $MnO_2 + 4H^+ + 2Br^- \rightarrow Mn^{2+} + Br_2 + 2H_2O$

(b) $CH_4 + 2O_2 \rightarrow CO_2 + 2H_2O$

(c) $5Fe^{2+} + MnO_4^- + 8H^+ \rightarrow 5Fe^{3+} + Mn^{2+} + 4H_2O$

13-6 **(a)** $2H_2O + Sn^{2+} \rightarrow SnO_2 + 4H^+ + 2e^-$

(b) $2H_2O + CH_4 \rightarrow CO_2 + 8H^+ + 8e^-$

(c) $e^- + Fe^{3+} \rightarrow Fe^{2+}$

(d) $6H_2O + I_2 \rightarrow 2IO_3^- + 12H^+ + 10e^-$

(e) $e^- + 2H^+ + NO_3^- \rightarrow NO_2 + H_2O$

13-7   **(a)**

$$S^{2-} \rightarrow S + 2e^- \quad | \times 3$$
$$\underline{3e^- + 4H^+ + NO_3^- \rightarrow NO + 2H_2O} \quad | \times 2$$
$$3S^{2-} + 8H^+ + 2NO_3^- \rightarrow 3S + 2NO + 4H_2O$$

(b) $I_2 + 2S_2O_3^{2-} \rightarrow 2I^- + S_4O_6^{2-}$

(c) $3SO_3^{2-} + ClO_3^- \rightarrow Cl^- + 3SO_4^{2-}$

13-8   **(a)**

$$4H_2O + Mn^{2+} \rightarrow MnO_4^- + 8H^+ + 5e^- | \times 2$$
$$\underline{2e^- + 6H^+ + BiO_3^- \rightarrow Bi^{3+} + 3H_2O} \quad | \times 5$$
$$2Mn^{2+} + 5BiO_3^- + 14H^+ \rightarrow 2MnO_4^- + 5Bi^{3+} + 7H_2O$$

(b)

$$2H_2O + SO_2 \rightarrow SO_4^{2-} + 4H^+ + 2e^- | \times 5$$
$$\underline{10e^- + 12H^+ + 2IO_3^- \rightarrow I_2 + 6H_2O} \quad | \times 1$$
$$2IO_3^- + 5SO_2 + 4H_2O \rightarrow I_2 + 5SO_4^{2-} + 8H^+$$

(d) $P_4 + 10HOCl + 6H_2O \rightarrow 4H_3PO_4 + 10Cl^- + 10H^+$

(f) $4Zn + 10H^+ + NO_3^- \rightarrow 4Zn^{2+} + NH_4^+ + 3H_2O$

13-9   **(a)** $2OH^- + SnO_2^{2-} \rightarrow SnO_3^{2-} + H_2O + 2e^-$

(b) $6e^- + 4H_2O + 2ClO_2^- \rightarrow Cl_2 + 8OH^-$

(c) $6OH^- + Si \rightarrow SiO_3^{2-} + 3H_2O + 4e^-$

(d) $4OH^- + Al \rightarrow Al(OH)_4^- + 3e^-$

13-10   **(a)**

$$8OH^- + S^{2-} \rightarrow SO_4^{2-} + 4H_2O + 8e^- | \times 1$$
$$\underline{I_2 + 2e^- \rightarrow 2I^-} \quad | \times 4$$
$$S^{2-} + 8OH^- + 4I_2 \rightarrow SO_4^{2-} + 8I^- + 4H_2O$$

(b) $8MnO_4^- + 8OH^- + I^- \rightarrow 8MnO_4^{2-} + IO_4^- + 4H_2O$

(c) $BiO_3^- + SnO_2^{2-} + 2H_2O \rightarrow Bi^{3+} + SnO_3^{2-} + 4OH^-$

(f) $2CrI_3 + 64OH^- + 27Cl_2 \rightarrow$
$$2CrO_4^{2-} + 6IO_4^- + 32H_2O + 54Cl^-$$

13-11 Cd^{2+} belongs below Zn^{2+} and above Ni^{2+}. It seems to have the same strength as Fe^{2+}.

13-12   Between F_2 and Cl_2

13-13 (a) Yes (b) N.R. (c) Yes (d) N.R. (e) Yes

13-14 (a) $CuCl_2(aq) + Fe(s) \rightarrow FeCl_2(aq) + Cu(s)$
(A coating of Cu forms on the nails.)

(b) N.R.

(e) $2Al(s) + 6H_2O \rightarrow 2Al(OH)_3(s) + 3H_2(g)$

(f) N.R.

(g) $Cu^{2+}(aq) + H_2(g) \rightarrow Cu(s) + 2H^+(aq)$

13-16 $Fe + 2H_2O \rightarrow$ N.R.
$Fe + 2H^+ \rightarrow Fe^{2+} + H_2$

13-18 The Daniell cell is more powerful. The greater the separation between the oxidizing agent and the reducing agent in the table, the more powerful the voltaic cell.

13-20 The strongest oxidizing agent is reduced the easiest. Thus the reduction of Ag^+ (i.e., $Ag^+ + e^- \rightarrow Ag$) occurs first. This procedure can be used to purify silver.

Chapter 14

14-1 (a) As the temperature increases, the frequency of collisions between molecules increases as well as the average energy of the collisions. Both contribute to the increased rate of reaction.

(b) The cooking of eggs initiates a chemical reaction which occurs more slowly at lower temperatures.

(c) The average energy of colliding molecules at room temperature is not sufficient to initiate a reaction between H_2 and O_2.

(d) A higher concentration of O_2 increases the rate of combustion.

(e) Metabolism generates energy. At a lower temperature this chemical reaction slows and therefore the wasp slows.

14-2 (a) $K_{eq} = \dfrac{[COCl_2]}{[CO][Cl_2]}$ (b) $K_{eq} = \dfrac{[CO_2][H_2]^4}{[CH_4][H_2O]^2}$

14-3 (a) $K_{eq} = \dfrac{[O_3]^2}{[O_2]^3}$ (b) $K_{eq} = \dfrac{[NO_2]^2}{[N_2][O_2]^2}$

14-4 $K_{eq} = 0.34$

14-6 $K_{eq} = \dfrac{[CO_2][H_2]^4}{[CH_4][H_2O]^2} = \dfrac{[2.20/30][4.00/30]^4}{[6.20/30][3.00/30]^2} = 0.0112$

14-7 (a) $[H_2] = [I_2] = 0.30$

  (b) $[H_2] = 0.30; [I_2] = 0.50$

  (c) $[HI] = 0.60 - 0.20 = 0.40; [H_2] = 0.10; [I_2] = 0.10$

14-8 (a) $[N_2] = 0.90; [H_2] = 0.70$

  (b) $[N_2] = 0.15; [H_2] = 0.45$

14-9 (a) The concentration of O_2 that reacts is

  $0.25 \text{ mol of NH}_3 \times \dfrac{5 \text{ mol of } O_2}{4 \text{ mol of NH}_3} = 0.31 \text{ mol of } O_2$

  (b) $[NH_3] = 0.75; [O_2] = 0.69; [NO] = 0.25; [H_2O] = 0.38$

  (c) $K_{eq} = \dfrac{[NO]^4[H_2O]^6}{[NH_3]^4[O_2]^5} = \dfrac{[0.25]^4[0.38]^6}{[0.75]^4[0.69]^5}$

14-11 $[PCl_5] = 0.28$

14-13 $[H_2O] = 0.26$

14-14 (a) Right (b) Left (c) Left (d) Right (e) Right (f) No effect (g) Decrease (h) Increase the yield but greatly decrease the rate of formation of NH_3

14-16 (a) Increase (b) Increase (c) Increase (d) decrease (e) Decrease (f) No effect

14-17 (a) Endothermic (b) Cooler (c) No effect, since there are the same number of moles of gas on both sides of the equation.

Chapter 15

15-1 The concentration of ions in pure water is extremely small (i.e., 10^{-7} M).

15-2 (a) $[OH^-] = 6.67 \times 10^{-12}$ (b) $[H_3O^+] = 3.88 \times 10^{-8}$

15-3 $[H_3O^+] = 0.0250; [OH^-] = 4.00 \times 10^{-13}$

15-5 $[H_3O^+][OH^-] = [10^{-2}][10^{-2}] = 10^{-4} \neq K_w$. H_3O^+ reacts with OH^- (forming H_2O) until the concentration of each ion drops to 10^{-7} M.

15-6 (a) Acidic (b) Basic (c) Acidic

15-7 (a) Acidic (b) Basic

15-8 (a) pH = 4.00 (acidic) (c) pH = -0.18 (acidic) (e) pH = 2.66 (acidic)

15-9 (a) 1.0×10^{-3} (b) 2.9×10^{-4} (c) 1.0×10^{-6}

15-10 pH = 1.12

15-12 pH = 13.56

15-13 (a) Strongly acidic (b) Strongly acidic (c) Weakly acidic (d) Strongly basic (e) Neutral

15-14 Ammonia, eggs, rain water, grape juice, vinegar, sulfuric acid

15-15 (a) $K_a = \dfrac{[H_3O^+][OBr^-]}{[HOBr]}$ (b) $K_b = \dfrac{[NH_4^+][OH^-]}{[NH_3]}$

(e) $K_a = \dfrac{[H_3O^+][H_2PO_4^-]}{[H_3PO_4]}$ (f) $K_b = \dfrac{[H_2N(CH_3)_2^+][OH^-]}{[HN(CH_3)_2]}$

15-16 (a) $[HOCN] = 0.20 - 0.0062 = 0.19$

(b) $K_a = \dfrac{[H_3O^+][OCN^-]}{[HOCN]} = \dfrac{(6.2 \times 10^{-3})(6.2 \times 10^{-3})}{(0.19)} = 2.0 \times 10^{-4}$

(c) pH = 2.21

15-17 (a) $HX + H_2O \rightleftharpoons H_3O^+ + X^-$; $K_a = \dfrac{[H_3O^+][X^-]}{[HX]}$

(b) $[H_3O^+] = 0.100 \times 0.58 = 0.058 = [X^-]$
$[HX] = 0.58 - 0.058 = 0.52$

(c) $K_a = 6.5 \times 10^{-3}$

(d) pH = 1.24

15-20 $K_b = 6.6 \times 10^{-6}$

15-21 $[H_3O^+] = 0.300 - 0.277 = 0.023$; pH = 1.64; $K_a = 1.9 \times 10^{-3}$

15-22 $[H_3O^+] = 3.2 \times 10^{-5}$; pH = 4.49

15-23 **(a)** $[H_3O^+] = 1.0 \times 10^{-5}$; pH = 5.00

(b) No. The decrease in H_3O^+ is less than 10-fold (about one-third).

15-25 pH = 11.32

15-27 **(a)** HS^- **(b)** H_3O^+ **(c)** OH^-

15-28 **(a)** $F^- + H_2O \rightleftharpoons HF + OH^-$

(b) $SO_3^{2-} + H_2O \rightleftharpoons HSO_3^- + OH^-$

(c) $H_2N(CH_3)_2^+ + H_2O \rightleftharpoons HN(CH_3)_2 + H_3O^+$

(d) Ca^{2+} does not hydrolyze

15-29 **(a)** Acidic; $H_2S + H_2O \rightleftharpoons HS^- + H_3O^+$

(b) Basic: $ClO^- + H_2O \rightleftharpoons HClO + OH^-$

(c) Neutral

(d) Basic: $NH_3 + H_2O \rightleftharpoons NH_4^+ + OH^-$

(e) Acidic: $HN_2H_4^+ + H_2O \rightleftharpoons N_2H_4 + H_3O^+$

(f) Acidic: $HOBr + H_2O \rightleftharpoons H_3O^+ + OBr^-$

(g) Basic: $CaO + H_2O \rightarrow Ca(OH)_2 \rightarrow Ca^{2+} + 2OH^-$

(h) Basic: $Ba(OH)_2 \rightarrow Ba^{2+} + 2OH^-$

15-30 LiOH strongly basic, pH = 13.0; $SrBr_2$ neutral, pH = 7.0; KOCl weakly basic, pH = 10.2; NH_4Cl weakly acidic, pH = 5.2; HI strongly acidic, pH = 1.0. The concentration is 0.10 M.

15-31 There is no unionized HCl present in solution to maintain a near constant concentration of H_3O^+. Also, the Cl^- ion does not react with H_3O^+ in aqueous solution.

15-32 **(a)** A buffer is formed, since HNO_2 is a weak acid and NO_2^- is its conjugate base.

(b) A buffer is formed, since NH_3 is a weak base and NH_4^+ is its conjugate acid.

(c) No buffer, HNO_3 is a strong acid.

(d) No buffer, NO_3^- is not the conjugate base of HNO_2.

(e) A buffer is formed, since HOBr is a weak acid and OBr^- is its conjugate base.

(i) A buffer is formed, since $H_2PO_4^-$ can act as a weak acid and HPO_4^{2-} is its conjugate base.

Chapter 16

16-1 (a)

$$H-\overset{\overset{\displaystyle H}{|}}{\underset{\underset{\displaystyle H}{|}}{C}}-\ddot{B}\ddot{r}:$$

(b)

$$\overset{\displaystyle H}{\underset{\displaystyle H}{>}}C=C=C\overset{\displaystyle H}{\underset{\displaystyle H}{<}}$$

(c)

$$\overset{\displaystyle H}{\underset{\displaystyle H}{>}}C=\overset{\overset{\displaystyle H}{|}}{\underset{\underset{\displaystyle H}{|}}{C}}-\overset{\overset{\displaystyle H}{|}}{\underset{\underset{\displaystyle H}{|}}{C}}-\overset{\overset{\displaystyle H}{|}}{\underset{\underset{\displaystyle H}{|}}{C}}-H$$

$$H-\overset{\overset{\displaystyle H}{|}}{\underset{\underset{\displaystyle H}{|}}{C}}-\overset{\overset{\displaystyle H}{|}}{\underset{}{C}}=\overset{\overset{\displaystyle H}{}}{\underset{}{C}}-\overset{\overset{\displaystyle H}{|}}{\underset{\underset{\displaystyle H}{|}}{C}}-H$$

$$\begin{array}{c}H \quad H \\ | \quad | \\ H-C-C-H \\ | \quad | \\ H-C-C-H \\ | \quad | \\ H \quad H \end{array}$$

(d)

$$H-\overset{\overset{\displaystyle H}{|}}{\underset{\underset{\displaystyle H}{|}}{C}}-\overset{\overset{\displaystyle ..}{}}{\underset{\underset{\displaystyle H}{|}}{N}}-H$$

16-2 $CH_3-CH_2-CH_2-CH_2-CH_2-CH_3$ $\qquad$ $CH_3-CH_2-CH_2-\underset{\underset{\displaystyle CH_3}{|}}{CH}-CH_3$

$CH_3-CH_2-\underset{\underset{\displaystyle CH_3}{|}}{CH}-CH_2-CH_3$ $\qquad$ $CH_3-CH_2-\overset{\overset{\displaystyle CH_3}{|}}{\underset{\underset{\displaystyle CH_3}{|}}{C}}-CH_3$ $\qquad$ $CH_3-\underset{\underset{\displaystyle CH_3}{|}}{CH}-\underset{\underset{\displaystyle CH_3}{|}}{CH}-CH_3$

16-3 (a) No (no oxygen in second compound) (b) No (unequal number of carbons)

16-4 (a) Six (*hex-*) (b) Eight (*oct-*) (c) Nine (*non-*)

16-5 (a) CH_3—, methyl; CH_3CH_2—, ethyl (b) $(CH_3)_2CH$—, isopropyl
(c) CH_3CH_2—, ethyl; $(CH_3)_3C$—, *t*-butyl

16-6 (a) $\left(\begin{array}{c} CH_2-CH \\ | \\ CH_3 \end{array}\right)$ Polypropylene

(b) $\left(\begin{array}{c} CH_2-CH \\ | \\ F \end{array} \right)$ Polyvinylfluoride

16-7 **(a)** $-C\!=\!C-$, alkene; and $\overset{\overset{\displaystyle O}{\|}}{C}-H$, aldehyde

(b) $-NH_2$, amine; and $-\underset{\underset{\displaystyle O}{\|}}{C}-$, ketone

(c) $-O-$, ether; and $-\underset{\underset{\displaystyle O}{\|}}{C}-NH_2$, amide

(d) $-C\!\equiv\!C-$, alkyne; and $-OH$ alcohol

16-8 **(a)** In an ether the O is between two C's. In an alcohol the O is between a C and a H.
(b) In an aldehyde the carbonyl (i.e., C=O) is between a C and a H. In a ketone the carbonyl is between two C's.
(c) In an amine the NH_2 group is attached to a hydrocarbon group. In an amide the NH_2 group is attached to a carbonyl group.

16-9 **(a)** Ketone **(b)** Alkene **(c)** Alkane **(d)** Aldehyde **(e)** Aromatic

Index

Abscissa, 404
Absolute zero, 188, 192
Accelerators, particle, 60, 61
Accuracy, 20
Acetic acid:
  ionization of, 234, 241
  structure of, 338
Acetone, 336
Acetylene, 326
Acid anhydride, 247
Acid-base reactions, 236
Acidic solutions, 298
  in terms of pH, 300
Acidosis, 310
Acid rain, 248
Acid salt, 238
Acids:
  Arrhenius, 234, 298
  Brønsted-Lowry, 243
  ionization of, 235
  ionization constants of, 303
  nomenclature of, 141
  polyprotic, 238
  properties of, 233
  strong, 240
  weak, 240, 301
Actinide elements, 93
Addition:
  review of, 348
  with scientific notation, 375
Air:
  composition of, 181
  pollution of, 4, 172
Alchemists, 2
Alcohols, 330
Aldehydes, 335
Algebra:
  review of, 359
  test of, 357
Algebra equations:
  operations on, 359
  from proportionalities, 369
  from word problems, 363
Alkali metals, 95
Alkaline earth metals, 96
Alkaloids, 335
Alkalosis, 310
Alkanes, 317

Alkenes, 323
Alkylation, 322
Alkyl groups, 320
Alkynes, 326
Allotropes, 97
Alloys, 96
Alpha particles, 54
Aluminum, 97
Amides, 337
Amines, 333
Ammonia:
  as Brønsted-Lowry base, 243
  formation of, 164
  ionization of in water, 236, 242
  ionization constant of, 305
  Lewis structure of, 119
Amphoterism, 244
Amu, 51
Angstroms, 100
Anhydrides, 247
Anions, 48
  table of monatomic, 137
  table of polyatomic, 139
Anode, 268
Antihistamines, 335
Antilogs, 399
Aromatic compounds, 328
Arrhenius, Svante, 243
Aspirin, 339
Atmosphere:
  composition of, 181
  pressure of, 183
Atom, 45
  Bohr's model of, 74
  radius of, 100, 102
  structure of, 50
Atomic bomb, 64
Atomic mass unit, 51, 151
Atomic number, 51
Atomic radius, 100
  periodic trends of, 102
Atomic weight, 53
Atomscope, 46
Aufbau principle, 79
Autoionization, 296
Avogadro's law, 198
Avogadro's number, 146

Balances, laboratory, 26
Balancing equations, 163
  ion-electron method in acidic solutions, 258
  ion-electron method in basic solutions, 260
  oxidation state method, 265
Barometer, 182
Base anhydride, 249
Bases:
  Arrhenius, 234, 298
  Brønsted-Lowry, 243
  ionization constants of, 304
  properties of, 233
  strong, 242
  weak, 242, 301
Basic solution, 298
  in terms of pH, 300
Battery, 268
  lead-acid, car, 252, 269
  nickel-cadmium, 270
Becquerel, H., 54
Benzene, 328
Beta particles, 55
Bimolecular collisions, 278
Binary acids, nomenclature of, 141
Binary compounds:
  covalent bonding of, 118
  ionic bonding of, 113
  metal-nonmetal nomenclature, 137
  nonmetal-nonmetal nomenclature, 140
Biochemistry, 97
Bismuth, 98
Blood:
  pH of, 308
  as solution, 211
Bohr groups, 93
Bohr model, 74
Bohr, Niels, 70
Boiling point, 9
Bonds:
  chemical, 47
  covalent, 115, 117
  ionic, 113
  multiple covalent, 120
  polarity of, 126
Borazine:
  empirical formula of, 155
  percent composition, 154
Boric acid, aqueous equilibrium, 299
Boron, 97
Boyle, Robert, 184
Boyle's law, 184
  result of kinetic theory, 192
Bromine, 100
Brønsted-Lowry acids and bases, 243

Buffer solutions, 308
Buoyancy, 37
Buret, 26
Butane, 317

Calcium carbide, reaction with water, 326
Calcium carbonate:
  Lewis structure of, 124
  percent composition, 154
  reaction with acid, 234
Cancellation in division, 350
Carbon, 97
Carbonated water, *see* Carbonic acid
Carbon dioxide:
  as acid anhydride, 247
  percent composition, 153
  polarity of, 129
Carbonic acid, 234, 247
Carbon tetrachloride, polarity of, 130
Carbonyl group, 335
Carboxyl group, 337
Carboxylic acid, 337
Carcinogen, 329
Catalysts, 290
  surface, 291
Catalytic converter, 291
Cathode, 268
Cations, 48
Caustic potash, *see* Potassium hydroxide
Caustic soda, *see* Sodium hydroxide
Celsius scale, 39
Centi-, 28
Centrifugal force, 74
Chain reaction, nuclear, 62
Chalcogen, 98
Characteristic, of a logarithm, 398
Charles, Jacque, 187
Charles' law, 188
  result of kinetic theory, 193
Chemical bond, *see* Bonds, chemical
Chemical change, 13
Chemical energy, 15
  conversion to electrical energy, 271
Chemical equations, 163. *See also* Equations
Chemical kinetics, 277
Chemical properties, 13
Chemical reactions, 2, 11, 12
  acid-base, 236
  combination, 166
  combustion, 166
  decomposition, 166
  double displacement, 167, 219, 237
  redox, 253
  single replacement or substitution, 167, 262

see also Reactions
Chemistry, 6
    study of, 3
Chlorine, 100, 253
Cobalt-60 treatment, 59
Coefficient:
    in chemical equations, 163
    in scientific notation, 24, 373
Collision theory, 278
Combination reactions, 166
Combined gas law, 189
Combustion reactions, 166
Common names, 133, 143
Compounds:
    basic structure of, 10
    formulas of, 47
    identification of, 157
    solubility of, 216
Concentrated solution, 222
Concentration, 222
    effect on equilibrium, 289
    molarity, 223
    percent by weight, 222
Conductor, 212
Conjugate acid-base pairs, 243
Control rods, 64
Conversion factors, 29, 36, 380
Conversions:
    procedure for, 34, 380
    between temperature scales, 40, 188
    between units, 34
Copper:
    basic structure, 46
    metal, 2
    ore, 1
Copper sulfate, 280
Coulomb's law, 101
Covalent bonding, 115
Cracking, 322
Critical mass, 63
Crude oil, see Petroleum
Crystals, 92
    growing of, 280
Cyclopropane, 320

Dacron, 341
Dalton, John, 3, 45
Dalton's law, 196
Daniell cell, 267
Decomposition reactions, 166
Density, 35
    calculation of, 36
    as conversion factor, 36
    table of, 36

Derivatives, 328
Detonation, 281
Diamagnetism, 85
Diamond, 98
Diatomic elements, 47
Diethyl ether, 333
Diffusion of gases, 194
Dilute solution, 222
Dilution of solutions, 224
Dimensional analysis, see Unit-factor method
Dinitrogen oxide, empirical formula of, 155
Dipole:
    bond, 128
    interactions, 214, 215
    molecular, 129
Dissociation of acids, 235
Distillation, 9
Division, 350
    of fractions, 351
    of scientific notation, 376
d orbitals, 77
    shapes of, 88
Dot structures, see Lewis dot structures
Double bond, 120
Double displacement reactions, 167, 219, 237
Dry cell, 270
Dynamic equilibrium, 240. See also Equilibrium

Einstein's law, 16
Electrical energy, 15
    conversion to chemical energy, 267
Electricity, 212
    from chemical reaction, 267
    from nuclear power, 64
Electrodes, 213, 268
    inert, 270
Electrolysis, 271
Electrolyte, 213
    weak, 213
Electrolytic cells, 270
Electron, 50
    box diagrams of, 86
    energy levels of, 74
    notation, 78
    from a nucleus, 55
    shielding, 100
    spin, 85
Electron affinity, 105
    table of, 106
Electronegativity, 128
    table of, 128
Electron notation, 78
    and periodic table, 83
Electron shielding, 100

Electroplating, 271
Electrostatic interactions, 49
    dipole-dipole, 214
    hydrogen bonding, 214
    ion-dipole, 215
    ion-ion, 216
Elements, 10
    Bohr groups, 93
    groups of, 70, 93
    as individual atoms, 46
    as molecules, 47
    physical properties of, 92
    periods of, 93
    symbols and names of, 11
    synthesis of, 61
Empirical formula, 155, 158
Endothermic reaction, 14
Energy, 14
    chemical, 15, 267
    conservation of, 15
    electrical, 15, 267
    heat, 15
    kinetic, 15
    light, 15, 72
    mechanical, 15
    minimum for reaction, 281
    nuclear, 16, 62, 65
    potential, 15
    relation to mass, 16
English system, 26
    relationship to metric, 29
Equations:
    balancing redox, 255, 258
    balancing simple, 163
    interpretation of, 163
    molecular, 219
    net ionic, 220
    total ionic, 218 , 219
Equilibrium, 240, 278
    and gaseous reactions, 278
    in pure water, 296
    and reaction to stress, 288
    and saturated solutions, 280
    and weak acids and bases, 301
Equilibrium constant, 284
    expression, 284
    magnitude of, 298
    tables of 303, 304
Ester, 337
Ethane, 318
Ether, 333
Ethyl alcohol, 331
    fermentation of, 332
Ethylene, 323

Ethylene glycol, 331
Exact numbers, 23
Excited state, 75
Exothermic reactions, 14
    and equilibrium constant, 290
Exponential notation, *see* Scientific notation
Exponents, 24, 349, 373

Factor form of division, 381
Fahrenheit scale, 39
Families of elements, *see* Groups
Fermentation, 332
Filtration, 8
Fire, 1
Fission, 62
Fluorine, bonding in, 116
f orbitals, 77
Force of a gas, 183
Formaldehyde, 336
Formic acid, 338
Formula:
    of compounds, 48
    condensed, 317
    empirical, 155, 158
    of ionic compounds, 48
    molecular, 156, 158
Formula weight, 151
Fractions:
    construction of, 354
    division of, 352
    expressed as percent, 355
    multiplication of, 351
Fuel cell, 270
Functional group, 330
Fusion, nuclear, 65
    control of, 66

Galvanic cell, 267
Gamma rays, 55
Gases, 6, 181
    compressibility of, 194
    see also Gas laws
Gas laws, 182
    Avogadro's, 198
    Boyle's, 184
    Charles', 187
    combined, 189
    Dalton's, 196
    Gay-Lussac's, 190
    Graham's, 194
    ideal, 199
    kinetic theory, 192
Gasohol, 332
Gasoline, 322
Gay-Lussac's law, 190

result of kinetic theory, 193
Glycerol, 331
Graduated cylinder, 26
Graham's law, 194
Grain alcohol, 331
Gram, 27
Graphite, 98
Graphs, 404
Greek prefixes, 141
Ground state, 75
Groups, 70, 93
  IA, 95
  IIA, 96
  IIIA, 96
  IVA, 97
  VA, 98
  VIA, 98
  VIIA, 99
  O, 110

Haber process, 293
Hahn, Otto, 62
Half-life, 56
Half-reactions, 254, 258
Halogens, 99
Heat energy, 15
  measurement of, 38
Helium, nucleus of, 54
Hetero atoms, 317
Heterogeneous matter, 7
Homogeneous matter, 7
Homologous series, 318
Hund's rule, 86
Hybrid structures, 125
Hydration, 215
Hydrazine, in aqueous equilibrium, 305
  Lewis structure of, 119
Hydrocarbons, 317
  alkanes, 317
  alkenes, 323
  alkynes, 326
  aromatic, 328
  in petroleum, 322
  unsaturated, 323
Hydrochloric acid:
  bonding in, 118
  as Brønsted-Lowry acid, 244
  in water, 234
Hydrocyanic acid:
  in aqueous solution, 302
  in buffer solution, 309
Hydrofluoric acid, ionization of, 240, 281
Hydrogen, 96, 100
  bonding in, 117

reaction with iodine, 277
Hydrogen bomb, 66
Hydrogen bonding, 214
Hydrogen peroxide, 163
Hydrolysis, 306
Hydrometer, 37
Hydronium ion, 235
Hydroxide ion, 139
Hydroxyl group, 330
Hypochlorous acid, aqueous equilibrium, 301, 304

Ice, structure of, 214
Ideal gas law, 199
Infrared light, 73
Inner transition elements, 93
Insoluble compounds, 216
Insulator, 212
Inverse relationships, 369
  in conjugate acid-base pairs, 244
  graph of, 407
  between hydronium and hydroxide ion concentrations, 297
  in oxidizing and reducing agents, 264
Iodine:
  basic structure of, 47
  properties of, 102
  reaction with hydrogen, 277
Ion-electron method, 258
Ionic compounds, 49, 113
  solution in water, 215
Ionic equations, 218
  net ionic, 220
  total ionic, 219
Ionic radii, 105
Ion interactions:
  ion-dipole, 215
  ion-ion, 216
Ionization, of acids, 235
  of atoms, 103
  constants, 303
  from radiation, 58
Ionization energy, 103
  periodic trends, 104
Ion product of water, 296
Ions, 48
  as acids and bases in water, 306
  formation of, 111
  monatomic, 52
  polyatomic, 115, 121, 139
  spectator, 220, 257
Iron, production of, 179
Iron(III) oxide, empirical formula of, 156
Isobutane, 317
Isomers, 316
Isopropanol, 332

Isotopes, 52
IUPAC, 135, 318

K_a, 301
K_b, 303
K_{eq}, 284
K_w, 296
Kelvin scale, 188
Ketones, 335
Kilo-, 28
Kilopascal, 27
Kinetic energy, 15
    of colliding molecules, 281
    and temperature, 192
Kinetic theory of gases, 192
    and Boyle's law, 192
    and Charles' law, 193
    and Dalton's law, 196
    and Gay-Lussac's law, 193
    and Graham's law, 194
    and reaction rates, 278

Lanthanide elements, 93
Lattice, 115
Lattice energy, 115
Laughing gas, *see* Dinitrogen oxide
Lavoisier, Antoine, 181
Law:
    Avogadro's, 198
    Boyle's, 184
    Charles', 187
    Coulomb's, 101
    combined gas, 189
    of conservation of energy, 15
    of conservation of mass, 14
    Dalton's, 196
    Einstein's, 16
    gas, 182
    Gay-Lussac's, 190
    Graham's, 194
    Ideal gas, 199
    rate, 283
Lead-acid battery, 252, 269
Le Châtelier's principle, 288
    and buffer solutions, 309
    and concentration, 289
    and pressure, 290
    and temperature, 290
    and volume, 290
Lewis dot structures, 110
    of carbon compounds, 315
    of molecules, 118
    of polyatomic ions, 121
    of representative elements, 111
    rules for writing, 121

Lewis, G.N., 111
Light:
    classification of, 73
    energy of, 15, 72
    infrared, 72
    speed of, 72
    ultraviolet, 72
    visible, 72
    wavelength of, 72
Limestone, *see* Calcium carbonate
Limiting reactant, 175
    determination of, 176
Linear relationships, algebra equations, 405
    graphs of, 404
Liquid state, 6
Liter, 27
Logarithms:
    review of, 397
    table of common, 400
Lye, *see* Sodium hydroxide

Magnesium, 97
Magnetism:
    diamagnetism, 85
    of molecules, 116
    paramagnetism, 85
Malachite, 2
Mantissa, of logarithm, 398
Mass, 25
    conservation of, 14
Mass number, 52
Mathematics:
    review of, 347
    test of, 346
Matter, 6
    heterogeneous, 7
    homogeneous, 7
    pure, 8, 158
    states of, 6
Measurements, 18
    standards for, 18
Mechanical energy, 15
Mechanism of a reaction, 277
Melmac, 337
Meltdown, 65
Melting point, 9
Mendeleev, Dimitri, 91
Mercury, 93
    in a barometer, 182
Metabolism, 276
Metalloids, 92
Metallurgy, 2
Metal oxides, as bases, 249
Metals, 92
    active, 70, 267

coinage, 70
  with one oxidation state, 138
  properties of, 92, 104
  with variable oxidation states, 138
Metathesis reactions, 219
Meter, 27
Methane, 318, 321
  as potential Brønsted-Lowry acid, 246
  Lewis structure of, 119
Methyl alcohol, 330
  as fuel, 176
  in water solution, 218
Metric measurements, 27, 28
  prefixes, 28
  relationships to English units, 29
  symbols, 27
Meyer, Lothar, 91
Milli-, 28
Mixtures, 7
  of gases, 195
  properties of, 9
Molar gas constant, 200
Molarity, 223
  and stoichiometry, 227
Molar mass:
  of compounds, 151
  of elements, 146
Molar number, 146
Molar volume, 202
Mole, 27, 146
Molecular equation, 219
Molecular formula, 156, 158
Molecular structure, 159
Molecular velocity, 194
Molecular weight, *see* Formula weight
Molecules, 47
Mole fraction, 196
Monomer, 323
Multiplication, 349
  of fractions, 351
  of scientific notation, 376
Muriatic acid, *see* Hydrochloric acid

Naming of compounds, 134
Natural gas, 321
Net ionic equation, 220
Neutralization reactions, 236
Neutral solution, 297
  in terms of pH, 300
Neutron, 50
Neutron activation, 62
Nickel, reduction of, 263
Nitric acid, 245
Nitric oxide, Lewis structure of, 125
Nitrogen, bonding in, 120

Nitrogen trichloride, Lewis structure of, 122
Nitroglycerin, 332
Nitrous acid:
  in aqueous equilibrium, 302
  as weak acid, 246
Noble gases, 92
  electron notation of, 80
Nomenclature:
  of acids, 141
  of metal-nonmetal binary compounds, 137
  of organic compounds, 318
  of polyatomic ions, 139
  by the Stock method, 137
Nonconductors, 212
Nonelectrolytes, 213
Nonlinear graphs, 406
Nonmetal oxides, as acids, 247
Nonmetals, 92
Nonpolar bond, 127
Nonpolar molecule, 129
Nuclear energy, 16, 62, 65
Nuclear equations, 54
Nuclear reactions, 59
Nuclear reactor, 64
Nucleons, 50
Nucleus, 50
Nylon, 340

Ocean:
  of air, 181
  of water, 211
Octane rating, 322
Octet rule, 110
Orbitals, 84
  shapes of, 87
Orbits, *see* Shells
Ordinate, 404
Organic chemistry, 97, 314
Orlon, 328
Oxidation, 253
Oxidation number, 135
Oxidation-reduction reactions, 253. *See also* Redox reactions
Oxidation state, 135
Oxidizing agents, 253
  table of comparative strengths, 266
Oxyacids, 142
Oxyanions, 139
Oxygen, 99, 181
  bonding in, 121
Ozone, 99

Paramagnetism, 85
  of molecules, 116
Partial pressure, 196
Particle accelerators, 60, 61

Pauli exclusion principle, 85
Pauling, Linus, 128
Percent, 355
  by weight solute, 222
Percent composition, 153
Percent purity, 174
Percent yield, 173
Periodic table, 11, 91
  Bohr groups, 93
  electronic basis, 70
  history of, 91
  periods, 93
  reading of, 54
Periodic trends:
  atomic radius, 100
  bonding, 109
  electron affinity, 105
  electronegativity, 128
  ionization energy, 103
Periods, 93
Petrochemicals, 323
Petroleum, refining of, 321
pH, 299
  of blood, 310
  in buffer solutions, 308
  of common substances, 301
Phase, 7
  gas, 6, 181
  liquid, 6
  solid, 6
Phenacetin, 339
Phenol, 329
Phosphoric acid, neutralization of, 238
Phosphorus, 98
Phosphorus trifluoride, Lewis structure of, 119
Photosynthesis, 14
Physical change, 13
Physical property, 12
Physical science, 19
Plutonium, fission of, 64
pOH, 299
Polar covalent bond, 127
Polarity:
  of bonds, 126
  of molecules, 129
Pollution:
  of air, 4
  of water, 172, 248
Polyatomic ions, 115
  bonding in, 121
  nomenclature of, 139
Polyethylene, 323
Polymers, 323, 340
  table of, 327
Polyprotic acid, 237

  neutralization of, 237
p orbitals, 77
  shapes of, 87
Potassium hydroxide, 236
Potential energy, 15
Precipitate, 218
Precision, 20
Prefixes, 27, 141
Pressure, 183
  correction factor, 186
  effect on gaseous equilibria, 290
  of a gas, 183
  units of, 184
Priestly, Joseph, 181
Prism, 72, 73
Problem solving, 28, 380
Products, 163
Prometheus, 1, 11
Proof, alcohol, 332
Propane, 318
Properties, 9
  of gases, 194
  of organic compounds, 315
Proportionalities, 369
  constant of, 369
  direct, 369
  inverse, 369
Proton, 50
Pseudo-noble gas structure, 112
Pure substances, 8, 158
Pyrite, 172

Quantized energy, 74
Quantum number:
  orbital, 84
  principal, shell, 77
  spin, 85
  subshell, 77

R, 200
Radiation, 54
  effects of, 57
  types of, 54, 55
Radioactive decay, 55
  rates of, 56
  series, 57
Radioactive isotope, 54
Radioactivity, *see* Radioactive decay
Radionuclides, 54
  half-life of, 56
Rare earths, 93
Rate constants, 283
Rate law, 283
Rates of reactions, 277
  effect of catalyst, 290

effect of concentration, 281
effect of temperature, 281
Reactants, 163
in excess, 175
limiting, 175
Reaction factor, 169
Reaction mechanism, 277
Reactions, 2, 11, 12
endothermic, 14, 290
exothermic, 14, 290
nuclear, 59
rates of, 277
reversible, 278
spontaneous, 267
see also Chemical reactions
Real gas, 201
Reciprocal, 352
Redox equations, 253
balancing of, 255, 258
Redox reactions, 253
electrolytic, 270
prediction of, 265
voltaic, 267
Reducing agents, 253
table of comparative strengths, 266
Reduction, 253
Refinery, 321
Reforming, 322
Representative elements, 93
IA, alkali metals, 95
IIA, alkaline earth metals, 96
IIIA, 96
IVA, 97
VA, 98
VIA, chalcogens, 98
VIIA, halogens, 99
bond formation of, 110
Lewis dot structures of, 111
Resonance structures, 125
Resultant force, 129
Reversible reactions, 278
Richter scale, 300
Roots:
in nomenclature, 137
of numbers, 353
in scientific notation, 378
Rounding off, 22
Rust, *see* Iron(III) oxide
Rutherford, Lord, 50, 59, 70

Salt, 236
Salt bridge, 268
Scientific method, 262
Scientific notation, 23
addition and subtraction of, 375

multiplication and division of, 376
powers and roots of, 378
review of, 373
and significant figures, 25
standard form for, 373
test of, 372
Second, 27
Semiconductors, 97
Shells, 76
Shielding, effects of, 100
Significant figures, 20
in mathematical operations, 21
role of zero, 21
Silver, tarnishing of, 176
Single replacement reactions, 167, 262
S.I. units, 27
Slope of a line, 405
Smog, 4
Sodium, 96, 253
Sodium chloride:
basic structure of, 49
preparation from elements, 253
preparation from neutralization, 236
solution equilibrium, 280
Sodium hydroxide, 236
in aqueous solution, 242
Solid state, 6
Solubility, rules for, 220
Soluble compounds, 216
Solute, 212
Solutions, 8, 212
acidic, 298
aqueous, 211
basic, 298
buffer, 308
concentrated, 222
conductivity of, 212
dilute, 222
formation of, 215
saturated, 280
Solvent, 212
s orbitals, 77
shapes of, 87
Specific gravity, 38
Spectator ions, 220, 257
Spectrum, 71
continuous, 71
discrete, line, 73
emission, of elements, 73
Spontaneous reactions, 267
Standard temperature and pressure, 190
States of matter, 6
Stock method, 137
Stoichiometry, 168
and gases, 204

general procedure, 171
mole-mole, 169
mole-weight, 169
and solutions, 227
weight-number, 170
weight-weight, 170
STP, 190
Strassmann, Fritz, 62
Strong acids, 240
Strong bases, 242
Structural formula, 278, 316
Subshells, 77
Substituent, 319
Subtraction, 348
— with scientific notation, 375
Sulfur:
  chemical properties of, 13
  physical properties of, 12, 99
Sulfur dioxide:
  in acid rain, 248
  bonding in, 120
  resonance structures of, 126
Sulfuric acid:
  in aqueous solution, 234
  composition of, 152
  Lewis structure of, 124
  neutralization of, 237
  production of, 288
Sulfur trioxide, equilibrium of, 288
Symbols of elements, 11
Synthesis, gas, 331
Systematic names, 135

Temperature, 38
  effect on equilibrium constant, 290
  correction factor, 189
Temperature scales:
  Celsius, 38
  Fahrenheit, 38
  Kelvin, 188
Tetraphosphorus decoxide, molecular formula of, 157
Theoretical yield, 173
Thermometer, 38
Thomson, J.J., 50
Three Mile Island, 65, 210
Toluene, 329
Torr, 184
Torricelli, Evangelista, 182
Transition elements, 93
Transmutation, 59
Triple bond, 120

Ultraviolet light, 72

Unit-factor method, 28, 380
  one step problems, 31, 385
  two or more step problems, 32, 390
Units of measurement:
  English, 29
  manipulation in calculations, 383
  S.I. or metric, 27
Unity factor, 30, 382
Unsaturated hydrocarbons, 323
Uranium, fission of, 62

Valence electrons, 110
Velocity of molecules, 194
Vinegar, 234
Vinyl chloride, 328
Visible light, 72
Volume of gases:
  as a correction factor, 187
  and pressure, 184
  and quantity, 198
  and temperature, 187
Voltaic cell, 267

Water 2, 69, 211
  dipole-dipole interactions in, 214
  formation of, 163
  formula of, 47
  importance of, 211
  ion product of, 296
  Lewis structure of, 119
  polarity of, 131, 214
  solvent properties of, 214
Wavelength, 72
Wave mechanics, 76
Weak acids, 240
Weak bases, 242, 301
Weak electrolytes, 213
  and covalent molecules, 217
Weight, 25
Weight percent, 153, 222
Wood alcohol, 330

X-rays, 72

Zinc:
  chemical properties of, 13
  physical properties of, 12
  reaction with acids, 235
  reduction of, 262, 265
Zinc sulfide:
  chemical properties of, 13
  physical properties of, 12